# Studies in Systems, Decision and Control

Volume 120

**Series editor**

Janusz Kacprzyk, Polish Academy of Sciences, Warsaw, Poland

The series "Studies in Systems, Decision and Control" (SSDC) covers both new developments and advances, as well as the state of the art, in the various areas of broadly perceived systems, decision making and control- quickly, up to date and with a high quality. The intent is to cover the theory, applications, and perspectives on the state of the art and future developments relevant to systems, decision making, control, complex processes and related areas, as embedded in the fields of engineering, computer science, physics, economics, social and life sciences, as well as the paradigms and methodologies behind them. The series contains monographs, textbooks, lecture notes and edited volumes in systems, decision making and control spanning the areas of Cyber-Physical Systems, Autonomous Systems, Sensor Networks, Control Systems, Energy Systems, Automotive Systems, Biological Systems, Vehicular Networking and Connected Vehicles, Aerospace Systems, Automation, Manufacturing, Smart Grids, Nonlinear Systems, Power Systems, Robotics, Social Systems, Economic Systems, and others. Of particular value to both the contributors and the readership are the short publication timeframe and the world-wide distribution and exposure which enables both a wide and rapid dissemination of research output.

More information about this series at http://www.springer.com/series/13304

Marcin Szuster · Zenon Hendzel

# Intelligent Optimal Adaptive Control for Mechatronic Systems

 Springer

Marcin Szuster
Department of Applied Mechanics
  and Robotics, Faculty of Mechanical
  Engineering and Aeronautics
Rzeszow University of Technology
Rzeszow
Poland

Zenon Hendzel
Department of Applied Mechanics
  and Robotics, Faculty of Mechanical
  Engineering and Aeronautics
Rzeszow University of Technology
Rzeszow
Poland

ISSN 2198-4182          ISSN 2198-4190   (electronic)
Studies in Systems, Decision and Control
ISBN 978-3-319-88663-3          ISBN 978-3-319-68826-8   (eBook)
https://doi.org/10.1007/978-3-319-68826-8

Printed on acid-free paper

This Springer imprint is published by Springer Nature
The registered company is Springer International Publishing AG
The registered company address is: Gewerbestrasse 11, 6330 Cham, Switzerland

*From* Marcin Szuster

*To my wife Sylwia, for supporting me all the way.*

.

.

.

*From* Zenon Hendzel

*To my grandsons Julian, Jan, Dominik, ...*
*To learn is not to know.*

# Contents

# Chapter 1
# Introduction

From the earliest days, man has desired to create machines that would reflect on the human condition i.e. machines that would demonstrate autonomy, capacity for movement and that would learn and adapt to changing environmental conditions. For many centuries it was not possible for man to build machines and devices that would have at least some of these characteristics. It was not until the 20th century that the rapid development of knowledge in fields such as automatic control, computer science, electronics and manufacturing processes allowed for the construction of robots i.e. machines with complex mechanical structures supplied with appropriate control software that could perform certain tasks previously done by humans. Further progress continued in such areas as robotics and mechatronics i.e. disciplines that are concerned with the mechanics, design, control and operation of robots. Current development of science and technology challenges the scientific community to provide innovative engineering solutions and encourages undertaking research on optimal solutions for mechatronic systems which include, inter alia, wheeled mobile robots and robotic manipulators. The issues relating to mechatronics are of interdisciplinary nature and require the knowledge of multiple disciplines. In particular, wheeled mobile robots are nonlinear, nonholonomic mechatronic systems composed of interacting mechanical, electrical, electronic components and software.

The class of robots that perform behaviors with a high degree of autonomy includes wheeled mobile robots, whose primary task is to execute a collision-free motion in their surrounding environment. Such robots can be used for inspection, patrol, maintenance, search and transportation tasks or for exploration of unknown environments e.g. other planets [4]. The last few years have seen a significant growth of interest in mobile robotics and increased research on wheeled mobile robots and legged robots. This is due to the improvement of mechanical components (sub-assemblies) used for building robots and development of microprocessor-based software systems that enable precise control of complex mechanical systems and reduce design and manufacturing costs of such devices. These factors are driving the increased interest

© Springer International Publishing AG 2018
M. Szuster and Z. Hendzel, *Intelligent Optimal Adaptive Control for Mechatronic Systems*, Studies in Systems, Decision and Control 120, https://doi.org/10.1007/978-3-319-68826-8_1

in personal mobile robots and service robots that have an enormous development potential and offer a wide range of applications. There is demand for mobile devices that could replace humans at tedious, strenuous, dangerous tasks or tasks which pose health risks, and would provide assistance to elderly or disabled persons.

Robotic manipulators are the second class of robots which are used for other applications. They resemble the human arm in structure and in many of its functions. Manipulators are increasingly used to replace humans on repetitive and arduous tasks such as operation of hydraulic presses and production lines. Due to their high precision of movement and repeatability, manipulators are used in advanced mechanical processing of machine parts or assembly operations e.g. assembly of electronic circuits. Not only does the use of robotic manipulators ensure high performance and repeatability of tasks carried out but it also leads to increased efficiency and, in consequence, reduction of production costs. Industrial manipulators are frequently used in high-volume production, particularly in automotive, electro-mechanical and chemical industries, and environments where human exposure to toxic and explosive gases occurs.

The early development of wheeled mobile robots dates back to the 1950s with the establishment of the first fully automated factories in the United States where mobile robots were utilized to automate manufacturing tasks such as material transfer and handling. However, the work on further development of robots was discontinued due to control-related difficulties, in particular, insufficient control techniques. It was not until the 1970s when major progress in automation of transport occurred, driven by growing demand for robotic solutions in electronics, pharmaceutical and automotive industries. The adopted solutions were cost-effective, which spurred interest in robotization of transport operations and increased research efforts. Construction of the first servo-controlled electric-powered teleoperator in the 1940s marked the starting point for development of robotic manipulators [23]. The first programmable robotic arm was created in 1954. This followed years of rapid development of new robotic manipulator designs and expansion of their capabilities and areas of application.

The increased interest in robotics has resulted in a large volume of scientific publications on robot design, kinematics and dynamics, control and trajectory generation.

One of the key elements of a robot is the control system which, together with the sensory system, determines the capabilities and usability of a device. Currently, complex robot control systems are mostly software programs executed by a processing unit linked to a sensory system and actuators of a robot. Given the nature of operations, a complex robot control system uses, inter alia, control algorithms for trajectory generation and execution.

A desired robot trajectory is executed by the tracking control system that generates control signals making the selected point attached to a robot follow the movement of a virtual point according to predefined parameters of motion.

Initially, tracking control systems employed simple controllers that consisted of proportional (P), derivative (D) and integral (I) elements. However, they did not deliver the required control performance with regard to the systems described by nonlinear dynamic equations i.e. robotic manipulators and wheeled mobile robots. Therefore, other types of control systems have been introduced. They consist of the aforementioned controllers and additional algorithms for the generation of control signals, where object dynamics is taken into account. Such control systems comprise robust systems that include parametric uncertainties related to the modeling of controlled object dynamics, and adaptive systems that can adjust the parameters of a control algorithm to variable robot operating conditions.

Over the last twenty years, the rapid development of artificial intelligence methods such as artificial neural networks, fuzzy logic systems and genetic algorithms, coupled with the development of microprocessor technologies, has enabled the design of more complex control systems. Artificial neural networks are the most commonly used controller structures in robotics, as they are adaptable and can be used for any nonlinear mapping. Fuzzy logic systems are also found useful, as they allow for development of control systems for complex objects on the basis of all available scientific information. Control systems that use artificial intelligence methods have the ability to adjust systems' parameters to variable operating conditions of a device and can perform complex tasks, thereby demonstrating a high level of autonomy and machine intelligence. Application of artificial neural networks to control algorithms often involves the use of learning with a teacher methods. The downside of such methods is that the reference signal to be mapped by the network needs to be predefined. Another approach to artificial neural networks involves the recently developed reinforcement learning methods. These methods take inspiration from animal and human behavior, where learning process consists in interaction with the environment and assessment of environment's response to a particular action given a defined objective function. Neural dynamic programming algorithms are examples of reinforcement learning algorithms.

Development of trajectory generation algorithms for mobile robots is also inspired by the study of animal locomotion. Such observations have led to the development of methods for trajectory planning e.g. behavioral control. This approach allows for the design of trajectory generation systems with a simple structure and high degree of autonomy that results from their capacity for adaptation to unknown environmental conditions when carrying out complex tasks. The popularity of these issues is due to the fact that such algorithms have practical applications and are used for a variety of tasks e.g. reaching a target point in an unknown environment, investigation of contaminated areas, exploration of planets, or other human-defined tasks such as mowing, patrolling, area search, cleaning. Artificial intelligence methods, in particular reinforcement learning algorithms, demonstrate considerable potential for application in trajectory generation systems of mobile robots, which is reflected in current research.

Generation and execution of mobile robots' trajectory are among the essential areas of robotics. So far, no universal methods have been devised to solve these highly complex problems. Solutions of the trajectory generation and tracking control problems presented in the scientific publications are merely of academic nature, and only a few have been implemented. Most of the proposed methods for trajectory generation algorithms are offline solutions that have no real applications. Current research efforts focus on development of practical mechatronic methods for motion planning and control of wheeled mobile robots and robotic manipulators.

Despite the rapid development of algorithms for trajectory generation and execution, there are still problems to be addressed. These include the execution of robots' planned motion with regard to the dynamics of a controlled object, and trajectory planning supported with the operation of the sensory system and its application in behavioral control so as to execute complex tasks.

## 1.1   Artificial Intelligence and Neural Networks

The term "artificial intelligence"(AI) was coined by John McCarthy in 1956 [21], to define an interdisciplinary field of science whose primary goal of research is access and incorporation of human intelligence into machines. There are many different definitions of artificial intelligence in the available literature. According to the definition proposed by R.J. Schalkoff artificial intelligence seeks to solve problems by emulating natural behavior and cognitive processes of humans with the use of computer programs [21]. Artificial intelligence methods include, in particular, fuzzy logic systems, evolutionary algorithms and artificial neural networks.

Research into evolutionary algorithms [11, 21] is nature-inspired. In other words, it consists in the imitation of natural processes, where every living organism has its own unique genetic material that stores information about inherited traits. In reproduction, the offspring inherits a complete set of traits encoded in genes. However, in the phase of chromosomal crossover i.e. exchange of genetic material between maternal and paternal chromosomes, mutations may occur which results in the offspring inheriting traits from both parents as well as having individual ones that differ from those found in either parent. Should such set of traits facilitate the organism's adaptation to environmental conditions, it will pass on the genetic material to its offspring. This approach has been used for solving optimization problems, where a population of individuals is identified as a pool of potential solutions to a given problem. The degree of their environmental adaptation (so-called fitness) is measured by means of an adequate fitness function. These individuals exchange their genetic material i.e. chromosomal crossover and mutations occur, and thus generate new solutions that are then selected in a fitness-based process to maximize values of the fitness function. Hence, successive generations of descendants (offspring) offer solutions that converge towards the optimal solution.

Development of fuzzy logic systems was initially undertaken in 1965 with the introduction of fuzzy sets proposed by Lotfi Zadeh [11, 12, 21], which in an ill-defined way associate numerical values with linguistic variables such as "very big", "big", "small". Fuzzy logic operations resemble human reasoning which is rather approximate than precise in nature. The area of mathematics that deals with fuzzy information is called "fuzzy set theory". Its key element is fuzzy logic which is used in fuzzy control and modeling.

Artificial neural network approach seeks alternative solutions to various computational problems through application of knowledge related to the brain structure and brain functions of living organisms [11, 12, 16, 17, 21, 24]. It is an interdisciplinary field linked to automatic control, biocybernetics, electronics, applied mechanics and medicine as well.

The study of nervous system activity of living organisms has contributed to the structure design and development of methods for artificial neural networks, in particular computational techniques, where in a problem-solving process the network's output can be configured through learning, as opposed to an algorithm which is written in form of a program. These methods can be applied to problems where the answer is not known and where traditional algorithms fail due to difficulties and constraints. This is particularly the case for correlation analysis tasks such as diagnostics, classification, forecasting and recognition. Neural networks operate based on parallel distributed processing (connectionism) where a large number of single and simple units (neurons), each of which demonstrates slow activity, perform complex computational tasks as they process data simultaneously. Information processing is spread, thus a partial damage to the structure of network may not have a negative impact on its functioning. Other characteristics of neural networks include its capacity for generalization and approximation of nonlinear mapping, and low sensitivity to errors. That is why these algorithms are used in input-output relations (where relationships are not fully known) and also for solving identification and optimization problems, where input data are inaccurate or partially incorrect. Due to their capabilities, neural networks are commonly applied in different areas of knowledge e.g. robotics, where they are used for the modeling and control of the systems described by nonlinear dynamic equations - such as robots. The simplest single-layer neural network with a single output consists of a weight vector (where weights adjust in the learning process) and a neuron activation function vector. The cross product of these two vectors is the output of the network.

## 1.2 Learning with a Critic

Development of the academic field related to artificial neural networks dates back to 1940s when McCulloch and Pitts [14] published the first works that introduced a systematic description of biological neural networks, which was followed by formulation of the perceptron model. The term "perceptron" was first used in 1957 by Frank Rosenblatt [20] to describe the machine he designed. It was in part an electro-mechanical and electronic device that recognized alphanumeric characters and used an innovative learning algorithm for system programming. The rapid development of neural networks was virtually put on hold for over a decade due to the 1969 publication by Minsky and Papert [15] which highlights the limitations of perceptrons. It was found, however, that these limitations did not concern neural networks with nonlinear activation functions. Moreover, the development of learning methods for multilayer neural networks based on backward propagation of errors, which took place in the early 1980s, led to further dynamic development of neural networks. It was established that learning with a teacher method did not apply to all cases, due to lack of the reference signal. New concepts were developed i.e. unsupervised learning and learning with a critic [1, 3, 16, 22]. Learning with a critic is a kind of supervised learning, which is also called reinforcement learning. It consists in interaction with the environment by taking actions and evaluating the effects of these actions. The main advantage is that there is no requirement for the output values to be presented; it shall be sufficient to determine whether the undertaken action produces relevant results in terms of desired system behavior. Given the action undertaken by the system has a positive effect with regard to the adopted evaluation criterion, a similar behavior tendency for analogous future situations is strengthened (i.e. reinforced), otherwise it is weakened. Learning with a critic requires the use of an adaptive algorithm, referred to as critic, that evaluates the current control strategy. Learning with a critic is more versatile in application than a typical learning with a teacher method, as in the learning with a critic approach the values of desired output signals may be unknown. However, this method requires a properly defined system for evaluating the undertaken actions in terms of a desired input-output behavior. The practical implementation of this neural network learning method is more complex compared to supervised learning. There are many different algorithms based on the learning with a critic method e.g. neural dynamic programming algorithms [1, 5, 6, 18, 19], Q-learning algorithm [1, 3, 7, 9], R-learning [1] and SARSA [1, 3].

## 1.3  Scope of Study

The purpose of this book is to prove the viability of reinforcement learning techniques and in particular the family of techniques known as adaptive dynamic programming, in optimal control of discrete and continuous mechatronic systems. This work discusses the problem of tracking control of dynamic objects such as wheeled mobile robot and robotic manipulator. It also addresses the issue of a defined trajectory generation with regard to a wheeled mobile robot moving in an unknown environment with static obstacles.

   This type of a sequential decision making process may be resolved by applying the theory of dynamic programming. Dynamic programming, which is used in optimal control, is an alternative method to the calculus of variations. It was developed in the 1950s by R. Bellman [2]. This method can be used for solving optimal control problems with regard to nonlinear, non-stationary systems. Despite its elegant mathematical form, the practical applications of dynamic programming are somewhat limited. For this procedure to be applied, a description of the controlled system is required, and with the large dimension of the state vector or the large number of discrete steps "curse of dimensionality" occurs.

   Adaptive dynamic programming method is applied by defining an iterative sequence of value functions and control signals, and is based on the principle of optimality, temporal difference learning, and uses two parametric structures (actor and critic). It comprises an approximate solution to the Hamilton–Jacobi–Bellman equation [13].

The extension of this subject is the differential game theory, construed as an element of decentralized control, widely used for complex mechatronic systems. The application of neural dynamic programming allows for the solution of two-player zero-sum game.

The theory of two-player zero-sum differential games and H∞ control follow from the solution to the Hamilton–Jacobi–Isaacs (HJI) equation [25], which is a generalization of the Hamilton–Jacobi–Bellman equation.

The issue of tracking control of wheeled mobile robots or robotic manipulators is a difficult one, due to the nonlinear character of the equations that describe the dynamics of such devices, whose mathematical model parameters might be unknown and time-varying, depending on the operating conditions.

In mechanics, tracking control relates to solving an inverse dynamics problem in order to calculate the control signal that would compensate nonlinearity of the controlled object. The theory of robust and adaptive systems was used in the synthesis of wheeled mobile robot tracking control algorithms. Stability of the proposed solutions is ensured by Lyapunov stability theory.

The problem of generating collision-free trajectories, in an unknown environment with static obstacles, consists in designing a feasible trajectory of an arbitrary point on a wheeled mobile robot in real-time, based on environment feedback provided by the robot's sensory system. This requires the acquisition of signals from proximity sensors and proper interpretation of information to gauge the distance of obstacles, and subsequently the generation of a feasible trajectory to be executed by a robot that ensures the performance of the adopted task.

The designed tracking control and trajectory generation algorithms were numerically tested in Matlab-Simulink computing environment, with the use of a mobile robot model, robotic manipulator model and a measurement test environment. The experimental verification of algorithms' performance was carried out on lab stations dedicated for motion analysis of a wheeled mobile robot and robotic manipulator.

The station for motion analysis of a wheeled mobile robot consists of the Pioneer 2-DX robot with a power supply and a PC equipped with dSpace DS1102 DSP controller board, dSpace Control Desk experiment management software and Matlab - Simulink computing package. The station for motion analysis of a robotic manipulator consists of the Scorbot-ER 4pc manipulator with a power supply and a PC equipped with dSpace DS1104 DSP controller board, dSpace Control Desk and Matlab-Simulink software. The stations are adapted for scientific research such as rapid prototyping of control systems for dynamic objects and their verification with the use of physical objects controlled in real-time. The control system is represented in form of code, written in Matlab-Simulink computing environment, which is compiled into object code compatible with dSpace Control Desk software that controls the experiment being run on dSpace DSP controller board. Once the experiment is launched, the board generates control signals in real-time, which is dictated by the adopted algorithm. The signals are generated based on the available measurement signals from the controlled object's sensory system. Simultaneously, data acquisition continues. This approach serves as a rapid and convenient research method with regard to the designed algorithms.

In most cases, the process of experimental research on designed algorithms requires adjustments to the values of certain control parameters and further numerical tests once the parameters are tuned.

The quality of motion generated with the proposed tracking control algorithms for a wheeled mobile robot, with the application of neural dynamic programming methods, has been compared with the quality of the motion generated with the use of a PD controller, adaptive control algorithm and neural control system, which are described in publications [8, 10]. Furthermore, this book focuses on tracking control algorithms for the robotic manipulator with the application of selected neural dynamic programming methods. To evaluate the quality of tracking control, performance indexes have been defined based on the tracking errors relating to the desired trajectory.

This book discusses several important aspects of mechatronics. Chapter 2 includes a description of controlled systems i.e. a two-wheeled mobile robot and a robotic manipulator. The kinematics and dynamics of a control object is presented with example solutions to inverse kinematics and inverse dynamics problems. Chapter 3 presents intelligent control methods for nonlinear systems illustrated with an example of neural control algorithm. It also provides basic information on artificial neural networks in terms of the implementation of tracking control algorithms for dynamic systems. Neural networks are further explained in the chapters to follow. Chapter 4 examines optimal control methods for dynamic systems such as Bellman's dynamic programming, linear-quadratic regulator and Pontryagin's maximum principle. It also provides an example of discrete-time linear system control with the use of optimal methods. Chapter 5 discusses learning methods for intelligent systems such as supervised learning, learning with a critic and learning without a teacher. Chapter 6 describes selected neural dynamic programming algorithms, in particular methods requiring knowledge of the mathematical model of a controlled object, such as heuristic dynamic programming algorithm, dual heuristic dynamic programming algorithm and globalised dual heuristic dynamic programming algorithm, and methods that do not require such knowledge i.e. action-dependent heuristic dynamic programming algorithm. Chapter 7 presents the following tracking control algorithms for a wheeled mobile robot: the PD controller, adaptive control algorithm, neural control algorithm, control algorithm that uses neural dynamic programming structures - in the configuration of heuristic dynamic programming, dual-heuristic dynamic programming, globalised dual-heuristic dynamic programming and action-dependent heuristic dynamic programming. This chapter also presents control algorithms for a robotic manipulator such as: the PD controller, control algorithm that uses neural dynamic programming structures - in the configuration of dual-heuristic dynamic programming and globalised dual-heuristic dynamic programming. Descriptions of individual control algorithms are illustrated with numerical test results which are derived from tracking control simulations of a dynamic object. Furthermore, this chapter provides an example of the use of neural dynamic programming algorithms in behavioral control of a wheeled mobile robot. Chapter 8 discusses methods for continuous nonlinear systems control, in particular classic reinforcement learning and its approximation, and reinforcement learning based on the actor-critic structure.

Descriptions of individual control algorithms are supplemented with numerical test results that are based on control simulations of a dynamic object with the use of the presented method. The following chapter focuses on two-player zero-sum differential games and H∞ control. The issues addressed in this book were used for the control of mobile robot drive unit and for the tracking control of a wheeled mobile robot. The results obtained in the simulation tests are provided. Chapter 10 includes a description of the laboratory stations used for the verification tests of control algorithms and discusses the experiments run with the use of the control systems presented in Chap. 7. Chapter 11 presents the summary of findings and conclusions from the conducted research.

# References

1. Barto, A., Sutton, R.: Reinforcement Learning: an Introduction. MIT Press, Cambridge (1998)
2. Bellman, R.: Dynamic Programming. Princeton University Press, New York (1957)
3. Cichosz, P.: Learning Systems. (in Polish) WNT, Warsaw (2000)
4. Fahimi, F.: Autonomous Robots - Modeling, Path Planning, and Control. Springer, New York (2009)
5. Ferrari, S.: Algebraic and Adaptive Learning in Neural Control Systems. Ph.D. Thesis, Princeton University, Princeton (2002)
6. Ferrari, S., Stengel, R.F.: An adaptive critic global controller. In: Proceedings of American Control Conference, vol. 4, pp. 2665–2670. Anchorage, Alaska (2002)
7. Gaskett, C., Wettergreen, D., Zelinsky, A.: Q-learning in continous state and action spaces. Lect. Notes Comput. Sci. **1747**, 417–428 (1999)
8. Giergiel, J., Hendzel, Z., Zylski, W.: Modeling and Control of Wheeled Mobile Robots. (in Polish) Scientific Publishing PWN, Warsaw (2002)
9. Hagen, S., Krose, B.: Neural Q-learning. Neural. Comput. Appl. **12**, 81–88 (2003)
10. Hendzel, Z., Trojnacki, M.: Neural Network Control of Mobile Wheeled Robots. (in Polish) Rzeszow University of Technology Publishing House, Rzeszow (2008)
11. Jamshidi, M., Zilouchian, A.: Intelligent Control Systems Using Soft Computing Methodologies. CRC Press, London (2001)
12. Kecman, V.: Learning and Soft Computing. MIT Press, Cambridge (2001)
13. Lewis, F.L., Vrabie, D., Syroms, V.L.: Optimal Control, 3rd edn. Wiley, New Jersey (2012)
14. McCulloch, W.S., Pitts, W.: A logical calculus of the ideas immanent in nervous activity. Bull. Math. Biophys. **5**, 115–133 (1943)
15. Minsky, M., Papert, S.: Perceptrons: an Introduction to Computational Geometry. MIT Press, Cambridge (1969)
16. Osowski, S.: Neural Networks. (in Polish) Warsaw University of Technology Publishing House, Warsaw (1996)
17. Osowski, S.: Neural Networks - an Algorithmic Approach. (in Polish) WNT, Warsaw (1996)
18. Powell, W.B.: Approximate Dynamic Programming: Solving the Curses of Dimensionality. Willey-Interscience, Princeton (2007)
19. Prokhorov, D., Wunch, D.: Adaptive critic designs. IEEE Trans. Neural Netw. **8**, 997–1007 (1997)
20. Rosenblatt, F.: The perceptron: a probabilistic model for information storage and organization in the brain. Psychol. Rev. **65**, 386–408 (1958)
21. Rutkowski, L.: Methods and Techniques of Artificial Intelligence. (in Polish) Scientific Publishing PWN, Warsaw (2005)

22. Si, J., Barto, A.G., Powell, W.B., Wunsch, D.: Handbook of Learning and Approximate Dynamic Programming. IEEE Press Wiley-Interscience, Hoboken (2004)
23. Spong, M.W., Vidyasagar, M.: Robots Dynamics and Control. (in Polish) WNT, Warsaw (1997)
24. Tadeusiewicz, R.: Neural Networks. (in Polish) AOWRM, Warsaw (1993)
25. Van Der Schaft, A.: $L_2$ - Gain and Passivity Techniques in Nonlinear Control. Spriner, Berlin (1996)

# Chapter 2
# Object of Research

This chapter describes the objects of research i.e. a two-wheeled mobile robot and a robotic manipulator with three degrees of freedom. It presents the structure, components and basic parameters of the controlled systems. The chapter discusses the kinematics of both robots and provides a simulation for solving the inverse kinematics problem. The desired trajectory of a given system was generated, based on a pre-planned path and velocity values of a selected point attached to a robot. The generated trajectories were used for numerical tests and verification analyses of tracking performance with regard to the controlled systems. The chapter discusses also the dynamics of both controlled systems and provides a simulation for solving the inverse dynamics problem for a given trajectory generated earlier. This resulted in waveforms of control signals for a given trajectory.

## 2.1 Two-Wheeled Mobile Robot

Pioneer 2-DX wheeled mobile robot (WMR) is a two-wheeled robot intended for laboratory use. It is equipped with the so-called caster, which is a third self-adjusting supporting wheel, whose dynamics was not taken into consideration, and is assumed to be negligible. Mobile robots with a similar structure are known as two-wheeled robots since they have two drive wheels. It is to be noted that in the last few years new robots have been introduced that have no extra supporting wheel.

The Pioneer 2-DX model [1] is shown in Fig. 2.1a, while its dimensional representation is shown in Fig. 2.1b.

The model is composed of:

- drive wheels 1 and 2,
- caster 3,
- frame 4.

© Springer International Publishing AG 2018
M. Szuster and Z. Hendzel, *Intelligent Optimal Adaptive Control for Mechatronic Systems*, Studies in Systems, Decision and Control 120, https://doi.org/10.1007/978-3-319-68826-8_2

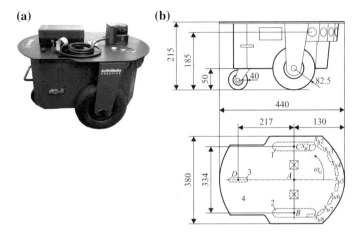

**Fig. 2.1** **a** Pioneer 2-DX wheeled mobile robot. **b** Pioneer 2-DX schematic diagram; main physical dimensions

Pioneer 2's total mass is $m_{DX} = 9$ [kg], and its maximum capacity (payload) is $m_{Lx} = 20$ [kg]. Drive wheels 1 and 2 are made of rubber and their main physical dimensions are: radius $r = r_1 = r_2 = 0.0825$ [m] and width $h_{w1} = h_{w2} = 0.037$ [m].

The maximum velocity of point $A$, which is attached to the Pioneer 2-DX WMR, is $v_{mx} = 1.6$ [m/s]. The WMR is equipped with a sensory system consisting of 8 ultrasonic sensors, affixed to the circumference on the front of the frame. The sensors are denoted by $s_{u1}, \ldots, s_{u8}$ as marked in Fig. 2.1b. In the programming phase, the maximum measuring range of rangefinders was limited to $d_{mx} = 4$ [m]. If the distance between an obstacle and the WMR is greater than the adopted limit, a default value $d_{mx}$ is assumed. The minimum measuring distance from an obstacle is $d_{mn} = 0.4$ [m]. If the distance between an obstacle and the WMR is less than $d_{mn}$, a default value $d_{mn}$ is assumed. Considering the above, the distance measurements made by the ultrasonic sensors fall in the range $d_{si} \in \langle 0.4, 4.0 \rangle$[m], $i = 1, \ldots, 8$. The deviations of ultrasonic sensors' axes from the frame's axis of symmetry are as follows: $\omega_{s1} = 90°$, $\omega_{s2} = 50°$, $\omega_{s3} = 30°$, $\omega_{s4} = 10°$, $\omega_{s5} = -10°$, $\omega_{s6} = -30°$, $\omega_{s7} = -50°$, $\omega_{s8} = -90°$.

An analysis of the WMRs motion is linked to the kinematics and dynamics of these objects which is further discussed in the subchapters below.

### 2.1.1   Description of the Kinematics of a Mobile Robot

The issues related to the kinematics of the WMR include, inter alia, forward and inverse kinematics problems. A problem-solving approach to forward kinematics consists in the determination of the WMR's position and orientation relative to a stationary frame of reference, where the motion parameters of the drive units and the geometry of the system are known. To solve the inverse kinematics problem, the angular parameters of drive wheels rotation need to be determined, given that a predefined path and velocity of a selected point of the system are known. An analysis of the inverse kinematics problem is carried out to determine the trajectory of the WMR that is to be executed by the tracking control system [7, 21].

Selection of a characteristic WMR's point that is to follow the desired path depends on the vehicle's navigation system and the design of the control system. The desired path for a selected WMR's point is represented by rectilinear and curvilinear segments [19] so as to ensure the completion of a defined task. There are many factors that affect the path design such as the type of tasks to be carried out, the structure and technical capabilities of the WMR. Section 7.8 provides a synthesis of the WMR's trajectory generation layer of the hierarchical control system. The layer generates control signals that are necessary to determine the trajectory for the WMR's selected point. This process is carried out on an ongoing basis during the performance of tasks such as "goal seeking with obstacle avoidance", which is a combination of two task types: "goal seeking" and "keep to the center of the empty space/obstacle avoidance".

The kinematics of a selected point of the system is analyzed by means of kinematic equations that may be established through application of different methods e.g. Denavit–Hartenberg convention [19] which employs homogenous coordinates and transformation matrices. Another approach is to use the classic methods applied in mechanics, in particular an analytical description of motion that is based on the parametric equations of motion [21].

In the following considerations it is assumed that all the WMR's components are perfectly rigid and the motion is executed on a flat horizontal surface. The WMR motion is described based on a model (see Fig. 2.2) corresponding to the structure of the Pioneer 2–DX robot used in the verification of solutions that are presented later in this work. The investigated WMR is a nonholonomic system with two degrees

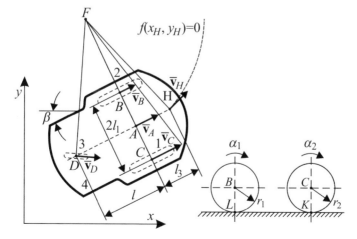

**Fig. 2.2** Schematic diagram of velocities of individual points on wheeled mobile robot's frame

of freedom. Two independent variables were used for motion description i.e. drive wheels 1 and 2 rotation angles, denoted by $\alpha_1$ and $\alpha_2$, respectively. In the case where angular velocity vectors of the WMR drive wheels take the same values, the robot's frame performs translational motion, whereas if angular velocity vectors of individual wheels have different values, the robot's frame is in plane motion (i.e. it moves in the plane (surface) of motion defined by axes $xy$).

In the analysis of the inverse kinematics problem, it was assumed that the selected point $H$ on the WMR moves on a given path with a desired velocity. The execution of trajectory is ensured by defining appropriate angular velocities for the rotation of the WMR drive wheels 1 and 2. It is also assumed that the WMR wheels roll without slipping and the velocity vector of the characteristic point $A$ lies in the plane parallel to the plane of motion and is directed perpendicularly to a line segment bounded by points $B$ and $C$ [7, 21].

In the case where the velocities of points $B$ and $C$ are such that $v_B > v_C$, the WMR's frame is in the plane motion and point $F$ shown in Fig. 2.2 is the instantaneous center of rotation of the frame.

Projections of the WMR's point $A$ velocity vector on stationary coordinate system axes $x$ and $y$ satisfy the relation

$$\dot{y}_A = \dot{x}_A \tan(\beta) \ , \tag{2.1}$$

where $\beta$ – instantaneous angle of rotation of the WMR's frame.

It follows from Eq. (2.1) that point $A$ velocity vector is bounded by nonholonomic constraints. Based on the geometry of the WMR the relation between points $H$ and $A$ in the $xy$ system can be determined

$$x_H = x_A + l_3 \cos(\beta) \ , \tag{2.2}$$

$$y_H = y_A + l_3 \sin(\beta) \ . \tag{2.3}$$

Differentiating the above relations with respect to time, we get

$$\dot{x}_H = \dot{x}_A + l_3 \dot{\beta} \sin(\beta) \ , \tag{2.4}$$

$$\dot{y}_H = \dot{y}_A + l_3 \dot{\beta} \cos(\beta) \ . \tag{2.5}$$

Given that $v_A$ is the value of point $A$ velocity vector, then values of vector projections on the coordinate system axes $xy$ are as follows

$$\dot{x}_A = v_A \cos(\beta) \ , \tag{2.6}$$

$$\dot{y}_A = v_A \sin(\beta) \ . \tag{2.7}$$

Plugging relations (2.6) and (2.7) into the set of Eqs. (2.4) and (2.5), we obtain

$$\dot{x}_H = v_A \cos{(\beta)} + l_3\dot{\beta}\sin{(\beta)} \; , \tag{2.8}$$

$$\dot{y}_H = v_A \sin{(\beta)} + l_3\dot{\beta}\cos{(\beta)} \; . \tag{2.9}$$

It is assumed that point $H$ moves in the $xy$ plane, on a path that is analytically expressed as

$$f(x_H, y_H) = 0 \; , \tag{2.10}$$

where by differentiating (2.10) with respect to time, we get

$$\dot{f}(x_H, y_H) = 0 \; . \tag{2.11}$$

Given the velocity of point $A$ and the desired path of point $H$, the set of Eqs. (2.8), (2.9) and (2.11) allows for the calculation of the change in the values of the following parameters

$$
\begin{aligned}
x_H &= x_H(t) \; , \\
y_H &= y_H(t) \; , \\
\beta &= \beta(t) \; .
\end{aligned} \tag{2.12}
$$

Relations that describe the velocities of points $B$ and $C$, when written as vectors, take the following form

$$
\begin{aligned}
\overline{v}_C &= \overline{v}_A + \overline{v}_{CA} \; , \\
\overline{v}_B &= \overline{v}_A + \overline{v}_{BA} \; ,
\end{aligned} \tag{2.13}
$$

and in the scalar form

$$
\begin{aligned}
v_C &= v_A + v_{CA} \; , \\
v_B &= v_A + v_{BA} \; ,
\end{aligned} \tag{2.14}
$$

where

$$
\begin{aligned}
v_{CA} &= \dot{\beta}l_1 \; , \\
v_{BA} &= -\dot{\beta}l_1 \; .
\end{aligned} \tag{2.15}
$$

It is assumed that the WMR drive wheels roll without slipping, hence the following relations are satisfied

$$
\begin{aligned}
v_B &= \dot{\alpha}_1 r_1 \; , \\
v_C &= \dot{\alpha}_2 r_2 \; ,
\end{aligned} \tag{2.16}
$$

where $r_1 = r_2 = r$ – radiuses of the WMR drive wheels.

By considering Eqs. (2.14)–(2.16), the angular velocities $\dot{\alpha}_1$ and $\dot{\alpha}_2$ of individual WMR drive wheels can be written as

$$
\begin{bmatrix} \dot{\alpha}_1 \\ \dot{\alpha}_2 \end{bmatrix} = \frac{1}{r} \begin{bmatrix} 1 & l_1 \\ 1 & -l_1 \end{bmatrix} \begin{bmatrix} v_A \\ \dot{\beta} \end{bmatrix} \; . \tag{2.17}
$$

The inverse form of relation (2.17) takes the form

$$\begin{bmatrix} v_A \\ \dot{\beta} \end{bmatrix} = \frac{r}{2} \begin{bmatrix} 1 & 1 \\ l_1^{-1} & l_1^{-1} \end{bmatrix} \begin{bmatrix} \dot{\alpha}_1 \\ \dot{\alpha}_2 \end{bmatrix} . \tag{2.18}$$

Kinematic relations described by Eqs. (2.17) and (2.18) are considered later in this work. To solve the inverse kinematics problem the values of the following variables need to be calculated

$$\begin{aligned}
x_H &= x_H(t) , \\
y_H &= y_H(t) , \\
\beta &= \beta(t) , \\
\dot{\alpha}_1 &= \dot{\alpha}_1(t) , \\
\dot{\alpha}_2 &= \dot{\alpha}_2(t) .
\end{aligned} \tag{2.19}$$

by means of Eqs. (2.8), (2.9), (2.11) and (2.19), taking into account the desired path for point $H$ and the velocity of point $A$ $v_A$.

Given that distance $l_3 = 0$, two trajectories of point $A$ were generated relative to the desired paths. The first being a loop-shaped path with loop radius $R = 0.75$ [m], and the second being an 8-shaped path consisting of two combined loops, each of radius $R = 0.75$ [m].

### 2.1.1.1  Simulation of Loop-Shaped Path Inverse Kinematics

The WMR's point $A$ motion on the desired loop-shaped path, from the initial position at point $S$ to the desired final position at point $G$, is divided into five characteristic phases [7]:

(a) Start-up (motion on a rectilinear path):

$$v_A = \frac{v_A^*}{t_r - t_p} (t - t_p) , \quad t_p \leq t < t_r , \quad \dot{\alpha}_1 = \dot{\alpha}_2 = \frac{v_A}{r} , \quad \dot{\beta} = 0 ,$$

where $v_A$ – linear velocity of point $A$, $t$ – time, $v_A^*$ – maximum desired linear velocity of point $A$, $t_p$ – (motion) initial time, $t_r$ – end-time of start-up, $\dot{\alpha}_1, \dot{\alpha}_2$ – desired angular velocities of the WMR drive wheels rotation, $r$ – drive wheel radius, $\dot{\beta}$ – instantaneous angular velocity of the WMRs frame rotation.

(b) Motion with the desired velocity, when $v_A = v_A^* = \text{const.}$:

$$\dot{\alpha}_1 = \dot{\alpha}_2 = \frac{v_A}{r} , \quad t_r \leq t < t_{c1} , \quad \dot{\beta} = 0 ,$$

where $t_{c1}$ – start-time of curvilinear motion.

(c) Motion in a circular path with radius $R$, when $v_A = v_A^* = \text{const.}$, $R = 0.75$ [m]:

$$\dot{\alpha}_1 = \frac{v_A}{r} + l_1\dot{\beta}, \quad \dot{\alpha}_2 = \frac{v_A}{r} - l_1\dot{\beta}, \quad t_{c1} \leq t < t_{c2},$$

where $t_{c2}$ – end-time of curvilinear motion, $l_1$ – length that results from the WMR's geometry.

(d) Curve exit including the transition period, followed by motion on a rectilinear path with constant velocity, when $v_A = v_A^* = \text{const.}$:

$$\dot{\alpha}_1 = \dot{\alpha}_{p1} - \left(\dot{\alpha}_{p1} - \frac{v_A}{r}\right)\left(1 - e^{-\varsigma t}\right), \quad \dot{\alpha}_2 = \dot{\alpha}_{p2} - \left(\frac{v_A}{r} - \dot{\alpha}_{p2}\right)\left(1 - e^{-\varsigma t}\right),$$
$$t_{c2} \leq t < t_h,$$

where $t_h$ – braking start-time, $\varsigma$ – transition curves approximation constant, $\dot{\alpha}_{p1}$, $\dot{\alpha}_{p2}$ – values of wheels' angular velocities at the beginning of the transition period. Introducing approximation allows for the execution of system's motion with smooth changes being applied to parameters such as velocity and acceleration.

(e) Braking:

$$v_A = v_A^* - \frac{v_A^*}{t_k - t_h}(t - t_h), \quad t_h \leq t < t_k, \quad \dot{\alpha}_1 = \dot{\alpha}_2 = \frac{v_A}{r}, \quad \dot{\beta} = 0,$$

where $t_k$ – end-time.

Equal lengths of time were adopted for the braking and the start-up phase. The start-time values for the remaining phases are respectively: $t_p = 3$ [s], $t_r = 5$ [s], $t_{c1} = 7.5$ [s], $t_{c2} = 20.5$ [s], $t_h = 23$ [s], $t_k = 25$ [s]. The maximum linear velocity value of point $A$ on the WMR is $v_A^* = 0.4$ [m/s]. To eliminate discontinuities in angular accelerations of trajectory with a triangular velocity profile, an approximation of the velocity profile was generated by means of the following relation

$$v_A = v_A^* \cdot \frac{1}{\left(1 + \exp^{-c_1(t-t_1)}\right)\left(1 + \exp^{c_1(t-t_2)}\right)}, \quad (2.20)$$

where $c_1 = 12$ [1/s] – sigmoid functions slope coefficients, $t_1$ – motion mean start-time, $t_1 = 0.5(t_r + t_p) = 4$ [s], $t_2$ – braking mean start-time, $t_2 = 0.5(t_k + t_h) = 24$ [s]. The diagrams of the assumed triangular velocity profile and its approximation are shown in Fig. 2.3a, b, respectively.

Based on the desired velocity profile (Fig. 2.3b), the desired motion path of point $A$ on the WMR (Fig. 2.4a) and the geometry of the WMR, the inverse kinematics problem was solved, thereby producing angular variables related to the WMR drive wheels rotation. Figure 2.4b shows the diagram of rotation angles $\alpha_1$ and $\alpha_2$ [rad] of the WMR drive wheels, Fig. 2.4c presents the angular velocities $\dot{\alpha}_1$ and $\dot{\alpha}_2$ [rad/s], whereas in Fig. 2.4d the angular accelerations $\ddot{\alpha}_1$ and $\ddot{\alpha}_2$ [rad/s$^2$] are presented.

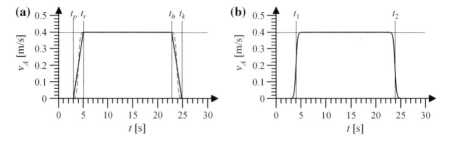

**Fig. 2.3** **a** WMR's point A triangular velocity profile, **b** approximation of point A triangular velocity profile

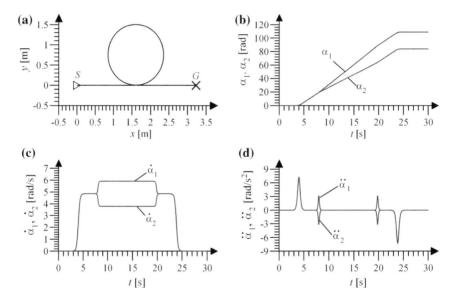

**Fig. 2.4** **a** WMR's point A desired motion path, **b** WMR drive wheels rotation angles $\alpha_1$ and $\alpha_2$, **c** angular velocities $\dot{\alpha}_1$ and $\dot{\alpha}_2$, **d** angular accelerations $\ddot{\alpha}_1$ and $\ddot{\alpha}_2$

### 2.1.1.2 Simulation of 8-Shaped Path Inverse Kinematics

The WMR's point A motion on the desired 8-shaped path, from the initial position at point S to the desired final position at point G, is divided into two characteristic phases (I and II), each of which consists in a "drive" on a loop-shaped path. Motion phases are separated by a stop at point P. Each phase consists of nine characteristic sub-phases of motion:

(a) Start-up (motion in a rectilinear path):

$$v_A = \frac{v_A^*}{t_{r1} - t_{p1}} \left( t - t_{p1} \right) , \quad t_{p1} \le t < t_{r1} , \quad \dot{\alpha}_1 = \dot{\alpha}_2 = \frac{v_A}{r} , \quad \dot{\beta} = 0 ,$$

where $t_{p1}$ – initial time of motion in the first phase of motion, $t_{r1}$ – start-up end-time in the first phase of motion.

(b) Motion with the desired velocity, when $v_A = v_A^* = \text{const.}$:

$$\dot\alpha_1 = \dot\alpha_2 = \frac{v_A}{r} , \quad t_{r1} \le t < t_{c1} , \quad \dot\beta = 0 ,$$

where $t_{c1}$ – start-time of the curvilinear motion in the first phase of motion.

(c) Motion along a circular path with radius $R$, when $v_A = v_A^* = \text{const.}, R = 0.75 \, [\text{m}]$:

$$\dot\alpha_1 = \frac{v_A}{r} + l_1\dot\beta , \quad \dot\alpha_2 = \frac{v_A}{r} - l_1\dot\beta , \quad t_{c1} \le t < t_{a1} ,$$

where $t_{a1}$ – start-time of decrease in velocity (deceleration) in the curvilinear motion.

(d) Deceleration in the motion along a circular path with radius $R$:

$$\dot\alpha_1 = \frac{v_A^* - \left(v_A^* - v_A^{**}\right) \cdot (t - t_{a2})}{r \, (t_{a2} - t_{a1})} + l_1\dot\beta , \quad \dot\alpha_2 = \frac{v_A^* - \left(v_A^* - v_A^{**}\right) \cdot (t - t_{a2})}{r \, (t_{a2} - t_{a1})} - l_1\dot\beta ,$$

$$t_{a1} \le t < t_{a2} ,$$

where $t_{a2}$ – end-time of deceleration in the curvilinear motion, $v_A^{**}$ – minimum desired linear velocity of the selected point $A$ on the WMR.

(e) Motion along a circular path with radius $R$, when $v_A = v_A^{**} = \text{const.}, R = 0.75 \, [\text{m}]$:

$$\dot\alpha_1 = \frac{v_A}{r} + l_1\dot\beta , \quad \dot\alpha_2 = \frac{v_A}{r} - l_1\dot\beta , \quad t_{a2} \le t < t_{a3} ,$$

where $t_{a3}$ – start-time of increase in velocity (acceleration) in the curvilinear motion.

(f) Acceleration in the motion on a circular path with radius $R$:

$$\dot\alpha_1 = \frac{v_A^* - \left(v_A^* - v_A^{**}\right) \cdot (t_{a3} - t)}{r \, (t_{a3} - t_{a4})} + l_1\dot\beta , \quad \dot\alpha_2 = \frac{v_A^* - \left(v_A^* - v_A^{**}\right) \cdot (t_{a3} - t)}{r \, (t_{a3} - t_{a4})} - l_1\dot\beta ,$$

$$t_{a3} \le t < t_{a4} ,$$

where $t_{a4}$ – end-time of acceleration in the curvilinear motion.

(g) Motion on a circular path with radius $R$, when $v_A = v_A^* = \text{const.}, R = 0.75 \, [\text{m}]$:

$$\dot\alpha_1 = \frac{v_A}{r} + l_1\dot\beta , \quad \dot\alpha_2 = \frac{v_A}{r} - l_1\dot\beta , \quad t_{a4} \le t < t_{c2} .$$

where $t_{c2}$ – end-time of the curvilinear motion in the first phase of motion.

(h) Curve exit including the transition period, followed by motion on a rectilinear path with constant velocity, when $v_A = v_A^* = \text{const.}$:

$$\dot\alpha_1 = \dot\alpha_{p1} - \left(\dot\alpha_{p1} - \frac{v_A}{r}\right)\left(1 - e^{-\varsigma t}\right) , \quad \dot\alpha_2 = \dot\alpha_{p2} - \left(\frac{v_A}{r} - \dot\alpha_{p2}\right)\left(1 - e^{-\varsigma t}\right) ,$$

$$t_{c2} \le t < t_{h1} ,$$

where $t_{h1}$ – braking start-time in the first phase of motion, $\varsigma$ – transition curves approximation constant, $\dot{\alpha}_{p1}, \dot{\alpha}_{p2}$ – values of wheels' angular velocities at the beginning of the transition period in the first phase of motion.

(i) Braking:

$$v_A = v_A^* - \frac{v_A^*}{t_{k1} - t_{h1}} (t - t_{h1}) , \quad t_{h1} \leq t < t_{k1} , \quad \dot{\alpha}_1 = \dot{\alpha}_2 = \frac{v_A}{r} , \quad \dot{\beta} = 0 .$$

The second characteristic period of the WMR's point $A$ motion on a loop-shaped path consists of identical motion phases.

Equal lengths of time were adopted for the braking and the start-up phase, equal lengths of time were assumed for the decrease and increase in velocities in the curvilinear motion phase. The start-time values for the subsequent phases are respectively: $t_{p1} = 3$ [s], $t_{r1} = 5$ [s], $t_{c1} = 7.5$ [s], $t_{a1} = 10$ [s], $t_{a2} = 11$ [s], $t_{a3} = 13.85$ [s], $t_{a4} = 14.85$ [s], $t_{c2} = 18$ [s], $t_{h1} = 18.4$ [s], $t_{k1} = 20.4$ [s], $t_{p2} = 24.6$ [s], $t_{r1} = 26.6$ [s], $t_{c3} = 22.2$ [s], $t_{b1} = 30.1$ [s], $t_{b2} = 31.1$ [s], $t_{b3} = 34.1$ [s], $t_{b4} = 35.1$ [s], $t_{c4} = 32.8$ [s], $t_{h2} = 40$ [s], $t_{k2} = 42$ [s]. The maximum velocity value of point $A$ on the WMR is $v_A^* = 0.4$ [m/s], whereas the minimum is $v_A^{**} = 0.3$ [m/s]. To eliminate the discontinuities in angular accelerations of trajectory with a triangular velocity profile, an approximation of the velocity profile was generated by means of the following relation

$$v_A = v_A^* \cdot \sum_{m=0}^{1} \frac{1}{\left(1 + exp^{-c_1(t-t_{1+4m})}\right)\left(1 + exp^{c_1(t-t_{4+4m})}\right)} +$$
$$- \left(v_A^* - v_A^{**}\right) \cdot \sum_{m=0}^{1} \frac{1}{\left(1 + exp^{-c_2(t-t_{2+4m})}\right)\left(1 + exp^{c_2(t-t_{3+4m})}\right)} , \quad (2.21)$$

where $c_1 = 6$ [1/s], $c_2 = 12$ [1/s] – sigmoid functions slope coefficients, $t_1$ – start-up's mean start-time in the first phase of motion, $t_1 = 0.5\left(t_{r1} + t_{p1}\right) = 4$ [s], $t_2$ – mean start-time of deceleration in the curvilinear motion in the first phase of motion, $t_2 = 0.5(t_{a1} + t_{a2}) = 10.5$ [s], $t_3$ – mean start-time of acceleration in the curvilinear motion, $t_3 = 0.5(t_{a3} + t_{a4}) = 14.35$ [s], $t_4$ – braking mean start-time, $t_4 = 0.5(t_{k1} + t_{h1}) = 19.4$ [s], $t_5$ – start-up's mean start-time in the second phase of motion, $t_5 = 0.5\left(t_{r2} + t_{p2}\right) = 25.6$ [s], $t_6$ – mean start-time of deceleration in the curvilinear motion in the second phase of motion, $t_6 = 0.5(t_{b1} + t_{b2}) = 30.6$ [s], $t_7$ – mean start-time of acceleration in the curvilinear motion, $t_7 = 0.5(t_{b3} + t_{b4}) = 33.6$ [s], $t_8$ – braking mean start-time, $t_8 = 0.5(t_{k2} + t_{h2}) = 41$ [s]. The diagrams of the assumed triangular velocity profile and its approximation are shown in Fig. 2.5a, b, respectively.

Based on the desired velocity profile (Fig. 2.5b), the desired motion path of point $A$ on WMR (Fig. 2.6a) and the geometry of the Pioneer 2-DX, the inverse kinematics problem was solved, thereby producing angular variables of the WMR drive wheels rotation. Figure 2.6b shows the diagram of rotation angles $\alpha_1$ and $\alpha_2$ [rad] of the WMR

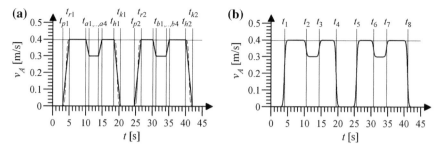

**Fig. 2.5** **a** WMR point $A$ triangular velocity profile, **b** approximation of point $A$ triangular velocity profile

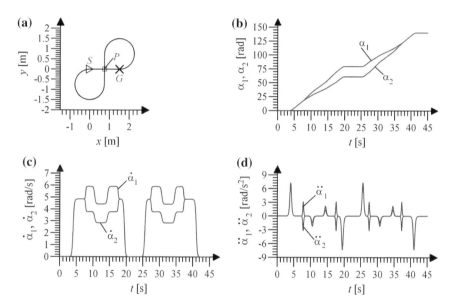

**Fig. 2.6** **a** WMR point $A$ desired motion path, **b** WMR drive wheels rotation angles $\alpha_1$ and $\alpha_2$, **c** angular velocities $\dot{\alpha}_1$ and $\dot{\alpha}_2$, **b** angular accelerations $\ddot{\alpha}_1$ and $\ddot{\alpha}_2$

drive wheels, Fig. 2.6c presents the angular velocities $\dot{\alpha}_1$ and $\dot{\alpha}_2$ [rad/s], whereas in Fig. 2.6d the angular accelerations $\ddot{\alpha}_1$ and $\ddot{\alpha}_2$ [rad/s$^2$] are presented.

The generated solutions to the inverse kinematics problem were applied in the quality analysis of tracking performance with regard to particular control systems, as the WMR's desired trajectory.

## 2.1.2 Description of the Dynamics of a Mobile Robot

The analysis of issues related to the WMR's dynamics consists in seeking correlations between the system's parameters of motion and the mechanisms that cause the motion. The forward dynamics problem is to determine the system's parameters of motion resulting from the generalized forces, whose values are known, whereas the inverse dynamics problem is solved by determining the values of the generalized forces that occur during the WMR's motion along the desired trajectory, where the values of the parameters of motion are known [21].

The analysis of the WMR's dynamics results in a mathematical model represented by ordinary differential equations that may be used for further analysis, identification of parameters and synthesis of control systems [11]. The model that describes the WMR's dynamics provides merely an approximation of real system's behavior and does not take into account every system-related event, due to the computational complexity and difficulty in measuring certain quantities.

The WMRs are complex mechanical systems with nonholonomic constraints and a nonlinear dynamic description. The dynamics of such systems can be described by means of Lagrange equations with multipliers [7, 21] to obtain dynamic equations of motion with the right-hand side in implicit form. Once reduced to equations of motion with nonholonomic constraints, they allow for the solution of the WMR's forward and inverse dynamics problems. The use of Lagrange equations with multipliers to describe the WMR's dynamics allows to determine the values of friction forces acting at the contact point between the WMR drive wheels and the surface, and thereby to determine whether the desired trajectory can be executed in the rolling without slipping motion. The disadvantage of this approach is in the computational complexity and difficulty in determining the values of Lagrange coefficients, therefore a transformation is applied that allows for decoupling multipliers from the driving moments. The WMR's dynamics may be also described by means of Maggi's mathematical formalism [7, 8, 10]. Thus obtained model, which has a convenient computational form, allows for the calculation of driving moments values for the WMR drive wheels.

Maggi's equations were selected to describe the WMR's dynamics. The motion of dynamic systems expressed by means of Maggi's equations in generalized coordinates is defined as follows

$$\sum_{j=1}^{h} C_{i,j} \left[ \frac{d}{dt} \left( \frac{\partial E}{\partial \dot{q}_j} \right) - \frac{\partial E}{\partial q_j} \right] = \Theta_i \,, \tag{2.22}$$

where $i = 1, \ldots, g$, $g$ – number of independent parameters of the system in generalized coordinates $q_j$, equal to the number of system's degrees of freedom, $j = 1, \ldots, h$, $h$ – number of generalized coordinates, $E = E(\dot{\mathbf{q}})$ – kinetic energy of the system, $C_{i,j}$ – coefficients. The diagram of the WMR is shown in Fig. 2.7.

All generalized velocities are written in the following form

$$\dot{q}_j = \sum_{i=1}^{g} C_{i,j} \dot{e}_i + G_j \,, \tag{2.23}$$

where $\dot{e}_i$ – kinetic parameters of the system in generalized coordinates, $G_j$ – coefficients. The right hand-sides of Eq. (2.22) are the coefficients of variations $\delta \dot{e}_i$

**Fig. 2.7** Diagram of the WMR

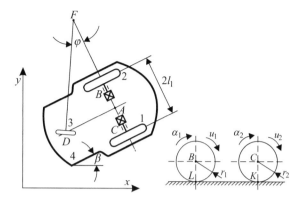

in expressions for virtual work of system's external forces. These coefficients are defined as follows [10]

$$\sum_{i=1}^{g} \Theta_i \delta e_i = \sum_{i=1}^{g} \delta e_i \sum_{j=1}^{h} C_{i,j} Q_{Fj} , \qquad (2.24)$$

where $\mathbf{Q}_F$ – vector of generalized forces, given the driving moments of the WMR wheels 1 and 2, and rolling resistance. Thus formulated Maggi's equations are used to describe the WMR point $A$ motion. It is assumed that robot's motion is executed in a single plane. In order to clearly determine its position and orientation it is necessary to provide the position of point $A$ i.e. coordinates $x_A$, $y_A$ related to the orientation of the frame defined by the instantaneous angle of rotation $\beta$, as well as the rotation angles of drive wheels 1 and 2, denoted by $\alpha_1$ and $\alpha_2$, respectively. Vectors of generalized coordinates and generalized velocities are as follows

$$\mathbf{q} = [x_A, y_A, \beta, \alpha_1, \alpha_2]^T , \qquad (2.25)$$

$$\dot{\mathbf{q}} = \left[\dot{x}_A, \dot{y}_A, \dot{\beta}, \dot{\alpha}_1, \dot{\alpha}_2\right]^T . \qquad (2.26)$$

In the analysis of the values of the external forces acting on the examined dynamic system, it is essential to include the dry friction forces which occur in the tangent plane of contact between the drive wheels and the surface. These forces are shown in Fig. 2.8, where $\overline{\mathbf{T}}_{1\tau}$, $\overline{\mathbf{T}}_{2\tau}$ – circumferential dry friction forces, $\overline{\mathbf{T}}_{1n}$, $\overline{\mathbf{T}}_{2n}$ – transverse dry friction forces [7].

The considered WMR is a system with two degrees of freedom. The movement of WMR is analyzed in the $xy$ plane. Given that independent coordinates are represented by drive wheels rotation angles $\alpha_1$ and $\alpha_2$, then the arrangement of velocity vectors with regard to points $A$, $B$ and $C$ implies the following equations for velocities

**Fig. 2.8** Friction forces
acting on the WMR drive
wheels

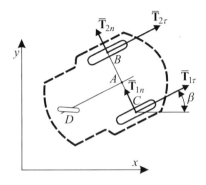

$$\dot{x}_A = \frac{r}{2} (\dot{\alpha}_1 + \dot{\alpha}_2) \cos (\beta) \ ,$$

$$\dot{y}_A = \frac{r}{2} (\dot{\alpha}_1 + \dot{\alpha}_2) \sin (\beta) \ , \qquad (2.27)$$

$$\dot{\beta} = \frac{r}{2l_1} (\dot{\alpha}_1 - \dot{\alpha}_2) \ .$$

Based on relation (2.23) generalized velocities are written as follows

$$\dot{q}_1 = \dot{x}_A = \frac{r}{2} (\dot{\alpha}_1 + \dot{\alpha}_2) \cos (\beta) = \frac{r}{2} (\dot{e}_1 + \dot{e}_2) \cos (\beta) = C_{1,1}\dot{e}_1 + C_{2,1}\dot{e}_2 + G_1 \ ,$$

$$\dot{q}_2 = \dot{y}_A = \frac{r}{2} (\dot{\alpha}_1 + \dot{\alpha}_2) \sin (\beta) = \frac{r}{2} (\dot{e}_1 + \dot{e}_2) \sin (\beta) = C_{1,2}\dot{e}_1 + C_{2,2}\dot{e}_2 + G_2 \ ,$$

$$\dot{q}_3 = \dot{\beta} = \frac{r}{2l_1} (\dot{\alpha}_1 - \dot{\alpha}_2) = \frac{r}{2l_1} (\dot{e}_1 - \dot{e}_2) = C_{1,3}\dot{e}_1 + C_{2,3}\dot{e}_2 + G_3 \ ,$$

$$\dot{q}_4 = \dot{\alpha}_1 = \dot{e}_1 = C_{1,4}\dot{e}_1 + C_{2,4}\dot{e}_2 + G_4 \ ,$$

$$\dot{q}_5 = \dot{\alpha}_2 = \dot{e}_2 = C_{1,5}\dot{e}_1 + C_{2,5}\dot{e}_2 + G_5 \ ,$$

$$(2.28)$$

where coefficients $C_{i,j}$ and $G_j$ have the following values

$$\left.\begin{array}{lll} C_{1,1} = \dfrac{r}{2} \cos (\beta) \ , & C_{2,1} = \dfrac{r}{2} \cos (\beta) \ , & G_1 = 0 \ , \\[2mm] C_{1,2} = \dfrac{r}{2} \sin (\beta) \ , & C_{2,2} = \dfrac{r}{2} \sin (\beta) \ , & G_2 = 0 \ , \\[2mm] C_{1,3} = \dfrac{r}{2l_1} \ , & C_{2,3} = -\dfrac{r}{2l_1} \ , & G_3 = 0 \ , \\[2mm] C_{1,4} = 1 \ , & C_{2,4} = 0 \ , & G_4 = 0 \ , \\[2mm] C_{1,5} = 0 \ , & C_{2,5} = 1 \ , & G_5 = 0 \ . \end{array}\right\} \qquad (2.29)$$

Given the generalized forces described by relation (2.24), and the values of coefficients $C_{i,j}$ (2.29), the WMR's dynamic equations can be written as

$$\Theta_1 = C_{1,1} Q_{F1} + C_{1,2} Q_{F2} + C_{1,3} Q_{F3} + C_{1,4} Q_{F4} + C_{1,5} Q_{F5} = u_1 - N_1 f_1 \ ,$$

$$\Theta_2 = C_{2,1} Q_{F1} + C_{2,2} Q_{F2} + C_{2,3} Q_{F3} + C_{2,4} Q_{F4} + C_{2,5} Q_{F5} = u_2 - N_2 f_2 \ .$$

$$(2.30)$$

In the present case, Maggi's equations take the following form

$$
\begin{aligned}
&(2m_1 + m_4)\left(\frac{r}{2}\right)^2 (\ddot{\alpha}_1 + \ddot{\alpha}_2) + 2m_4\left(\frac{r}{2l_1}\right)^2 rl_2\,(\dot{\alpha}_2 - \dot{\alpha}_1)\,\dot{\alpha}_2 + I_{z1}\ddot{\alpha}_1 + \\
&+ \left(2m_1 l_1^2 + m_4 l_2^2 + 2I_{x1} + I_{z4}\right)\frac{r}{2l_1}(\ddot{\alpha}_1 - \ddot{\alpha}_2) = u_1 - N_1 f_1\,, \\
&(2m_1 + m_4)\left(\frac{r}{2}\right)^2 (\ddot{\alpha}_1 + \ddot{\alpha}_2) + 2m_4\left(\frac{r}{2l_1}\right)^2 rl_2\,(\dot{\alpha}_1 - \dot{\alpha}_2)\,\dot{\alpha}_1 + I_{z2}\ddot{\alpha}_2 + \\
&- \left(2m_1 l_1^2 + m_4 l_2^2 + 2I_{x1} + I_{z4}\right)\frac{r}{2l_1}(\ddot{\alpha}_1 - \ddot{\alpha}_2) = u_2 - N_2 f_2\,,
\end{aligned}
\tag{2.31}
$$

where $m_1 = m_2$ – substitute masses of WMR drive wheels 1 and 2,
$m_4$ – substitute mass of the WMR's frame,
$I_{x1} = I_{x2}$ – substitute mass moments of inertia of drive wheels 1 and 2 that are determined relative to axes $x_1$ and $x_2$, which relate to the drive wheels, respectively,
$I_{z1} = I_{z2}$ – substitute mass moments of inertia of drive wheels 1 and 2 relative to drive wheels rotation axes,
$I_{z4}$ – substitute mass moment of inertia of the WMR's frame relative to axis $z_4$ that relates to the frame, assuming the axes of the reference frame bounded with a given WMR's part are the central principal axes of inertia,
$N_1$, $N_2$ – normal forces relative to wheels 1 and 2,
$f_1$, $f_2$ – rolling friction coefficients relative to respective WMR wheels,
$u_1$, $u_2$ – driving moments (control signals),
$l, l_1, l_2, h_1$ – respective lengths resulting from the geometry of the system,
$r_1 = r_2 = r$ – radiuses of respective wheels.

Maggi's equations (2.31) were applied to describe the WMR's dynamics. Their vector/matrix form is shown below [7, 8].

$$
\begin{aligned}
&\begin{bmatrix}
(2m_1 + m_4)\left(\frac{r}{2}\right)^2 + I_{z1} + \\
+ \left(2m_1 l_1^2 + m_4 l_2^2 + 2I_{x1} + I_{z4}\right)\frac{r}{2l_1}
& (2m_1 + m_4)\left(\frac{r}{2}\right)^2 + \\
- \left(2m_1 l_1^2 + m_4 l_2^2 + 2I_{x1} + I_{z4}\right)\frac{r}{2l_1}
\\[2ex]
(2m_1 + m_4)\left(\frac{r}{2}\right)^2 + \\
- \left(2m_1 l_1^2 + m_4 l_2^2 + 2I_{x1} + I_{z4}\right)\frac{r}{2l_1}
& (2m_1 + m_4)\left(\frac{r}{2}\right)^2 + I_{z2} + \\
+ \left(2m_1 l_1^2 + m_4 l_2^2 + 2I_{x1} + I_{z4}\right)\frac{r}{2l_1}
\end{bmatrix}
\begin{bmatrix} \ddot{\alpha}_1 \\ \ddot{\alpha}_2 \end{bmatrix} \\[3ex]
&+ \begin{bmatrix}
0, & 2m_4\left(\frac{r}{2l_1}\right)^2 rl_2\,(\dot{\alpha}_2 - \dot{\alpha}_1) \\[2ex]
2m_4\left(\frac{r}{2l_1}\right)^2 rl_2\,(\dot{\alpha}_1 - \dot{\alpha}_2)\,, & 0
\end{bmatrix}
\begin{bmatrix} \dot{\alpha}_1 \\ \dot{\alpha}_2 \end{bmatrix} \\[3ex]
&+ \begin{bmatrix} N_1 f_1 \operatorname{sgn}(\dot{\alpha}_1) \\ N_2 f_2 \operatorname{sgn}(\dot{\alpha}_2) \end{bmatrix}
= \begin{bmatrix} u_1 \\ u_2 \end{bmatrix}.
\end{aligned}
\tag{2.32}
$$

The following parameters can be introduced to simplify the WMR's dynamic equations

$$a_1 = (2m_1 + m_4)\left(\frac{r}{2}\right)^2 ,$$

$$a_2 = \left(2m_1 l_1^2 + m_4 l_2^2 + 2I_{x1} + I_{z4}\right)\frac{r}{2l_1} ,$$

$$a_3 = I_{z1} = I_{z2} ,$$

$$a_4 = 2m_4\left(\frac{r}{2l_1}\right)^2 rl_2 ,$$   \qquad (2.33)

$$a_5 = N_1 f_1 ,$$

$$a_6 = N_2 f_2 ,$$

given the above, relation (2.32) can be written as

$$\mathbf{M}\ddot{\boldsymbol{\alpha}} + \mathbf{C}(\dot{\boldsymbol{\alpha}})\dot{\boldsymbol{\alpha}} + \mathbf{F}(\dot{\boldsymbol{\alpha}}) + \boldsymbol{\tau}_d(t) = \mathbf{u} , \qquad (2.34)$$

where $\mathbf{M}$ – inertia matrix of the WMR, $\mathbf{C}(\dot{\boldsymbol{\alpha}})\dot{\boldsymbol{\alpha}}$ – vector of moments of the centrifugal and Coriolis forces, $\mathbf{F}(\dot{\boldsymbol{\alpha}})$ – friction vector, $\boldsymbol{\tau}_d(t) = [\tau_{d1}(t), \tau_{d2}(t)]^T$ – vector of bounded disturbances - that also relates to the dynamics of the WMR's elements that are not modeled, $\mathbf{u}$ – control vector - that includes control signals in the form of driving moments of respective drive units, $\dot{\boldsymbol{\alpha}} = [\dot{\alpha}_1, \dot{\alpha}_2]^T$ – angular velocities vector of the WMR drive wheels rotation.

Matrices $\mathbf{M}$, $\mathbf{C}(\dot{\boldsymbol{\alpha}})$, and vector $\mathbf{F}(\dot{\boldsymbol{\alpha}})$ take the form

$$\mathbf{M} = \begin{bmatrix} a_1 + a_2 + a_3 , & a_1 - a_2 \\ a_1 - a_2 , & a_1 + a_2 + a_3 \end{bmatrix} ,$$

$$\mathbf{C}(\dot{\boldsymbol{\alpha}}) = \begin{bmatrix} 0 , & a_4(\dot{\alpha}_2 - \dot{\alpha}_1) \\ -a_4(\dot{\alpha}_1 - \dot{\alpha}_2) , & 0 \end{bmatrix} , \qquad (2.35)$$

$$\mathbf{F}(\dot{\boldsymbol{\alpha}}) = \begin{bmatrix} a_5 \mathrm{sgn}(\dot{\alpha}_1) \\ a_6 \mathrm{sgn}(\dot{\alpha}_2) \end{bmatrix} .$$

Maggi's equations make it possible to circumvent the procedure of decoupling multipliers in Lagrange equations, which is time-consuming for complex dynamic systems. Thus obtained relations allow for the solution of the forward and inverse dynamics problems of the WMR's dynamics.

Equation (2.34) may be written in a linear form with respect to parameters $\mathbf{a}$

$$\mathbf{Y}(\dot{\boldsymbol{\alpha}}, \ddot{\boldsymbol{\alpha}})^T \mathbf{a} + \boldsymbol{\tau}_d(t) = \mathbf{u} , \qquad (2.36)$$

where $\mathbf{Y}(\dot{\boldsymbol{\alpha}}, \ddot{\boldsymbol{\alpha}})$ – so-called regression matrix that takes the form

$$\mathbf{Y}(\dot{\boldsymbol{\alpha}}, \ddot{\boldsymbol{\alpha}}) = \begin{bmatrix} \ddot{\alpha}_1 + \ddot{\alpha}_2 , & \ddot{\alpha}_1 , & \ddot{\alpha}_1 - \ddot{\alpha}_2 , & \dot{\alpha}_2^2 - \dot{\alpha}_1\dot{\alpha}_2 , & \mathrm{sgn}(\dot{\alpha}_1) , & 0 \\ \ddot{\alpha}_1 + \ddot{\alpha}_2 , & \ddot{\alpha}_2 , & \ddot{\alpha}_2 - \ddot{\alpha}_1 , & \dot{\alpha}_1^2 - \dot{\alpha}_1\dot{\alpha}_2 , & 0 , & \mathrm{sgn}(\dot{\alpha}_2) \end{bmatrix}^T .$$
$$(2.37)$$

The nominal set of parameters of the WMR Pioneer 2-DX was adopted and denoted by $\mathbf{a}$, whereas the second set denoted by $\mathbf{a}_d$ refers to the parameters of the WMR carrying a load of $m_{RL} = 4$ [kg]. These parameters were applied to simulate

**Table 2.1** Adopted parameter values for the Pioneer2-DX WMR

| Parameter | Value (no load) (**a**) | Parameter | Value (load applied) (**a**$_d$) | Unit |
|---|---|---|---|---|
| $a_1$ | 0.1207 | $a_{d1}$ | 0.1343 | kgm$^2$ |
| $a_2$ | 0.0768 | $a_{d2}$ | 0.0945 | kgm$^2$ |
| $a_3$ | 0.037 | $a_{d3}$ | 0.037 | kgm$^2$ |
| $a_4$ | 0.001 | $a_{d4}$ | 0.001 | kgm$^2$ |
| $a_5$ | 2.025 | $a_{d5}$ | 2.296 | Nm |
| $a_6$ | 2.025 | $a_{d6}$ | 2.296 | Nm |

disturbances during the movement of the WMR. Both sets of parameters are presented in Table 2.1.

The model of forward dynamics of a mobile robot can be defined based on Eq. (2.34), thus

$$\ddot{\boldsymbol{\alpha}} = \mathbf{M}^{-1} \left[ \mathbf{u} - \mathbf{C} \left( \dot{\boldsymbol{\alpha}} \right) \dot{\boldsymbol{\alpha}} - \mathbf{F} \left( \dot{\boldsymbol{\alpha}} \right) - \boldsymbol{\tau}_d \left( t \right) \right] . \tag{2.38}$$

Equation (2.38) will be applied further in this work for the modeling of mobile robot's motion.

A continuous model of the WMR's dynamics (2.38) was discretized with the use of Euler's method of approximation of continuous derivative by a difference (forward rectangular rule). The parameters of motion were computed (in a simulation) or measured (in an experiment) in discrete-time periods $\boldsymbol{\alpha} \left( t_{\{k\}} \right), \dot{\boldsymbol{\alpha}} \left( t_{\{k\}} \right)$, where $t_{\{k\}} = kh$, $k$ – integer indicating an iteration step, $k = 1, \ldots, N$, $N$ – number of iteration steps, $h = t_{\{k+1\}} - t_{\{k\}}$ – time discretization parameter, $\boldsymbol{\alpha} \left( t_{\{k\}} \right)$ and $\boldsymbol{\alpha} \left( t_{\{k+1\}} \right)$ – WMR drive wheels rotation angles in discrete-time periods $t_{\{k\}}$ and $t_{\{k+1\}}$ (for steps $k$ and $k + 1$). Substituting $\mathbf{z}_1 = \boldsymbol{\alpha}$, $\mathbf{z}_2 = \dot{\boldsymbol{\alpha}}$ into (2.38), the following discrete description of the WMR's dynamics was obtained

$$\begin{aligned} \mathbf{z}_{1\{k+1\}} &= \mathbf{z}_{1\{k\}} + h\mathbf{z}_{2\{k\}} , \\ \mathbf{z}_{2\{k+1\}} &= \mathbf{z}_{2\{k\}} - h\mathbf{M}^{-1} \left[ \mathbf{C} \left( \mathbf{z}_{2\{k\}} \right) \mathbf{z}_{2\{k\}} + \mathbf{F} \left( \mathbf{z}_{2\{k\}} \right) + \boldsymbol{\tau}_{d\{k\}} - \mathbf{u}_{\{k\}} \right] , \end{aligned} \tag{2.39}$$

where $\mathbf{z}_{1\{k\}} = \left[ z_{11\{k\}}, z_{12\{k\}} \right]^T$ – vector of discrete drive wheels rotation angles that corresponds to the continuous vector $\boldsymbol{\alpha}$, $\mathbf{z}_{2\{k\}} = \left[ z_{21\{k\}}, z_{22\{k\}} \right]^T$ – vector of discrete angular velocities that corresponds to the continuous vector $\dot{\boldsymbol{\alpha}}$.

Difference equations of mobile robot's dynamics (2.39) that were derived by means of Maggi's equations will be used further in this work for the modeling of motion of a controlled object, while a discrete form of relation (2.34) was applied in the synthesis of robot's tracking control algorithms.

The structural properties of the WMR's dynamic model (2.34) are used in the synthesis of tracking control algorithms. These properties are presented below.

**Structural Properties of the WMR's Mathematical Model**

Description of the WMR's dynamics in the form of Eq. (2.34) meets the following assumptions [6, 11, 14]:

1. Inertia matrix $\mathbf{M}$ is symmetric and positive-definite, and meets the constraints

$$\sigma_{\min}(\mathbf{M})\,\mathbf{I} \le \mathbf{M} \le \sigma_{\max}(\mathbf{M})\,\mathbf{I}\,, \qquad (2.40)$$

   where $\mathbf{I}$ – identity matrix, $\sigma_{\min}(\mathbf{M})$, $\sigma_{\max}(\mathbf{M})$ – respectively, minimum and maximum strictly positive eigenvalue of the inertia matrix.

2. Matrix $\mathbf{C}(\dot{\alpha})$ is such that the matrix

$$\mathbf{S}(\dot{\alpha}) = \dot{\mathbf{M}} - 2\mathbf{C}(\dot{\alpha})\,, \qquad (2.41)$$

   is skew-symmetric. Thus the following relation

$$\boldsymbol{\xi}^{T}\mathbf{S}(\dot{\alpha})\,\boldsymbol{\xi} = \mathbf{0}\,, \qquad (2.42)$$

   where $\boldsymbol{\xi}$ – any given vector of appropriate dimension.

3. Vector of disturbances $\boldsymbol{\tau}_d(t)$ acting on the WMR is bounded so that $\|\tau_{dj}(t)\| \le b_{dj}$, where $b_{dj}$ is a positive constant, $j = 1, 2$.

4. The WMR's dynamic Eq. (2.34) can be written in a linear form with respect to the parameters, hence

$$\mathbf{M}\ddot{\alpha} + \mathbf{C}(\dot{\alpha})\,\dot{\alpha} + \mathbf{F}(\dot{\alpha}) + \boldsymbol{\tau}_d(t) = \mathbf{Y}(\dot{\alpha}, \ddot{\alpha})^{T}\,\mathbf{a} + \boldsymbol{\tau}_d(t) = \mathbf{u}\,, \quad (2.43)$$

   vector of parameters $\mathbf{a}$ shall consist of a minimum number of linearly independent elements.

### 2.1.2.1 Simulation of Inverse Dynamics Problem – The WMR's Motion on a Closed-Loop Path

In the simulation of the WMR's motion the robot's characteristic point $A$ follows the desired trajectory with a loop path. The inverse kinematics solution trajectory from Sect. 2.1.1.1 was applied. Based on the values of the desired WMR's parameters such as angular velocities ($\dot{\alpha}_1, \dot{\alpha}_2$) and angular accelerations ($\ddot{\alpha}_1, \ddot{\alpha}_2$), and according to Eq. (2.34) control signals $u_1$ and $u_2$ were generated. The control signals determine the WMR's motion on the desired path given the desired angular velocities and accelerations of the drive wheels. The nominal values of the WMR's model parameters $\mathbf{a}$ were adopted, in accordance with Table 2.1.

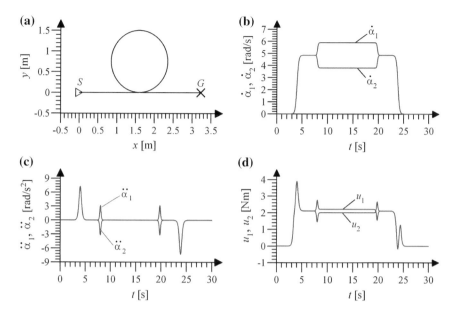

**Fig. 2.9** **a** Desired path of point $A$ on the WMR, **b** angular velocities $\dot{\alpha}_1$ and $\dot{\alpha}_2$, **c** angular accelerations $\ddot{\alpha}_1$ and $\ddot{\alpha}_2$, **d** control signals $u_1$, $u_2$

The diagram of the desired motion path of point $A$ is shown in Fig. 2.9a, desired angular velocities of the drive wheels ($\dot{\alpha}_1$, $\dot{\alpha}_2$) and angular accelerations ($\ddot{\alpha}_1$, $\ddot{\alpha}_2$) are illustrated in Fig. 2.9b, c, respectively. Figure 2.9d presents the generated control signals.

It can be noted that the waveforms shown in Fig. 2.9d directly correspond to the phases of motion related to the desired trajectory. At the initial time, the position of the WMR's point $A$ is at point $S$. In the acceleration period ($t \in < 3, 5 >$) the control signals are generated, thereby initializing the motion of robot's mass despite robot's innate inertia and resistance to motion. Hence, the highest values of control signals $u_1$ and $u_2$ are observed in this phase of motion. In the next phase, which is the constant motion on a rectilinear path ($t \in < 5, 7.5 >$), the values of angular accelerations are equal to zero, thus in accordance with Eq. (2.34) the values of control signals are determined merely by the modeled effects related to the resistance to motion. Another phase is the motion along a circular path (with radius $R = 0.75$ [m]) preceded by the curve entrance transition phase and completed with the curve exit transition phase. The curve entrance requires an increase in the angular velocity of wheel 1 and a decrease in the angular velocity of wheel 2 which results in the increase in the value of control signal $u_1$ and the decrease in the value of control signal $u_2$ in $t \in < 7.5, 8.5 >$. In the constant motion on a circular path ($t \in < 7.5, 20.5 >$) the control signals are affected by the vector of moments of the centrifugal and Coriolis forces, thus the different values of drive wheels control signals in this phase of motion. The curve exit phase is followed by another phase of constant motion on a rectilinear path ($t \in < 20.5, 23 >$). The final phase of motion is braking ($t \in < 23, 25 >$) and complete stop at point $G$, where the angular velocities tend to zero and the control signals go to zero as well.

### 2.1.2.2    Simulation of Inverse Dynamics Problem – The WMR's Motion on an 8-Shaped Path

In the simulation of the WMR's motion the robot's characteristic point $A$ moves on an 8-shaped path. The inverse kinematics solution trajectory from Sect. 2.1.1.2 was used. The simulation conditions applied are those from Sect. 2.1.2.1. The diagram of the defined motion path of point $A$ is shown in Fig. 2.10a, desired angular velocities of the drive wheels ($\dot{\alpha}_1$, $\dot{\alpha}_2$) and angular accelerations ($\ddot{\alpha}_1$, $\ddot{\alpha}_2$) are illustrated in Fig. 2.10b, c, respectively. Figure 2.10d presents the generated control signals.

The desired trajectory consists of two main periods. The first one consists in point $A$'s motion on a rectilinear path from its initial position at point $S$, followed by the motion along a curved path in the leftward direction, curve exit and the rectilinear motion with braking and stop at point $P$. The second main period of motion begins at point $P$ and consists in the rectilinear motion, followed by the motion on a curved path in the rightward direction and the rectilinear motion with braking and stop at point $G$. Each of these two main periods consists of analogous phases to those described in Sect. 2.1.2.1 except that during the motion on a curved path there is

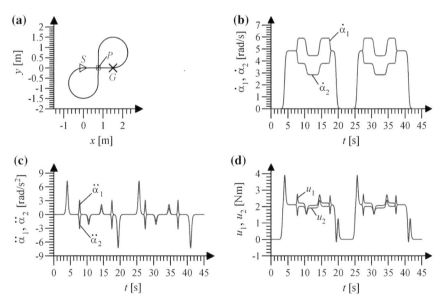

**Fig. 2.10  a** Desired path of point $A$ on the WMR, **b** angular velocities $\dot{\alpha}_1$ and $\dot{\alpha}_2$, **c** angular accelerations $\ddot{\alpha}_1$ and $\ddot{\alpha}_2$, **d** control signals $u_1$, $u_2$

an additional decrease in point $A$'s velocity from $v_A = 0.4 \left[\frac{m}{s}\right]$ to $v_A = 0.3 \left[\frac{m}{s}\right]$ in the periods $t \in< 10, 15 >$ [s] and $t \in< 30, 35 >$ [s]. The decrease in the values of angular velocities of drive wheels rotation in the aforementioned periods, results in the lower values of the control signals.

## 2.2 Robotic Manipulator

Scorbot–ER 4pc is a laboratory robotic manipulator (RM) with five degrees of freedom comprising revolute kinematic pairs of the arm and wrist. The robot was designed and developed to emulate an industrial robot, however its unique feature is the open mechanical structure that allows to observe the movement of the arm mechanisms and the open structure of the control system. Scorbot–ER 4pc [17] RM is shown in Fig. 2.11a, and its structure is illustrated by means of a schematic diagram in Fig. 2.11b.

The gripper is the robot's end-effector that can reach the desired position through the movement of the robot arm with three degrees of freedom. Another two degrees of freedom allow the gripper to be oriented arbitrarily and a servo motor allows closing of the gripper jaws. The robot is driven by drive units that comprise 12 [V] DC motors, gears and incremental encoders. Reduction gears were installed with ratios of $i_1 = i_2 = i_3 = 1 : 127.1$ in the drive units of links 1, 2 and 3, and ratios of $i_4 = i_5 = 1 : 65.5$ in the drive units of links 4 and 5. The motors that drive link 2 (upper arm), 3 (forearm) and the gripper are mounted in link 1 (base), thus they do not weigh down the links, whose masses and mass moments of inertia are relatively small. The power is transmitted by cogbelts, whose flexibility has a marginal impact on the robot arm movement given the position of the drive units in link 1. Robot's payload capacity is $m_L = 1$ [kg], including gripper, and the robot mass is $m_R = 11.5$ [kg].

The open structure of manipulator's control system along with dSpace measurement and diagnostic tools allow for experimental verification of different methods for modeling and control of RMs. Research on the synthesis of control algorithms with regard to the motion of dynamic systems requires knowledge of the system's

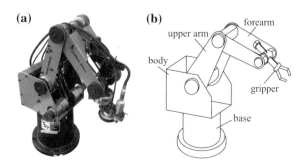

**Fig. 2.11** **a** Scorbot-ER 4pc robotic manipulator, **b** schematic diagram of the robotic manipulator

mathematical model that should be simple, namely, consider solely the most important effects. Therefore, the following assumptions are made [23]:

- analyzed kinematic chain of the manipulator has three degrees of freedom, gripper movements are negligible,
- manipulator links were modeled as rigid bodies,
- gripper was modeled as mass concentrated at point $C$, at the end of the robotic arm,
- flexibility and backlash in gears are negligible,
- dynamics of actuators was not considered, due to short time constants of motors,
- link 1 was modeled as a cylinder,
- links 2 and 3 were assumed to have axes of symmetry which intersect the axes of their respective joints.

### 2.2.1   Description of the Kinematics of a Robotic Manipulator

The issues related to the kinematics of robotic manipulators include forward and inverse kinematics problems. A problem-solving approach to forward kinematics consists in the determination of the tool center point (TCP) position and tool orientation within the manipulator's workspace, taking into account the joint coordinates (configuration coordinates). To solve the inverse kinematics problem joints' coordinates, velocities and accelerations must be determined based on the assumed motion of the tool within the workspace. From a mathematical point of view solving the inverse kinematics problem is more difficult than solving the problem of forward kinematics, however it is more important in terms of robotics. An analysis of the inverse kinematics problem is carried out to determine the trajectory which is to be executed by the tracking control system [23]. Except in extraordinary cases, solving the inverse kinematics problem is difficult, which is due to ambiguity of solutions, discontinuity in solutions and kinematic singularity.

This subchapter describes the kinematics of the Scorbot - ER 4pc RM by means of the Denavit–Hartenberg notation (D–H) [2, 9, 15, 16, 19, 20] which employs homogenous coordinates and transformation matrices [15, 16, 19]. The Scorbot - ER 4pc's arm has a joint kinematic structure (anthropomorphic). According to the adopted D-H notation, the $i$th coordinate frame is attached to each $i$th link of the manipulator's kinematic chain. The homogenous transformation matrix $\mathbf{A}_{i-1}^{i}$ transforms coordinates of a selected point from the $i$th coordinate system into the $i-1$ system. It is obtained by combining four transformations [19], according to the relation

$$\mathbf{A}_{i-1}^{i} = \mathbf{Rot}_{z,\Theta i}\,\mathbf{Trans}_{z,di}\,\mathbf{Trans}_{x,ai}\,\mathbf{Rot}_{x,\alpha i}\ , \tag{2.44}$$

where individual matrices of rotation and translation express the angle of rotation and displacement of the $i$th system to the $i-1$ system, $\mathbf{Rot}_{z,\Theta i}$ is the matrix of rotation through angle $\Theta_i$ about axis $z$, expressed by the relation

$$\mathbf{Rot}_{z,\Theta i} = \begin{bmatrix} \cos(\Theta_i) & -\sin(\Theta_i) & 0 & 0 \\ \sin(\Theta_i) & \cos(\Theta_i) & 0 & 0 \\ 0 & 0 & 1 & 0 \\ 0 & 0 & 0 & 1 \end{bmatrix}, \tag{2.45}$$

translation matrix $\mathbf{Trans}_{z,di}$ expresses translation along axis $z$ by distance $d_i$, according to the relation

$$\mathbf{Trans}_{z,di} = \begin{bmatrix} 1 & 0 & 0 & 0 \\ 0 & 1 & 0 & 0 \\ 0 & 0 & 1 & d_i \\ 0 & 0 & 0 & 1 \end{bmatrix}, \tag{2.46}$$

translation matrix $\mathbf{Trans}_{x,ai}$ expresses translation along axis $x$ by distance $a_i$

$$\mathbf{Trans}_{x,ai} = \begin{bmatrix} 1 & 0 & 0 & a_i \\ 0 & 1 & 0 & 0 \\ 0 & 0 & 1 & 0 \\ 0 & 0 & 0 & 1 \end{bmatrix}, \tag{2.47}$$

and rotation matrix $\mathbf{Rot}_{x,\alpha i}$ expresses rotation through angle $\alpha_i$ relative to axis $x$, according to the relation

$$\mathbf{Rot}_{x,\alpha i} = \begin{bmatrix} 1 & 0 & 0 & 0 \\ 0 & \cos(\alpha_i) & -\sin(\alpha_i) & 0 \\ 0 & \sin(\alpha_i) & \cos(\alpha_i) & 0 \\ 0 & 0 & 0 & 1 \end{bmatrix}, \tag{2.48}$$

where the parameters of the $i$th link of the kinematic chain are: $\Theta_i$ – joint's angle of rotation, $d_i$ – joint offset, $a_i$ - link length, $\alpha_i$ – link twist angle. The application of D-H notation requires the following selection of coordinate frames attached to the links matrix [19]:

- axis $x_i$ of the $i$th coordinate system is perpendicular to axis $z_{i-1}$ of the $i-1$ coordinate system,
- axis $x_i$ of the $i$th coordinate system intersects axis $z_{i-1}$ of the $i-1$ coordinate system,

where parameters $\Theta_i$, $d_i$, $a_i$ and $\alpha_i$ are unambiguously determined and satisfy the relation (2.44).

Finally, matrix $\mathbf{A}_{i-1}^{i}$ has the form

$$\mathbf{A}_{i-1}^{i} = \begin{bmatrix} \cos(\Theta_i) & -\sin(\Theta_i)\cos(\alpha_i) & \sin(\Theta_i)\sin(\alpha_i) & a_i\cos(\Theta_i) \\ \sin(\Theta_i) & \cos(\Theta_i)\cos(\alpha_i) & -\cos(\Theta_i)\sin(\alpha_i) & a_i\sin(\Theta_i) \\ 0 & \sin(\alpha_i) & \cos(\alpha_i) & d_i \\ 0 & 0 & 0 & 1 \end{bmatrix}. \tag{2.49}$$

**Fig. 2.12** Schematic
diagram of the Scorbot-ER
4pc robotic manipulator with
coordinate systems and
Devanit–Hartenberg
parameters

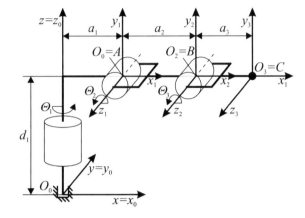

**Table 2.2** Scorbot-ER 4pc kinematic parameters relevant to Denavit–Hartenberg notation

| Czon | $\Theta_i$ [rad] | $a_i$ [m] | $d_i$ [m] | $\alpha_i$ [rad] |
|------|------------------|-----------|-----------|------------------|
| 1 | $\Theta_1$ | $l_1 = 0.026$ | $d_1 = 0.35$ | $\pi/2$ |
| 2 | $\Theta_2$ | $l_2 = 0.22$ | 0 | 0 |
| 3 | $\Theta_3$ | $l_3 = 0.22$ | 0 | 0 |

Figure 2.12 shows a schematic diagram of RM structure with kinematic parameters relevant to the D-H notation.

Relevant kinematic parameters of the links are presented in Table 2.2.

Homogenous transformation matrices of the Scorbot-ER 4pc RM [9, 23] take the form:

$$A_0^1(\Theta_1) = \begin{bmatrix} \cos(\Theta_1) & 0 & \sin(\Theta_1) & l_1\cos(\Theta_1) \\ \sin(\Theta_1) & 0 & -\cos(\Theta_1) & l_1\sin(\Theta_1) \\ 0 & 1 & 0 & d_1 \\ 0 & 0 & 0 & 1 \end{bmatrix}, \tag{2.50}$$

$$A_1^2(\Theta_2) = \begin{bmatrix} \cos(\Theta_2) & 0 & -\sin(\Theta_2) & l_2\cos(\Theta_2) \\ \sin(\Theta_2) & 0 & \cos(\Theta_2) & l_2\sin(\Theta_2) \\ 0 & 0 & 1 & 0 \\ 0 & 0 & 0 & 1 \end{bmatrix}, \tag{2.51}$$

$$A_2^3(\Theta_3) = \begin{bmatrix} \cos(\Theta_3) & 0 & -\sin(\Theta_3) & l_3\cos(\Theta_3) \\ \sin(\Theta_3) & 0 & \cos(\Theta_3) & l_3\sin(\Theta_3) \\ 0 & 0 & 1 & 0 \\ 0 & 0 & 0 & 1 \end{bmatrix}. \tag{2.52}$$

Considering the mechanical structure of the manipulator and the methods for measuring rotation angles of the Scorbot-ER 4pc manipulator's links, a new vector of configuration coordinates was adopted

$$\mathbf{q} = \begin{bmatrix} q_1 \\ q_2 \\ q_3 \end{bmatrix} = \begin{bmatrix} \Theta_1 \\ \Theta_2 \\ \Theta_2 + \Theta_3 \end{bmatrix}, \tag{2.53}$$

and new homogenous transformation matrices defined by coordinates $q_i$ in the form

$$\mathbf{A}_0^1 (q_1) = \begin{bmatrix} \cos(q_1) & 0 & \sin(q_1) & l_1 \cos(q_1) \\ \sin(q_1) & 0 & -\cos(q_1) & l_1 \sin(q_1) \\ 0 & 1 & 0 & d_1 \\ 0 & 0 & 0 & 1 \end{bmatrix}, \tag{2.54}$$

$$\mathbf{A}_1^2 (q_2) = \begin{bmatrix} \cos(q_2) & 0 & -\sin(q_2) & l_2 \cos(q_2) \\ \sin(q_2) & 0 & \cos(q_2) & l_2 \sin(q_2) \\ 0 & 0 & 1 & 0 \\ 0 & 0 & 0 & 1 \end{bmatrix}, \tag{2.55}$$

$$\mathbf{A}_2^3 (q_2, q_3) = \begin{bmatrix} \cos(q_3 - q_2) & 0 & -\sin(q_3 - q_2) & l_3 \cos(q_3 - q_2) \\ \sin(q_3 - q_2) & 0 & \cos(q_3 - q_2) & l_3 \sin(q_3 - q_2) \\ 0 & 0 & 1 & 0 \\ 0 & 0 & 0 & 1 \end{bmatrix}, \tag{2.56}$$

that will be used further in this work to determine the manipulator Jacobian.

Transformation of the $j$th coordinate system to the $i - 1$ system is defined by matrix $\mathbf{T}_{i-1}^j \in \mathfrak{R}^{4x4}$ expressed in the form [20]:

$$\mathbf{T}_{i-1}^j = \prod_{k=i}^j \mathbf{A}_{k-1}^k (\mathbf{q}) , \tag{2.57}$$

whose general form is

$$\mathbf{T}_{i-1}^j = \left[ \begin{array}{c|c} \mathbf{R}_{i-1}^j & \mathbf{p}_{i-1}^j \\ \hline \mathbf{0} & 1 \end{array} \right], \tag{2.58}$$

where $\mathbf{R}_{i-1}^j \in \mathfrak{R}^{3x3}$ – matrix of rotation of the $j$th coordinate system relative to the $i - 1$ system, $\mathbf{p}_{i-1}^j \in \mathfrak{R}^{3x1}$ – vector of translation of the $j$th coordinate system relative to the $i - 1$ system.

The Jacobian matrix (in robotics also referred to as the Jacobian) is essential for the modeling, motion planning and RM control. Scientific publications provide several different types of the Jacobian [19, 20], this monograph only presents the analytical Jacobian [19, 20, 23].

The analytical Jacobian of the RM is derived from the so-called kinematic functions, where the joint coordinates (configuration coordinates) $\mathbf{q}$ are related to the task coordinates $\mathbf{y}$ of manipulator's link e.g. tool position and orientation in robot's task space.

The task coordinates are defined by the relation

$$\mathbf{y} = \mathbf{k}(\mathbf{q}) \ , \tag{2.59}$$

where $\mathbf{k}(\mathbf{q}) \in \mathfrak{R}^{mx1}$ – the so-called kinematic function, $\mathbf{q} \in \mathfrak{R}^{nx1}$ – vector of configuration coordinates, $m$ – dimension of manipulator's task space, $n$ – dimension of manipulator's configuration space.

The connection between the values of velocities in the configuration space and the values of joint velocities is expressed by means of the analytical Jacobian, according to the following relation

$$\dot{\mathbf{y}} = \mathbf{J}(\mathbf{q})\,\dot{\mathbf{q}} \ , \tag{2.60}$$

$\dot{\mathbf{y}} \in \mathfrak{R}^{mx1}$ – velocity vector in task coordinates, $\dot{\mathbf{q}} \in \mathfrak{R}^{nx1}$ – vector of configuration velocities, $\mathbf{J}(\mathbf{q}) \in \mathfrak{R}^{mxn}$ – analytical Jacobian expressed in the form

$$\mathbf{J}(\mathbf{q}) = \frac{\partial \mathbf{k}(\mathbf{q})}{\partial \mathbf{q}} \ . \tag{2.61}$$

The selection of manipulator's task coordinates $\mathbf{y}$ has a significant impact on the dimension and form of the Jacobian [19, 20, 23]. The analytical Jacobian of $n$-link manipulator consists of columns $\mathbf{J}_i$, $i = 1, 2, \ldots, n$, that is

$$\mathbf{J}(\mathbf{q}) = [\mathbf{J}_1 \ \ldots \ \mathbf{J}_i \ \ldots \ \mathbf{J}_n] \ , \tag{2.62}$$

where the $i$th column of the revolute joint Jacobian matrix is expressed by

$$\mathbf{J}_i = \begin{bmatrix} \mathbf{R}_{0\ 3^{th}\ col}^{i-1} \times (\mathbf{p}_0^n - \mathbf{p}_0^{i-1}) \\ \mathbf{R}_{0\ 3^{th}\ col}^{i-1} \end{bmatrix} \ , \tag{2.63}$$

and for a prismatic joint by

$$\mathbf{J}_i = \begin{bmatrix} \mathbf{R}_{0\ 3^{th}\ col}^{i-1} \\ \mathbf{0} \end{bmatrix} \ , \tag{2.64}$$

where $\mathbf{R}_{0\ 3^{th}\ col}^{i-1}$ – is the third column of the rotation matrix resulting from relation (2.58).

The analytical Jacobian of the Scorbot-ER 4pc was determined assuming a 6-dimensional task space, where point $C$ coordinates of the manipulator's kinematic chain $\mathbf{y} = [y_1 y_2 y_3 y_4 y_5 y_6]^T$ describe the position ($y_1$, $y_2$, $y_3$) and the orientation ($y_4$, $y_5$, $y_6$) of the end-effector with respect to the base coordinate system. By applying configuration coordinates (2.53), the analytical Jacobian (2.61) has the form

$$
\mathbf{J}(\mathbf{q}) =
\begin{bmatrix}
-(l_1 + l_2 \cos(q_2) + l_3 \cos(q_3)) \sin(q_1) & -l_2 \sin(q_2) \cos(q_1) & -l_3 \sin(q_3) \cos(q_1) \\
(l_1 + l_2 \cos(q_2) + l_3 \cos(q_3)) \cos(q_1) & -l_2 \sin(q_2) \sin(q_1) & -l_3 \sin(q_3) \sin(q_1) \\
0 & l_2 \cos(q_2) & l_3 \cos(q_3) \\
0 & 0 & \sin(q_1) \\
0 & 0 & -\cos(q_1) \\
1 & 0 & 0
\end{bmatrix}.
$$
(2.65)

Another important issue in the analysis of manipulator's kinematics is to determine singular configurations based on the analysis of the Jacobian [20, 23]. The theory indicates that if the manipulator's number of degrees of freedom is lower than the number of task space dimensions, the manipulator has only singular configurations. In this analysis, 6-dimensional task space was adopted for a manipulator with three coordinates in the configuration space (2.53). A manipulator thus defined allows the end-effector to either reach the desired position, in that case the end-effector's orientation depends on the position, or to reach the desired orientation, then the position depends on the orientation. The determination of singular configurations of the Scorbot-ER 4pc manipulator is described in detail in [23].

#### 2.2.1.1 Inverse Kinematics Problem

The following considerations are limited to the first three degrees of freedom of the kinematic chain. Under this assumption the Scorbot-ER 4pc RM is not capable of reaching any arbitrary position and orientation within the 6-dimensional task space. Further considerations deal with the position kinematics which comes down to the division of task coordinates into coordinates $\mathbf{y}_p = [y_1, y_2, y_3]^T$ that relate to the position of the end-effector and coordinates $\mathbf{y}_o = [y_4, y_5, y_6]^T$ that relate to the orientation [23].

A similar division may be made for the analytical Jacobian (2.65) where $\mathbf{J}_p \in \mathfrak{R}^{3 \times 3}$ is the part related to the position

$$
\mathbf{J}_p(\mathbf{q}) =
\begin{bmatrix}
-(l_1 + l_2 \cos(q_2) + l_3 \cos(q_3)) \sin(q_1) & -l_2 \sin(q_2) \cos(q_1) & -l_3 \sin(q_3) \cos(q_1) \\
(l_1 + l_2 \cos(q_2) + l_3 \cos(q_3)) \cos(q_1) & -l_2 \sin(q_2) \sin(q_1) & -l_3 \sin(q_3) \sin(q_1) \\
0 & l_2 \cos(q_2) & l_3 \cos(q_3)
\end{bmatrix},
$$
(2.66)

and $\mathbf{J}_o \in \mathfrak{R}^{3 \times 3}$ is the part related to the orientation of the manipulator's end-effector

$$
\mathbf{J}_o(\mathbf{q}) =
\begin{bmatrix}
0 & 0 & \sin(q_1) \\
0 & 0 & -\cos(q_1) \\
1 & 0 & 0
\end{bmatrix}.
$$
(2.67)

The relation between the velocities of end-effector's point $C$ and the joint velocities of the manipulator is given by

$$\dot{\mathbf{y}}_p = \mathbf{J}_p(\mathbf{q})\,\dot{\mathbf{q}}\ . \tag{2.68}$$

The problems of forward and inverse kinematics of the RM can be solved by means of Eq. (2.68). However, solving the inverse kinematics problem is difficult, which is due to the singularity of manipulator's kinematics as well as the ambiguity and discontinuity in solutions. When solved, it provides the desired trajectory that is necessary for the tracking control of a RM.

The following considerations concern solution of the inverse kinematics problem of a RM, given the desired motion path and velocity of the end-effector's point $C$. In the general case point $C$'s motion path is a curve, defined in the task space

$$\begin{cases} f_1(x_C, z_C) = 0\ , \\ f_2(x_C, y_C) = 0\ , \end{cases} \tag{2.69}$$

where $x_C$, $y_C$, $z_C$ – end-effector's point $C$ coordinates in the base coordinate system $xyz$, corresponding to the position coordinates $\mathbf{y}_p$. The velocity vector of point $C$ has the form

$$\mathbf{v}_C = \begin{bmatrix} \dot{x}_C \\ \dot{y}_C \\ \dot{z}_C \end{bmatrix} = \dot{\mathbf{y}}_p\ , \tag{2.70}$$

whose value is expressed by the relation

$$v_C = \sqrt{\dot{x}_C^2 + \dot{y}_C^2 + \dot{z}_C^2}\ . \tag{2.71}$$

The direction of the velocity vector of point $C$ must be tangent to the path of motion, thus the following relations must be satisfied

$$\begin{cases} \operatorname{grad} f_1(x_C, z_C)\,\mathbf{v}_C = 0\ , \\ \operatorname{grad} f_2(x_C, y_C)\,\mathbf{v}_C = 0\ , \end{cases} \tag{2.72}$$

which can be written as

$$\begin{cases} \begin{bmatrix} \dfrac{\partial f_1(x_C,z_C)}{\partial x_C} & \dfrac{\partial f_1(x_C,z_C)}{\partial y_C} & \dfrac{\partial f_1(x_C,z_C)}{\partial z_C} \end{bmatrix} \begin{bmatrix} \dot{x}_C \\ \dot{y}_C \\ \dot{z}_C \end{bmatrix} = 0\ , \\[20pt] \begin{bmatrix} \dfrac{\partial f_2(x_C,y_C)}{\partial x_C} & \dfrac{\partial f_2(x_C,y_C)}{\partial y_C} & \dfrac{\partial f_2(x_C,y_C)}{\partial z_C} \end{bmatrix} \begin{bmatrix} \dot{x}_C \\ \dot{y}_C \\ \dot{z}_C \end{bmatrix} = 0\ . \end{cases} \tag{2.73}$$

Given that $f_{1x} = \frac{\partial f_1(x_C,z_C)}{\partial x_C}$, $f_{1y} = \frac{\partial f_1(x_C,z_C)}{\partial y_C}$, $f_{1z} = \frac{\partial f_1(x_C,z_C)}{\partial z_C}$, $f_{2x} = \frac{\partial f_2(x_C,y_C)}{\partial x_C}$, $f_{2y} = \frac{\partial f_2(x_C,y_C)}{\partial y_C}$, $f_{2z} = \frac{\partial f_2(x_C,y_C)}{\partial z_C}$, solution of Eqs. (2.71) and (2.73) take the form

$$\begin{cases} \dot{x}_C = \mp v_C \dfrac{f_{1y}f_{2z}-f_{1z}f_{2y}}{\sqrt{\left(f_{1x}f_{2y}-f_{2x}f_{1y}\right)^2+\left(f_{1x}f_{2z}-f_{2x}f_{1z}\right)^2+\left(f_{1y}f_{2z}-f_{2y}f_{1z}\right)^2}} \, , \\[2mm] \dot{y}_C = \pm v_C \dfrac{f_{1x}f_{2z}-f_{1z}f_{2x}}{\sqrt{\left(f_{1x}f_{2y}-f_{2x}f_{1y}\right)^2+\left(f_{1x}f_{2z}-f_{2x}f_{1z}\right)^2+\left(f_{1y}f_{2z}-f_{2y}f_{1z}\right)^2}} \, , \\[2mm] \dot{z}_C = \mp v_C \dfrac{f_{1x}f_{2y}-f_{1y}f_{2x}}{\sqrt{\left(f_{1x}f_{2y}-f_{2x}f_{1y}\right)^2+\left(f_{1x}f_{2z}-f_{2x}f_{1z}\right)^2+\left(f_{1y}f_{2z}-f_{2y}f_{1z}\right)^2}} \, . \end{cases} \tag{2.74}$$

By solving the set of differential equations (2.74) including the initial conditions and pre-determined value of point $C$ velocity vector, it is possible to determine the values of point $C$ velocity vector components with respect to time, as well as the coordinates $x_C$, $y_C$ i $z_C$. Relation (2.68) can be transformed into

$$\dot{\mathbf{q}} = \left[\mathbf{J}_p\left(\mathbf{q}\right)\right]^{-1}\dot{\mathbf{y}}_p \, . \tag{2.75}$$

By solving differential equations (2.75) based on the pre-defined values of point $C$ velocity components, it is possible to determine the values of angular velocities and rotation angles of manipulator's links, with respect to time. Differentiating the values of angular velocities provides the values of angular accelerations. Thus obtained variables may constitute the desired reference trajectory to be tracked by the control system. To avoid singular configurations, for which no inverse of the Jacobian $\mathbf{J}_p\left(\mathbf{q}\right)$, exists, the end-effector's motion path must be properly planned.

### 2.2.1.2  Simulation of Inverse Kinematics Problem with a Semicircle Path

Following example presents a solution to the inverse kinematics problem with regard to the RM. The assumed path of motion of point $C$ attached to the end-effector has the shape of a semicircle, with center $E$ and radius $R$, and lies in the $xy$ plane.

Thus, it is assumed that there is no motion in the $z$-axis direction, hence the coordinate $z_C = \text{const.}$, and point $C$ velocity relative to $z$-axis is $\dot{z}_C = 0$. The assumed equation for point $C$ motion path is

$$(x_C - x_E)^2 + (y_C - y_E)^2 - R^2 = 0 \, , \tag{2.76}$$
$$z_C - d_1 = 0 \, ,$$

where $x_E$, $y_E$ – coordinates of point $E$ (center of the circle), $x_E = 0.36$ [m], $y_E = 0$ [m], $R$ – circle radius, $R = 0.1$ [m]. The assumed value of point $C$ velocity vector is expressed by the following relation

$$v_C = v_C^* \cdot \sum_{m=0}^{5} \frac{(-1)^m}{\left(1 + \exp^{-c_1(t-t_{1+2m})}\right)\left(1 + \exp^{c_1(t-t_{2+2m})}\right)} \, , \tag{2.77}$$

where $v_C^*$ – maximum velocity of end-effector's point $C$, $v_C^* = 0.08$ [m/s], $c_1$ – slope coefficient of unipolar sigmoid function, $c_1 = 10$ [1/s], $t \in < 0, 40 >$ [s], $t_{1+2m}$ –

mean start-time of $m$th period of acceleration, $t_{2+2m}$ – mean start-time of $m$th period of braking, $m = 0, \ldots, 5, t_1 = 4\,[\text{s}], t_2 = 7.93\,[\text{s}], t_3 = 10\,[\text{s}], t_4 = 13.93\,[\text{s}], t_5 = 16$ [s], $t_6 = 19.93\,[\text{s}], t_7 = 22\,[\text{s}], t_8 = 25.93\,[\text{s}], t_9 = 28\,[\text{s}], t_{10} = 31.93\,[\text{s}], t_{11} = 34$ [s], $t_{12} = 37.93\,[\text{s}]$.

The velocity profile described by relation (2.77) approximates the triangular velocity profile; it was applied to eliminate discontinuities in angular accelerations of trajectory with a triangular velocity profile. In the presented case $f_{1x} = 2\,(x_C - x_E)$, $f_{1y} = 2y_C, f_{1z} = f_{2x} = f_{2y} = 0, f_{2z} = 1$, and the set of Eq. (2.74) takes a simplified form

$$\begin{cases} \dot{x}_C = \mp v_C \dfrac{f_{1y}}{\sqrt{f_{1x}^2 + f_{1y}^2}}\,, \\[2mm] \dot{y}_C = \pm v_C \dfrac{f_{1x}}{\sqrt{f_{1x}^2 + f_{1y}^2}}\,, \\[2mm] \dot{z}_C = 0\,. \end{cases} \tag{2.78}$$

While solving the set of differential equations (2.78) the following initial conditions were assumed: $x_C\,(0) = 0.46\,[\text{m}], y_C\,(0) = 0\,[\text{m}]$, and while solving differential equations (2.75) the assumed initial values of manipulator arms' rotation angles were $q_1\,(0) = 0\,[\text{rad}], q_2\,(0) = 0.1653\,[\text{rad}], q_3\,(0) = -0.1653\,[\text{rad}]$. The adopted geometry of the Scorbot-ER 4pc RM corresponds to the parametric data included in Table 2.2.

The velocity values of the end-effector's point $C$ $v_C$, calculated from relation (2.77), are shown in Fig. 2.13a. The projection of the velocity vector on the axis tangent to the point's motion path takes either positive or negative values for individual motion periods. Each of the $m = 6$ periods consists of the acceleration phase, where $v_C$ changes from $v_C = 0\,[\text{m/s}]$ to $v_C = \pm v_C^*\,[\text{m/s}]$, the constant velocity phase, where $v_C = \pm v_C^* = \text{const.}$, and the braking phase, where the velocity of point $C$ tends to zero. The path of point $C$, calculated from differential equations (2.78), is shown in Fig. 2.13b. Point $C$ moves on a circular path with the assumed value of the velocity vector; with point $S$ being the initial position, marked with the triangle in Fig. 2.13b. Point $G$ is the point of reversal, where the end-effector's point $C$ comes to a halt, followed by the second phase, where point $C$ moves along the same path from point $G$ to point $S$. These two phases make a full motion cycle that is repeated three times within $t \in\, < 0, 40 >\,[\text{s}]$.

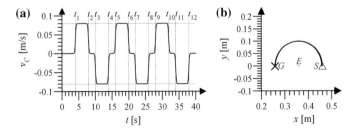

**Fig. 2.13** **a** Velocity profile of the end-effector's point $C$, **b** motion path of point $C$

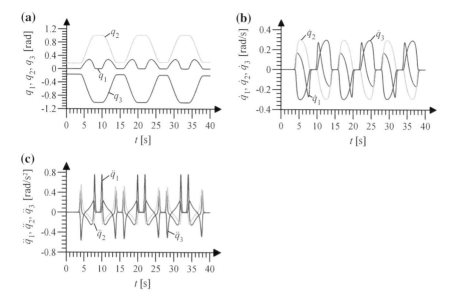

**Fig. 2.14**  **a** Manipulator's links rotation angles $q_1, q_2, q_3$, **b** angular velocities $\dot{q}_1, \dot{q}_2, \dot{q}_3$, **c** angular accelerations $\ddot{q}_1, \ddot{q}_2, \ddot{q}_3$

The case described above illustrates a typical path planning/path execution problem that is challenged by industrial manipulators in the pick and place process.

Based on the assumed path and point $C$ velocity, the trajectory in the robot's configuration space was determined, by solving the inverse kinematics problem i.e. by calculating joints' variables from differential equations (2.75) such as: rotation angles $q_1, q_2, q_3$ (Fig. 2.14a), angular velocities $\dot{q}_1, \dot{q}_2, \dot{q}_3$ (Fig. 2.14b) and angular accelerations $\ddot{q}_1, \ddot{q}_2, \ddot{q}_3$ (Fig. 2.14c).

Thus obtained inverse kinematics problem solution, in the form of joints' variables, may constitute the desired reference trajectory to be tracked by the Scorbot-ER 4pc tracking control system.

### 2.2.2  Description of the Dynamics of a Robotic Manipulator

Robotic manipulators are mechanical systems with a nonlinear dynamic description, which is due to the complexity of their structure. A dynamic model of such systems can be developed for example by adopting a classical mechanics approach or by means of the Lagrange's equations of the second kind. The basic issues relating to the modeling of robotic manipulator's dynamics are addressed in publications [2–4, 15, 18–20, 23].

The analysis of issues concerned with the dynamics of a RM consists in seeking correlations between the system's parameters of motion and the mechanisms that cause the motion. The forward dynamics problem is to determine the manipulator's kinematic parameters of motion resulting from the imposed forces, while the inverse dynamics problem is solved by determining the control variables required for the execution of the system's motion with regard to the assumed kinematic parameters.

The analysis of the RM's dynamics results in a mathematical model described with ordinary differential equations, with certain parameters and structure. The parameters of the model come in the form of coefficients in the differential equations of motion. However, it is difficult to determine their values, as they result from the geometry of the system, mass distribution or resistance to motion and thus can be defined in the process of parametric identification [3, 5, 12, 13, 18, 21–23]. Besides parametric identification the dynamic model may be used for the synthesis of the manipulator's tracking control systems. The model usually captures the most essential effects that occur in the manipulator's motion, due to the computational complexity and difficulty in measuring certain quantities. The dynamics of actuators is often deemed negligible and thus not considered, given the short time constants of the applied electric motors.

An important issue is the sensitivity of the mathematical model to changes in parameters [5, 21–23]. A sensitivity analysis allows to identify the impact of individual parameters on the accuracy of the mathematical model that can be simplified by isolating the parameters that have no significant influence on the model. The reduction of complexity allows the model to be applied in control based on the model performed in the real-time. A description of the Scorbot-ER 4pc dynamics model sensitivity analysis can be found in publications [22, 23].

This chapter presents a model of the dynamics of a three-degrees-of-freedom RM based on the Scorbot-ER 4pc robot. The model was derived using the Lagrange's equations of the second kind [22, 23]. The schematic diagram of the manipulator's kinematic chain is shown in Fig. 2.15.

The following denotations are used in Fig. 2.15: point $C$ – TCP, $|OO'| = d_1$, $|O'A| = l_1, |AB| = l_2, |BC| = l_3$ – dimensions resulting from the geometry of RM's kinematic chain, $u_1, u_2, u_3$ – links movement control signals. Point $O$ on the RM's

**Fig. 2.15** Schematic diagram of the Scorbot-ER 4pc robotic manipulator

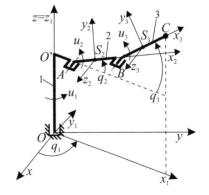

base is bounded by the stationary Cartesian coordinate system $xyz$. Manipulator's link 1 is bounded by the moving coordinate system $x_1 y_1 z_1$ with origin at point $O$, where axis $z_1$ is identical to axis $z$ of the base system. Link 1 rotates about axis $z_1$ through an angle $q_1$. Links 2 and 3 rotate through angles $q_2$ and $q_3$, respectively, about joints' axes whose directions are perpendicular to the $x_1 z_1$ plane and that pass through points $A$ and $B$. Link 3 rotation angle, $q_3$, is independent from link 2 rotation angle, $q_2$. Links' 2 and 3 resultant motion is composed of links' rotation motion with respect to joints and the transportation of link 2 imposed by link 1, as well as the transportation of link 3 imposed by links 1 and 2.

Given the assumptions set out in Sect. 2.2, a model of the RM's dynamics was derived by means of the Lagrange's equation of the second kind, thus

$$\frac{d}{dt}\left(\frac{\partial L}{\partial \dot{q}_j}\right) - \frac{\partial L}{\partial q_j} = Q_j , \tag{2.79}$$

$q_j$ – $j$th generalized coordinate, $Q_j$ – $j$th generalized force, $L$ – Lagrangian function is expressed by the relation

$$L = E - U , \tag{2.80}$$

where $E$ – kinetic energy of the system, $U$ – potential energy of the system. To formulate Lagrange's equations it is necessary to determine the kinetic and potential energy, and the generalized forces acting on the system. The kinetic energy of a multilink system is expressed by

$$E = \frac{1}{2}\dot{q}^T M(q)\dot{q} , \tag{2.81}$$

where $M(q)$ – inertia matrix, $M(q) \in \mathfrak{R}^{n \times n}$. Since no deformations are considered, the manipulator's potential results only from the mass potentials of individual links of the kinematic chain within Earth's gravitational field, and is expressed by a function of generalized coordinates. Thus, the Lagrangian function is written as

$$L = \frac{1}{2}\dot{q}^T M(q)\dot{q} - U(q) . \tag{2.82}$$

By substituting relation (2.82) to Lagrange's equation (2.79) and performing model sensitivity analysis together with all the necessary computations given in [22, 23], the following equation of manipulator's dynamics was obtained

$$M(q)\ddot{q} + C(q, \dot{q})\dot{q} + F(\dot{q}) + G(q) + \tau_d(t) = u , \tag{2.83}$$

where $q$ – vector of generalized coordinates, including manipulator's links rotation angles, $M(q)$ – inertia matrix, $C(q, \dot{q})\dot{q}$ – vector of moments resulting from centrifugal and Coriolis forces, $F(\dot{q})$ – resistance to motion vector, $G(q)$ – vector of moments resulting from gravitational forces, $\boldsymbol{\tau}_d(t)$ – vector of bounded disturbance moments, $\mathbf{u}$ – control signals vector.

Matrices $\mathbf{M}(\mathbf{q})$, $\mathbf{C}(\mathbf{q}, \dot{\mathbf{q}})$, and vectors $\mathbf{F}(\dot{\mathbf{q}})$, $\mathbf{G}(\mathbf{q})$, $\boldsymbol{\tau}_d(t)$, $\mathbf{u}$, take the form

$$
\mathbf{M}(\mathbf{q}) = \begin{bmatrix} M_{1,1} & 0 & 0 \\ 0 & p_6 & M_{2,3} \\ 0 & M_{3,2} & p_7 \end{bmatrix},
$$

$$
\mathbf{C}(\mathbf{q}, \dot{\mathbf{q}}) = \begin{bmatrix} -b\dot{q}_2 - c\dot{q}_3 & -b\dot{q}_1 & -c\dot{q}_1 \\ b\dot{q}_1 & 0 & C_{2,3} \\ c\dot{q}_1 & C_{3,2} & 0 \end{bmatrix},
$$

$$
\mathbf{F}(\dot{\mathbf{q}}) = \begin{bmatrix} p_8\dot{q}_1 + p_{11}\mathrm{sgn}(\dot{q}_1) \\ p_9\dot{q}_2 + p_{12}\mathrm{sgn}(\dot{q}_2) \\ p_{10}\dot{q}_3 + p_{13}\mathrm{sgn}(\dot{q}_3) \end{bmatrix},
$$

$$
\mathbf{G}(\mathbf{q}) = \begin{bmatrix} 0 \\ p_1 g \cos(q_2) \\ p_2 g \cos(q_3) \end{bmatrix}, \tag{2.84}
$$

$$
\boldsymbol{\tau}_d(t) = \begin{bmatrix} \tau_{d1} \\ \tau_{d2} \\ \tau_{d3} \end{bmatrix},
$$

$$
\mathbf{u}(t) = \begin{bmatrix} u_1 \\ u_2 \\ u_3 \end{bmatrix},
$$

where

$$
\begin{aligned}
M_{1,1} &= 2p_1 l_1 \cos(q_2) + 2p_2 (l_1 + l_2 \cos(q_2)) \cos(q_3) + \tfrac{1}{2} p_3 \cos(2q_2) + \\
&\quad + \tfrac{1}{2} p_4 \cos(2q_3) + p_5, \\
M_{2,3} &= M_{3,2} = p_2 l_2 \cos(q_3 - q_2), \\
b &= p_1 l_1 \sin(q_2) + p_2 l_2 \sin(q_2) \cos(q_3) + \tfrac{1}{2} p_3 \sin(2q_2), \\
c &= p_2 (l_1 + l_2 \cos(q_2)) \sin(q_3) + \tfrac{1}{2} p_4 \sin(2q_3), \\
C_{2,3} &= -p_2 l_2 \sin(q_3 - q_2) \dot{q}_3, \\
C_{3,2} &= p_2 l_2 \sin(q_3 - q_2) \dot{q}_2,
\end{aligned} \tag{2.85}
$$

where $\mathbf{p} = [p_1, \ldots, p_{13}]^T$ – vector of RM's parameters, which result from the geometry of the object, distribution of masses and resistance to motion, $g$ – gravitational acceleration, $g = 9.81 \left[\frac{m}{s^2}\right]$.

The values of individual parameters are expressed by coefficients grouped as follows

$$p_1 = m_2 l_{c2} + (m_3 + m_C) l_2 ,$$
$$p_2 = m_3 l_{c3} + m_C l_3 ,$$
$$p_3 = m_2 l_{c2}^2 + (m_3 + m_C) l_2^2 - I_{2xx} + I_{2yy} ,$$
$$p_4 = m_3 l_{c3}^2 + m_C l_3^2 - I_{3xx} + I_{3yy} ,$$
$$p_5 = I_{1yy} + \tfrac{1}{2} \left( I_{2xx} + I_{2yy} + I_{3xx} + I_{3yy} \right) + m_2 \left( l_1^2 + \tfrac{1}{2} l_{c2}^2 \right) +$$
$$+ m_3 \left( l_1^2 + \tfrac{1}{2} l_2^2 + \tfrac{1}{2} l_{c3}^2 \right) + m_C \left( l_1^2 + \tfrac{1}{2} l_2^2 + \tfrac{1}{2} l_3^2 \right) ,$$
$$p_6 = m_2 \left( l_2^2 + l_{c2}^2 \right) + m_C l_2^2 + I_{2zz} ,$$
$$p_7 = m_3 l_{c3}^2 + l_3^2 + I_{3zz} ,$$
$$p_8 = F_{v1} ,$$
$$p_9 = F_{v2} ,$$
$$p_{10} = F_{v3} ,$$
$$p_{11} = F_{C1} ,$$
$$p_{12} = F_{C2} ,$$
$$p_{13} = F_{C3} ,$$

$$(2.86)$$

where $m_i$ – manipulator's $i$th link mass, $m_C$ – gripper mass concentrated at TCP, $l_i$ – the $i$th link length, $l_{ci}$ – distance of the $i$th link's center of mass from the $i - 1$ link's end, $I_{ixx}$, $I_{iyy}$, $I_{izz}$ – mass moments of inertia of the $i$th link with respect to $x_i$, $y_i$, $z_i$ axes, $F_{vi}$ – the $i$th link viscous friction coefficient, $F_{Ci}$ – moment of dry friction, $i = 1, 2, 3$. The construction of the resistance to motion vector $\mathbf{F}(\dot{\mathbf{q}})$ results from the assumed approximation of the resistance to motion moments of each kinematic pair with the application of moments that result from the viscous friction ($F_{vi} \dot{q}_i$) and the dry friction ($F_{Ci} \operatorname{sgn}(\dot{q}_i)$).

The set of Eq. (2.83) can be written in a linear form with respect to parameters $\mathbf{p}$

$$\mathbf{Y}(\mathbf{q}, \dot{\mathbf{q}}, \ddot{\mathbf{q}})^T \mathbf{p} + \boldsymbol{\tau}_d(t) = \mathbf{u} , \qquad (2.87)$$

where $\mathbf{Y}(\mathbf{q}, \dot{\mathbf{q}}, \ddot{\mathbf{q}})$ – the so-called regression matrix, which takes the form

$$\mathbf{Y}(\mathbf{q}, \dot{\mathbf{q}}, \ddot{\mathbf{q}}) = \begin{bmatrix} Y_{11} & Y_{12} & Y_{13} & Y_{14} & \ddot{q}_1 & 0 & 0 & \dot{q}_1 & 0 & 0 & \operatorname{sgn}(\dot{q}_1) & 0 & 0 \\ Y_{21} & Y_{22} & Y_{23} & 0 & 0 & \ddot{q}_2 & 0 & 0 & \dot{q}_2 & 0 & 0 & \operatorname{sgn}(\dot{q}_2) & 0 \\ 0 & Y_{32} & 0 & Y_{34} & 0 & 0 & \ddot{q}_3 & 0 & 0 & \dot{q}_3 & 0 & 0 & \operatorname{sgn}(\dot{q}_3) \end{bmatrix}^T ,$$

$$(2.88)$$

where

$$Y_{11} = 2l_1 \cos(q_2) \ddot{q}_1 - 2l_1 \sin(q_2) \dot{q}_2 \dot{q}_1 ,$$
$$Y_{12} = 2 (l_1 + l_2 \cos(q_2)) (\cos(q_3) \ddot{q}_1 - \sin(q_3) \dot{q}_3 \dot{q}_1) - 2l_2 \sin(q_2) \cos(q_3) \dot{q}_2 \dot{q}_1 ,$$
$$Y_{13} = \tfrac{1}{2} \cos(2q_2) \ddot{q}_1 - \sin(2q_2) \dot{q}_2 \dot{q}_1 ,$$
$$Y_{14} = \tfrac{1}{2} \cos(2q_3) \ddot{q}_1 - \sin(2q_3) \dot{q}_3 \dot{q}_1 ,$$
$$Y_{21} = l_1 \sin(q_2) \dot{q}_1^2 + g \cos(q_2) ,$$
$$Y_{22} = l_2 \cos(q_3 - q_2) \ddot{q}_3 + l_2 \sin(q_2) \cos(q_3) \dot{q}_1^2 - l_2 \sin(q_3 - q_2) \dot{q}_3^2 ,$$
$$Y_{23} = \tfrac{1}{2} \sin(2q_2) \dot{q}_1^2 ,$$
$$Y_{32} = l_2 \cos(q_3 - q_2) \ddot{q}_2 + (l_1 + l_2 \cos(q_2)) \sin(q_3) \dot{q}_1^2 + l_2 \sin(q_3 - q_2) \dot{q}_2^2 + g \cos(q_3) ,$$
$$Y_{34} = \tfrac{1}{2} \sin(2q_3) \dot{q}_1^2 .$$

$$(2.89)$$

**Table 2.3**  Asummed parametric values of the Scorbot-ER 4pc robotic manipulator

| Parameter | Value (no load) (**p**) | Parameter | Value (load applied) (**p**$_d$) | Unit |
|-----------|-------------------------|-----------|----------------------------------|------|
| $p_1$ | 0.0065 | $p_{d1}$ | 0.0082 | kgm |
| $p_2$ | 0.0018 | $p_{d2}$ | 0.0041 | kgm |
| $p_3$ | 0.0113 | $p_{d3}$ | 0.0161 | kgm$^2$ |
| $p_4$ | 0.0064 | $p_{d4}$ | 0.0113 | kgm$^2$ |
| $p_5$ | 0.0114 | $p_{d5}$ | 0.0163 | kgm$^2$ |
| $p_6$ | 0.0113 | $p_{d6}$ | 0.0162 | kgm$^2$ |
| $p_7$ | 0.0065 | $p_{d7}$ | 0.0113 | kgm$^2$ |
| $p_8$ | 0.5276 | $p_{d8}$ | 0.5345 | Nms |
| $p_9$ | 0.5232 | $p_{d9}$ | 0.5340 | Nms |
| $p_{10}$ | 0.5235 | $p_{d10}$ | 0.5342 | Nms |
| $p_{11}$ | 0.0195 | $p_{d11}$ | 0.0221 | Nm |
| $p_{12}$ | 0.0182 | $p_{d12}$ | 0.0216 | Nm |
| $p_{13}$ | 0.0183 | $p_{d13}$ | 0.0217 | Nm |

A set of nominal parameters of the Scorbot-ER 4pc RM was assumed and denoted by **p**, whereas the second set of parameters denoted by **p**$_d$ relates to the manipulator with the gripper carrying a load of $m_L = 1$ [kg]. The parameters **p**$_d$ were applied to simulate disturbance during the robot's movement. Both sets of parameters are presented in Table 2.3.

The inverse dynamics model of the RM can be defined based on Eq. (2.83), thus

$$\ddot{\mathbf{q}} = \mathbf{M}^{-1}(\mathbf{q}) \left[ \mathbf{u} - \mathbf{C}(\mathbf{q}, \dot{\mathbf{q}})\, \dot{\mathbf{q}} - \mathbf{F}(\dot{\mathbf{q}}) - \mathbf{G}(\mathbf{q}) - \boldsymbol{\tau}_d(t) \right] . \qquad (2.90)$$

Equation (2.90) is used further in this work for the modeling of the RM's motion by means of the generated control signals.

By applying Euler's method of approximation of continuous derivative by a difference, and substituting $\mathbf{z}_1 = \mathbf{q}$, $\mathbf{z}_2 = \dot{\mathbf{q}}$ to (2.90), in the same manner as in Sect. 2.1.2, the following discrete description of the RM's dynamics was obtained

$$\begin{aligned}
\mathbf{z}_{1\{k+1\}} &= \mathbf{z}_{1\{k\}} + h\mathbf{z}_{2\{k\}} , \\
\mathbf{z}_{2\{k+1\}} &= \mathbf{z}_{2\{k\}} - h\mathbf{M}^{-1}\left(\mathbf{z}_{1\{k\}}\right) \left[ \mathbf{C}\left(\mathbf{z}_{1\{k\}}, \mathbf{z}_{2\{k\}}\right) \mathbf{z}_{2\{k\}} + \mathbf{F}\left(\mathbf{z}_{2\{k\}}\right) + \mathbf{G}\left(\mathbf{z}_{1\{k\}}\right) + \boldsymbol{\tau}_{d\{k\}} - \mathbf{u}_{\{k\}} \right] ,
\end{aligned}$$
$$(2.91)$$

where $\mathbf{z}_{1\{k\}} = \left[ z_{11\{k\}}, z_{12\{k\}}, z_{13\{k\}} \right]^T$ – vector of links' discrete angles of rotation corresponding to the continuous vector $\mathbf{q}$, $\mathbf{z}_{2\{k\}} = \left[ z_{21\{k\}}, z_{22\{k\}}, z_{23\{k\}} \right]^T$ – vector of discrete angular velocities that corresponds to the continuous vector $\dot{\mathbf{q}}$, $k$ – index of iteration steps, $h$ – time discretization parameter. Difference equations of RM's dynamics (2.91) that were derived by means of Lagrange's equations of the second kind are used further in this work for the modeling of motion of a controlled object, while a discrete form of relation (2.83) was applied in the synthesis of manipulator's TCP tracking control algorithms.

The structural properties of the RM's dynamic model (2.83) are used in the synthesis of tracking control algorithms. These properties are presented below.

**Structural Properties of the RM's Mathematical Model**
Description of the RM's dynamics in the form of Eq. (2.83) complies with the following assumptions [4, 18, 19]:
1. Inertia matrix $\mathbf{M}(\mathbf{q})$ is symmetric and positive-definite, and meets the conditions

$$\sigma_{min}(\mathbf{M}(\mathbf{q}))\,\mathbf{I} \le \mathbf{M}(\mathbf{q}) \le \sigma_{max}(\mathbf{M}(\mathbf{q}))\,\mathbf{I}, \tag{2.92}$$

   where $\mathbf{I}$ – identity matrix, $\sigma_{min}(\mathbf{M}(\mathbf{q}))$, $\sigma_{max}(\mathbf{M}(\mathbf{q}))$ – respectively, minimum and maximum strictly positive eigenvalue of the interia matrix.
2. Matrix $\mathbf{C}(\mathbf{q}, \dot{\mathbf{q}})$ is such that the matrix

$$\mathbf{S}(\mathbf{q}, \dot{\mathbf{q}}) = \dot{\mathbf{M}} - 2\mathbf{C}(\mathbf{q}, \dot{\mathbf{q}}), \tag{2.93}$$

   is skew-symmetric. Thus, the following relation holds

$$\boldsymbol{\xi}^T \mathbf{S}(\mathbf{q}, \dot{\mathbf{q}})\, \boldsymbol{\xi} = \mathbf{0}, \tag{2.94}$$

   where $\boldsymbol{\xi}$ – any given vector of appropriate dimension.
3. Vector of disturbances $\boldsymbol{\tau}_d(t)$ acting on the RM is bounded so that $\| \tau_{dj}(t) \| \le b_{dj}$, where $b_{dj}$ is a positive constant, $j = 1, 2, 3$.
4. The RM's dynamic Eq. (2.83) can be written in a linear form with respect to the parameters, hence

$$\mathbf{M}(\mathbf{q})\ddot{\mathbf{q}} + \mathbf{C}(\mathbf{q}, \dot{\mathbf{q}})\dot{\mathbf{q}} + \mathbf{F}(\dot{\mathbf{q}}) + \mathbf{G}(\mathbf{q}) + \boldsymbol{\tau}_d(t) = \mathbf{Y}(\mathbf{q}, \dot{\mathbf{q}}, \ddot{\mathbf{q}})^T \mathbf{p} + \boldsymbol{\tau}_d(t) = \mathbf{u}, \tag{2.95}$$

   vector of parameters $\mathbf{p}$ shall consist of a minimum number of linearly independent elements.
5. The following relation holds for any two $\mathbf{x}$ and $\mathbf{y}$ vectors

$$\mathbf{C}(\mathbf{q}, \mathbf{x})\,\mathbf{y} = \mathbf{C}(\mathbf{q}, \mathbf{y})\,\mathbf{x}. \tag{2.96}$$

6. Matrix $\mathbf{C}(\mathbf{q}, \dot{\mathbf{q}})$ meets the condition

$$\|\mathbf{C}(\mathbf{q}, \dot{\mathbf{q}})\| \le K_C \|\dot{\mathbf{q}}\|, \tag{2.97}$$

   where $K_C > 0$.

7. Vector $\mathbf{C}(\mathbf{q}, \dot{\mathbf{q}})\,\dot{\mathbf{q}}$ meets the condition

$$\|\mathbf{C}(\mathbf{q}, \dot{\mathbf{q}})\,\dot{\mathbf{q}}\| \le k_C \|\dot{\mathbf{q}}\|^2 , \tag{2.98}$$

where $k_C > 0$.
8. Vector $\mathbf{F}(\dot{\mathbf{q}})$ meets the condition

$$\|\mathbf{F}(\dot{\mathbf{q}})\| \le K_F \|\dot{\mathbf{q}}\| + k_F , \tag{2.99}$$

where $K_F > 0$, and $k_F > 0$.
9. Vector $\mathbf{G}(\mathbf{q})$ meets the condition

$$\|\mathbf{G}(\mathbf{q})\| \le k_G , \tag{2.100}$$

where $k_G > 0$.

### 2.2.2.1   Simulation of Inverse Dynamics Problem of a Robotic Manipulator

In the simulation of the manipulator's motion the robot's TCP moves along the desired circular segment path with center $E$ and radius $R = 0.1$ [m]. The inverse kinematics solution trajectory from Sect. 2.2.1.2 was used. The values of control signals $u_1$, $u_2$ and $u_3$ were generated based on manipulator's parameters of motion such as rotation angles ($\mathbf{q}$), angular velocities ($\dot{\mathbf{q}}$) and angular accelerations ($\ddot{\mathbf{q}}$), according to relation (2.83). The control signals cause the robotic manipulator's TCP motion on the determined path. The joint variables take their set values. The nominal values of manipulator's parameters $\mathbf{p}$ were assumed as defined in Table 2.3. The diagram of the defined TCP's motion path is shown in Fig. 2.16a, pre-determined rotation angles of manipulator's links ($q_1, q_2, q_3$) and angular velocities values ($\dot{q}_1, \dot{q}_2, \dot{q}_3$) are illustrated in Fig. 2.16b,c, respectively. Figure 2.16d presents the generated control signals $u_1$, $u_2$ and $u_3$.

The desired reference trajectory consists of several motion phases, which result from the TCP's defined motion path in a pick and place type of task, where point $S$ marks the TCP's initial position and point $G$ is the path's reversal point. The trajectory is executed such that the TCP's motion cycle (from point $S$ to point $G$ and back) is performed three times. Each time the manipulator's TCP stops at point $S$ and $G$ for the period $t = 2$ [s]. In the initial phase of the simulation ($t \in < 0, 4 >$), the signals controlling links 2 and 3 movement take non-zero values, though no manipulator movement occurs. This is because the model of dynamics (2.83) includes the vector of moments resulting from gravitational forces $\mathbf{G}(\mathbf{q})$ that must be balanced by control signals to ensure the performance of motion parameters. In the case of link 1, whose

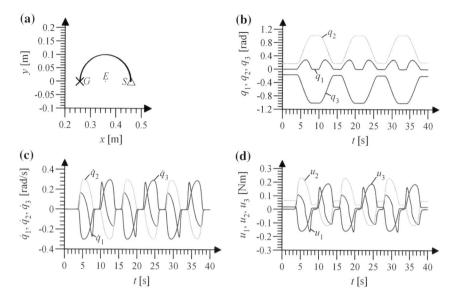

**Fig. 2.16** **a** Desired motion path of point $C$ attached to the manipulator's arm, **b** links rotation angles $q_1$, $q_2$, $q_3$, **c** angular velocities $\dot{q}_1$, $\dot{q}_2$, $\dot{q}_3$, **d** control signals $u_1$, $u_2$, $u_3$

axis of rotation is parallel to $z$-axis of the global coordinate system, there is no impact of the moments of gravitational forces upon the link movement. In the following motion phases, the values of control signals depend on the parameters of motion $\mathbf{q}$, $\dot{\mathbf{q}}$ and $\ddot{\mathbf{q}}$, where the dominant influence is exercised by the vector of moments resulting from the resistance to motion $\mathbf{F}(\dot{\mathbf{q}})$, whose values are the functions of links' angular velocities vector $\dot{\mathbf{q}}$.

# References

1. Active Media Robotics Pioneer 2/PeopleBot Operations Manual, Active Media, version 9, Peterborough (2001)
2. Angeles, J.: Fundamentals of Robotic Mechanical Systems: Theory, Methods, and Algorithms. Springer, New York (2007)
3. Canudas de Wit, C., Siciliano, B., Bastin, G.: Theory of Robot Control. Springer, London (1996)
4. Craig, J.J.: Introduction to Robotics. Prentice Hall, Upper Saddle River (2005)
5. Eykhoff, P.: Identification in Dynamical Systems. (in Polish) PWN, Warsaw (1980)
6. Fierro, R., Lewis F.L.: Control of a nonholonomic mobile robot using neural networks. IEEE Trans. Neural Netw. **9**(4), 589–600 (1998)
7. Giergiel, J., Hendzel, Z., Zylski, W.: Modeling and Control of Wheeled Mobile Robots. (in Polish) Scientific Publishing PWN, Warsaw (2002)
8. Giergiel, J., Zylski, W.: Description of motion of a mobile robot by Maggie's equations. JTAM **43**, 511–521 (2005)

9. Gierlak, P.: Analysis of the kinematics of the 5DOF manipulator. (in Polish) Scientific Letters of Rzeszow University of Technology, Mechanics **86**, 491–500 (2014)
10. Gutowski, R.: Analytical Mechanics. (in Polish) PWN, Warsaw (1971)
11. Hendzel, Z.: Tracking Control of Wheeled Mobile Robots. (in Polish) Rzeszow University of Technology Publishing House, Rzeszow (1996)
12. Hendzel, Z., Nawrocki, M.: Neural identification for Scorbot manipulator. (in Polish) Modelling. Eng. **36**, 135–142 (2008)
13. Hendzel, Z., Nawrocki, M.: Identification of the robot model parameters. (in Polish) Acta Mech. Autom. **4**, 69–73 (2010)
14. Hendzel, Z., Trojnacki, M.: Neural Network Control of Mobile Wheeled Robots. (in Polish) Rzeszow University of Technology Publishing House, Rzeszow (2008)
15. Kozowski, K., Dutkiewicz, P., Wrblewski, W.: Modeling and Control of Robots. (in Polish) Scientific Publishing PWN, Warsaw (2003)
16. Morecki, A., Knapczyk, J.: Basis of Robotics: Theory and Elements of Manipulators and Robots. WNT, Warsaw (1999)
17. SCORBOT-ER 4pc User's Manual, Eshed Robotec, version A, Rosh Haayin (1999)
18. Slotine, J.J., Li, W.: Applied Nonlinear Control. Prentice Hall, New Jersey (1991)
19. Spong, M.W., Vidyasagar, M.: Robot Dynamics and Control. (in Polish) WNT, Warsaw (1997)
20. Tcho, K., Mazur, A., Dulba, I., Hossa R., Muszyski, R.: Mobile Manipulators and Robots. (in Polish) Academic Publ. Company PLJ, Warsaw (2000)
21. Zylski, W.: Kinematics and Dynamics of Wheeled Mobile Robotos. (in Polish) Rzeszow University of Technology Publishing House, Rzeszow (1996)
22. Zylski, W., Gierlak, P.: Modelling of Movement of Selected Manipulator. (in Polish) Acta Mech. Autom. **4**, 112–119 (2010)
23. Zylski, W., Gierlak, P.: Tracking Control of Robotic Manipulator. (in Polish) Rzeszow University of Technology Publishing House, Rzeszow (2014)

# Chapter 3
# Intelligent Control of Mechatronic Systems

Among the variety of methods for mechatronic systems control, the intelligent control uses modern algorithms that compensate for the nonlinearity of controlled systems. These algorithms can adapt their parameters to variable operating conditions and comprise artificial intelligence methods such as artificial neural networks, and fuzzy logic algorithms. A distinction can be made between fuzzy control and neural control based on the type of artificial intelligence algorithms. This group of algorithms also includes the adaptive control algorithm, whose structure results from the model construction of a controlled object and whose parametric values can be adapted to ensure the required control performance.

## 3.1 Methods for Control of Nonlinear Systems

Due to its complex mechanical structure, from a mathematical point of view, the WMR is an object described with nonlinear dynamic equations and subjected to nonholonomic constraints, which affects the synthesis of a stable control law [6, 7]. The tracking control system can be construed as a motion execution layer of the WMR's hierarchical control system, where a trajectory is generated by a superior layer. The RM, however, is a mechatronic system subjected to holonomic constraints [34]. Its dynamics is described by means of nonlinear equations of motion, where some parameters may be unknown or change depending on the operating point of a device, which makes such an object difficult to control. Promoting fail-safe practices in the field of robotics requires development of effective tracking control algorithms for RMs and WMRs. The literature on the control of mechatronic systems provides different classes of control algorithms. This chapter focuses on a selected structure of nonlinear objects' control system.

Control systems of mechatronic objects were primarily composed solely of linear PID controllers (proportional-integral-derivative) based on error signals, its integral

© Springer International Publishing AG 2018
M. Szuster and Z. Hendzel, *Intelligent Optimal Adaptive Control for Mechatronic Systems*, Studies in Systems, Decision and Control 120, https://doi.org/10.1007/978-3-319-68826-8_3

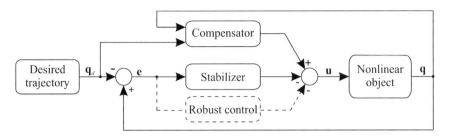

**Fig. 3.1** Schematic diagram of the structure of a nonlinear object tracking control system

and derivative [3, 7, 20, 30]. Application of a linear control law to objects described with nonlinear dynamic equations results in major errors in the execution of the desired trajectory. An advantage of the aforementioned algorithms is that they have simple structures and the knowledge of the mathematical model of a controlled object is not required. In contrast, their main disadvantage is that they do not provide for the compensation for nonlinearity of the controlled object, which results in low performance of the tracking motion.

To improve the quality of the tracking motion performed by nonlinear dynamic objects, a more elaborate control system shall be applied (see Fig. 3.1). The control system presented in the diagram consists of two basic elements, a compensator and a stabilizer. The compensator structure compensates for nonlinearity of a controlled object, and the stabilizer generates control actions that eliminate tracking errors resulting from incomplete compensation. An optional element of the control system is the robust control algorithm which ensures stability of the control system in a presence of major disturbances.

The concept of such a control system leads to a flexible structure, where performance of different control system components depends on the assumptions adopted at the stage of the control algorithm synthesis [6, 7, 34].

The process of synthesis of a tracking control algorithm for a nonlinear dynamic object, with the use of classic methods such as optimal control algorithms or different algebraic methods, often requires the knowledge of the mathematical model of a controlled object. It is not an easy task, and it causes inconvenience to select a relevant class of models, the most appropriate model structure within a given class and model coefficients. The accuracy of real object mapping i.e. creating a model representation of the effects that occur within a real object, is related to the complexity of a model. This affects the number of computations that are performed by a microprocessor system, which executes tasks of the control system.

The change in the properties of a controlled object may result from a variety of factors such as components obsolescence, variable operating conditions e.g. changes in resistance to motion as the WMR enters a surface with different parameters, changes in mass moments of inertia and in the values of gravity forces acting on the arm of a load lifting manipulator. These factors have an impact on the accuracy of a model under given operating conditions. For instance, the mass moment of inertia of the WMR's frame varies depending on the distribution of mass and mass value of the load carried. In practice, models that are applied to describe dynamic objects only approximate their properties, which is why they are called models with uncertainty [29]. The selection of an appropriate object model for a control algorithm is an inconvenient and difficult task. Therefore, efforts are made to devise methods for control system design that would not require a highly accurate model.

Adaptive control system is an example of such an algorithm [7, 10, 13, 29, 30], which is designed in such a way that it can independently modify its properties in real time, should there be a change of controlled object parameters. This feature allows for high tracking control accuracy in the event of major changes of controlled object parameters. Adaptive control algorithms belong to the class of direct algorithms, which are globally stable and, in the ideal case, ensure the execution of the tracking motion. In practical applications, where external interference occurs and the parameters of a controlled object may change, the lack of proper stimulation of the adaptive structure may lead to unwanted events which can be prevented by application of persistent excitation input signals [1, 16, 23, 29] or by adequate modification of the control algorithm [16, 23, 24, 31].

In respect of nonlinearities that occur in the description of nonlinear objects dynamics, the application of modern AI methods to tracking control algorithms has proved positive. These methods include, inter alia, artificial neural networks (NNs) that have become attractive tools used in modern control algorithms for nonlinear systems [3, 4, 7, 11, 13, 15, 18, 21, 22, 31]. This is due to the fact that NNs can approximate any nonlinear function, they have the ability to learn and can also adapt their parameters and structure. However, not all of the known NN models are suitable for application to motion control algorithms for dynamic objects. This is due to their structural complexity or the time needed for the adaptation of parameters. Neural tracking control systems mainly employ simple NNs due to hardware limitations and necessity to perform real-time computation. These networks include e.g. single-layer NNs with a fixed number of neurons in the hidden layer and values of neuron activation function that are easy to compute, as well as ontogenic NNs [4, 11, 17, 19, 22], which adapt the complexity of their structure to the problem that is being solved.

Reinforcement learning methods are currently becoming more common in terms of nonlinear object control. These comprise the neural dynamic programming algorithms [2, 5, 8, 9, 12, 26–28, 32, 33], which are a combination of the classical theory of optimization and modern AI methods such as NNs. They can be adapted to different classes of problems and allow for the optimization of control processes related to nonlinear objects e.g. RMs or WMRs. A broader description of the reinforcement learning concept is given in Sect. 5.2, while a detailed description of selected neural dynamic programming algorithms is presented in Chap. 6.

Adaptive methods are effectively used for nonlinear object control, where the operating conditions of a controlled object are subject to change. A neural control algorithm is an example of a method whose key feature is the application of structures that can adapt their parameters and adjust the generated control signal to the changing properties of a controlled object, which ensures high quality performance of the tracking motion. A neural control algorithm, together with examples of NN models that apply to it, is discussed below.

## 3.2   Neural Control

In neural control systems, object nonlinearity compensation is carried out by means of an NN. One of the NN's characteristics is its ability to approximate any nonlinear function. It can be demonstrated that an NN composed of two layers of neurons is capable of approximating a given continuous function with a predefined accuracy.

Adaptation or training of NN's parameters such as weights, can be performed with different methods e.g. gradient algorithms for learning with a teacher. NN learning methods are discussed in Chap. 5.

Weight adaptation is the process of updating weights in an NN according to the assumed adaptive algorithm. The process is carried out online with continuous stream of data into an NN. The concept of online weight update involves certain limitations with regard to the adopted NN structure, which may not be too complex as the output must be computed within a given time interval. Neural control algorithms are mostly based on linear neural networks with one layer of adaptive weights. Figure 3.2 shows a schematic diagram of a neural control algorithm.

Weight training is the process of adjusting weight values according to the adopted learning algorithm, to minimize error occurring in supervised learning, that follows the representation of all samples from the set of learning data. It is to be noted, that the number of trained NN's parameters may range from tens to several hundred, while the number of pattern representations from a learning set may amount to several hundred thousand, until the expected mapping quality is achieved.

When applying NNs to control algorithms it is essential to normalize the input signals $\mathbf{x}$ to the NN $\mathbf{x}_N = \kappa_{N*}\mathbf{x}$, $x_{Ni} \in \langle -1, 1 \rangle$, where $i = 1, \ldots, M$, $M$ – number of

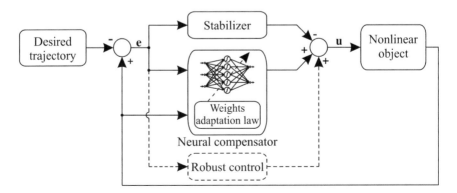

**Fig. 3.2**   General schematic diagram of a neural control system for a nonlinear object

**Fig. 3.3** General schematic
diagram of a three-layer NN
structure

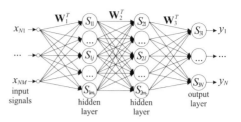

inputs to the NN, $\kappa_{N*}$ – diagonal design matrix of the NN's input scaling coefficients.
Figure 3.3 shows a general schematic of a three-layer Multi Input Multi Output
(MIMO) neural network. The structure of such an NN is very flexible and, depending
upon the type of neuron activation function, may have a specific architecture.

The presented NN consists of three layers:

- the first hidden layer with inputs $x_{Ni}$, outputs $S_{1j}$, and weights $W_{1i,j}$,
- the second hidden layer with inputs $S_{1j}$, outputs $S_{2l}$, and weights $W_{2j,l}$,
- the output layer with inputs $S_{2l}$, outputs $y_g$, and weights $W_{3l,g}$,

where $j, l, g$ – respective indexes, $j = 1, \ldots, m_1, l = 1, \ldots, m_2, g = 1, \ldots, N, m_1$
– number of neurons in the first hidden layer, $m_2$ – number of neurons in the second
hidden layer, $N$ – number of outputs from the NN, $\mathbf{S}_1, \mathbf{S}_2, \mathbf{S}_3$ – vectors of neuron
activation functions of respective layers, $\mathbf{W}_1, \mathbf{W}_2, \mathbf{W}_3$ – weight matrices. NNs that
are used for approximation may apply different types of neuron activation functions
such as continuous local functions (e.g. Gaussian functions, bicentral functions) or
non-local (e.g. linear functions, unipolar and bipolar sigmoid functions). There are
also NNs that implement discontinuous neuron activation functions e.g. unit step
function, however they are generally used for signal classification. For instance, the
$l$-th neuron of the second hidden layer is described with the equation

$$
S_{2l} = S \left( \sum_{j=1}^{m_1} W_{2j,l} S_{1j} \right) . \tag{3.1}
$$

Implementing unipolar sigmoid activation functions, the value of the $l$-th neuron of
the second layer is given by

$$
S_{2l} = \frac{1}{1 + \exp\left(-\beta_N \left(\mathbf{W}_{2:,l}^T \mathbf{S}_1 + W_{b2l}\right)\right)}, \tag{3.2}
$$

where $\beta_N$ – slope coefficient of sigmoid function at inflection point, $\mathbf{W}_{2:,l}$ – $l$-th
column of the matrix of input weights to the NN second layer, $W_{b2l}$ – extra weight
of the $l$-th neuron of the second layer, the so-called bias, resulting from the specific
nature of the applied neuron activation functions.

**Fig. 3.4** Values of unipolar
sigmoid functions

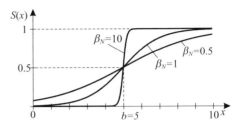

Figure 3.4 shows unipolar sigmoid neuron activation functions with respect to different $\beta_N$ coefficient values.

The synthesis of control systems employs Random Vector Functional Link NNs and NNs with Gaussian - type activation functions that are discussed in the subchapters to follow.

### 3.2.1  Random Vector Functional Link Neural Network

NN with functional extensions Random Vector Functional Link (RVFL) [7, 14, 21] is a single-layer, linear network with respect to the output layer weights $\mathbf{W}$, with fixed values of the input layer weights $\mathbf{D}_{N*}$, randomly selected in the initialization process, and with bipolar sigmoid neuron activation functions

$$S\left(\mathbf{x}_N\right) = \frac{2}{1 + \exp\left(-\beta_N\left(\mathbf{D}_{N*}^T\mathbf{x}_N + D_{Nb}\right)\right)} - 1 , \qquad (3.3)$$

where $D_{Nb}$ – additional weight associated with the neuron, the so-called bias. Notation (3.3) is simplified by introducing augmented vector of inputs to the RVFL $\mathbf{x}_N = \kappa_N\left[1, \mathbf{x}^T\right]^T$ where $\kappa_N$ is the augumented matrix $\kappa_{N*}$, thus $D_{Nb}$ is not considered in the description of neuron activation function.

The value of output from the RVFL NN is given by

$$y = \mathbf{W}^T\mathbf{S}\left(\mathbf{D}_N^T\mathbf{x}_N\right) , \qquad (3.4)$$

where $\mathbf{D}_N$ – augmented vector of fixed weights of the RVFL NN's input layer $\mathbf{D}_N = \left[D_{Nb}, \mathbf{D}_{N*}^T\right]^T$.

Figure 3.5 shows a schematic diagram of RVFL NN with one output.

RVFL NNs have a simple structure and are linear with respect to the output layer weights. The adjustment of RVFL NN's output layer weights is carried out by means of gradient methods, therefore network's adaptation process does not require significant computation overhead.

**Fig. 3.5**  General schematic diagram of an RVFL NN

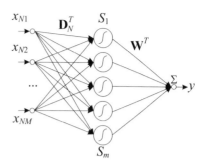

**Fig. 3.6**  Gaussian-type activation function

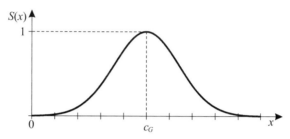

**Fig. 3.7**  General schematic diagram of an NN with Gaussian-type activation functions

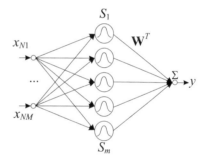

### 3.2.2  *Neural Network with Gaussian-Type Activation Functions*

A single-layer NN with Gaussian - type activation functions [25] is a linear network with respect to the output layer weights. One-dimensional Gaussian-type activation function is described by the relation

$$S\left(x_{N}\right)=\exp^{-\dfrac{\left(x_{N}-c_{G}\right)^{2}}{2r_{G}^{2}}}, \tag{3.5}$$

where $c_{G}$ – position of the center of the function, $r_{G}$ - width of the Gaussian function. Gaussian function is shown in Fig. 3.6.

The value of output from a single-layer NN with Gaussian-type activation functions can be written as

$$y = \mathbf{W}^T \mathbf{S}(\mathbf{x}_N) \ . \tag{3.6}$$

A schematic structure of an NN with Gaussian-type activation functions is shown in Fig. 3.7.

Single-layer NNs with Gaussian-type activation functions have a simple structure. However, Gaussian curves are local functions, thus an increase in the magnitude of the input vector results in a rapid growth of the number of NN's neurons. Therefore, NNs with Gaussian-type activation functions are used for approximation of functions with a small number of variables.

# References

1. Astrom, K.J., Wittenmark, B.: Adaptive Control. Addison-Wesley, New York (1979)
2. Barto, A., Sutton, R., Anderson, C.: Neuronlike adaptive elements that can solve difficult learning problems. IEEE Trans. Syst. Man Cybern. Syst. **13**, 834–846 (1983)
3. Burns, R.S.: Advanced Control Engineering. Butterworth-Heinemann, Oxford (2001)
4. Fabri, S., Kadirkamanathan, V.: Dynamic structure neural networks for stable adaptive control of nonlinear systems. IEEE Trans. Neural. Netw. **12**, 1151–1167 (1996)
5. Ferrari, S., Stengel, R.F.: An adaptive critic global controller. In: Proceedings of American Control Conference, vol. 4, pp. 2665–2670. Anchorage, Alaska (2002)
6. Giergiel, J., Hendzel, Z., ylski, W.: Kinematics, Dynamics and Control of Wheeled Mobile Robots in Mechatronic Aspect. (in Polish) Faculty IMiR AGH, Krakow (2000)
7. Giergiel, J., Hendzel, Z., ylski, W.: Modeling and Control of Wheeled Mobile Robots. (in Polish) Scientific Publishing PWN, Warsaw (2002)
8. Gierlak, P., Szuster, M., ylski, W.: Discrete Dual-heuristic Programming in 3DOF Manipulator Control. Lecture Notes in Artificial Intelligence. vol. 6114, 256–263 (2010)
9. Hendzel, Z.: An adaptive critic neural network for motion control of a wheeled mobile robot. Nonlinear Dyn. **50**, 849–855 (2007)
10. Hendzel, Z., Burghardt, A.: Behavioural Control of Wheeled Mobile Robots. (in Polish) Rzeszow University of Technology Publishing House, Rzeszow (2007)
11. Hendzel, Z., Szuster, M.: A dynamic structure neural network for motion control of a wheeled mobile robot. In: Rutkowski, L., Tadeusiewicz, R., Zadeh, L.A., Zurada, J. (eds.) Computational Inteligence: Methods ans Applications, pp. 365–376. EXIT, Warsaw (2008)
12. Hendzel, Z., Szuster, M.: Discrete neural dynamic programming in wheeled mobile robot control. Commun. Nonlinear. Sci. Numer. Simul. **16**, 2355–2362 (2011)
13. Hendzel, Z., Trojnacki, M.: Neural Network Control of Mobile Wheeled Robots. (in Polish) Rzeszow University of Technology Publishing House, Rzeszow (2008)
14. Igelnik, B., Pao, Y.-H.: Stochastic choice of basis functions in adaptive function approximation and the functional-link net. IEEE Trans. Neural. Netw. **6**, 1320–1329 (1995)
15. Jamshidi, M., Zilouchian, A.: Intelligent Control Systems Using Soft Computing Methodologies. CRC Press, London (2001)
16. Janecki, D.: The Role of Uniformly Excitation Signals in Adaptive Control Systems. (in Polish) IPPT PAS, Warsaw (1995)
17. Jankowski, N.: Ontogenic Neural Networks. (in Polish) Exit, Warsaw (2003)
18. Kecman, V.: Learning and Soft Computing. MIT Press, Cambridge (2001)
19. Kim, Y.H., Lewis, F.L.: A dynamical recurrent neural-network-based adaptive observer for a class of nonlinear systems. Automatica **33**, 1539–1543 (1997)

20. Kozowski, K., Dutkiewicz, P., Wrblewski, W.: Modeling and Control of Robots. (in Polish) Scientific Publishing PWN, Warsaw (2003)
21. Levis, F.L., Liu, K., Yesildirek, A.: Neural net robot controller with guaranted tracking performance. IEEE Trans. Neural. Netw. **6**, 703–715 (1995)
22. Liu, G.P.: Nonlinear Identification and Control. Springer, London (2001)
23. Niderliski, A., Mociski, J., Ogonowski, Z.: Adaptive Regulation. (in Polish) PWN, Warsaw (1995)
24. Ortega, R., Spong, M.W.: Adaptive motion control of rigid robots: a tutorial. Automatica **25**, 877–888 (1989)
25. Osowski, S.: Neural Networks - An Algorithmic Approach. (in Polish) WNT, Warsaw (1996)
26. Powell, W.B.: Approximate Dynamic Programming: Solving the Curses of Dimensionality. Princeton, Wiley-Interscience (2007)
27. Prokhorov, D., Wunch, D.: Adaptive critic designs. IEEE Trans. Neural. Netw. **8**, 997–1007 (1997)
28. Si, J., Barto, A.G., Powell, W.B., Wunsch, D.: Handbook of Learning and Approximate Dynamic Programming. IEEE Press, Wiley-Interscience (2004)
29. Slotine, J.J., Li, W.: Applied Nonlinear Control. Prentice Hall, New Jersey (1991)
30. Spong, M.W., Vidyasagar, M.: Robot Dynamics and Control. (in Polish) WNT, Warsaw (1997)
31. Spooner, J.T., Passio, K.M.: Stable adaptive control using fuzzy systems and neural networks. IEEE Trans. Fuzzy. Syst. **4**, 339–359 (1996)
32. Syam, R., Watanabe, K., Izumi, K.: Adaptive actor-critic learning for the control of mobile robots by applying predictive models. Soft. Comput. **9**, 835–845 (2005)
33. Visnevski, N., Prokhorov, D.: Control of a nonlinear multivariable system with adaptive critic designs. In: Proceedings of Artificial Neural Networks in Engineering. vol. 6, pp. 559–565 (1996)
34. Zylski, W., Gierlak, P.: Tracking Control of Robotic Manipulator. (in Polish) Rzeszow University of Technology Publishing House, Rzeszow (2014)

# Chapter 4
# Optimal Control Methods for Mechatronic Systems

A number of methods have been devised to deal with the problem of optimal control. These include the Bellman's dynamic programming, which is a very common method that allows for the algorithm synthesis of an optimal linear-quadratic regulator. Equally interesting are the methods for control law synthesis that use the Pontryagin's maximum principle. This chapter provides a brief description of the aforementioned methods and examples of their application with regard to optimal control of a discrete linear object of the first order.

## 4.1 Bellman's Dynamic Programming

The dynamic programming method (DP) was proposed by Richard Bellman in 1957 [1, 2, 5–7], as an alternative to the calculus of variations used in optimal control theory. This method allows for the determination of an optimal control law for a dynamic object.

A description of a nonlinear dynamic system is given by

$$\mathbf{x}_{\{k+1\}} = \mathbf{f}\left(\mathbf{x}_{\{k\}}, \mathbf{u}_{\{k\}}\right) , \tag{4.1}$$

where $\mathbf{x}_{\{k\}}$ – the object's state vector, $\mathbf{u}_{\{k\}}$ – the control vector, $k$ – the discrete-time step index, $k = 0, \ldots, n$.

A performance index was adopted, referred to as a finite-horizon value function of $n$ iteration steps

$$V_{\{k\}}\left(\mathbf{x}_{\{k\}}, \mathbf{u}_{\{k\}}\right) = \gamma^n \Phi\left(\mathbf{x}_{\{n\}}\right) + \sum_{k=0}^{n-1} \gamma^k L_{C\{k\}}\left(\mathbf{x}_{\{k\}}, \mathbf{u}_{\{k\}}\right) , \tag{4.2}$$

© Springer International Publishing AG 2018
M. Szuster and Z. Hendzel, *Intelligent Optimal Adaptive Control for Mechatronic Systems*, Studies in Systems, Decision and Control 120,
https://doi.org/10.1007/978-3-319-68826-8_4

where $\Phi\left(\mathbf{x}_{\{n\}}\right)$ – local cost value in the $n$-th iteration step, $L_{C\{k\}}\left(\mathbf{x}_{\{k\}}, \mathbf{u}_{\{k\}}\right)$ – local cost value in the $k$-th iteration step, $\gamma$ – discount factor of local cost values in subsequent iteration steps, $\gamma \in \langle 0, 1\rangle$. In further considerations it is assumed that $\gamma = 1$. Local cost $L_{C\{k\}}\left(\mathbf{x}_{\{k\}}, \mathbf{u}_{\{k\}}\right)$, also referred to as the reinforcement or the cost function, in the general case, depends on the object's state and control, and is given by

$$L_{C\{k\}}\left(\mathbf{x}_{\{k\}}, \mathbf{u}_{\{k\}}\right) = \frac{1}{2}\left(\mathbf{x}_{\{k\}}^T \mathbf{R} \mathbf{x}_{\{k\}} + \mathbf{u}_{\{k\}}^T \mathbf{Q} \mathbf{u}_{\{k\}}\right), \tag{4.3}$$

where $\mathbf{R}, \mathbf{Q}$ – design matrices with positive coefficients, that define the influence of the object's state $\mathbf{x}_{\{k\}}$ and control $\mathbf{u}_{\{k\}}$ on the cost function value. Step $k = n$ is the final stage of the iterative process, thus, by definition, there is no such control $\mathbf{u}_{\{n\}}$, that shall transfer the object to state $\mathbf{x}_{\{n+1\}}$. The local cost $\Phi\left(\mathbf{x}_{\{n\}}\right)$, based on (4.3), takes the form

$$\Phi\left(\mathbf{x}_{\{n\}}\right) = \frac{1}{2}\left(\mathbf{x}_{\{n\}}^T \mathbf{R}_E \mathbf{x}_{\{n\}}\right), \tag{4.4}$$

where $\mathbf{R}_E$ – positive-definite, diagonal design matrix. $\mathbf{R}$ and $\mathbf{R}_E$ matrices coefficients may differ, as more emphasis is placed on the reduction of the state vector value in the final $n$-th step of the process than in other iteration stages; in a specific case $\mathbf{R} = \mathbf{R}_E$.

In a general case of an infinite-time horizon $k \to \infty$, relation (4.4) takes the form

$$V_{\{k\}}\left(\mathbf{x}_{\{k\}}, \mathbf{u}_{\{k\}}\right) = \lim_{n\to\infty} \sum_{k=0}^{n} L_{C\{k\}}\left(\mathbf{x}_{\{k\}}, \mathbf{u}_{\{k\}}\right). \tag{4.5}$$

In the case where an object is in any physically allowable state $\mathbf{x}_{\{k\}}$, the value function is related to the cost of system's transition from the $k$-th step state to the $n$-th step state, which is given by [1, 3]

$$V_{\{k\}}\left(\mathbf{x}_{\{k\}}, \mathbf{u}_{\{k\}}, \ldots, \mathbf{u}_{\{n-1\}}\right) = L_{C\{k\}}\left(\mathbf{x}_{\{k\}}, \mathbf{u}_{\{k\}}\right) + V_{\{k+1\}}\left(\mathbf{x}_{\{k+1\}}, \mathbf{u}_{\{k+1\}}, \ldots, \mathbf{u}_{\{n-1\}}\right), \tag{4.6}$$

where $\mathbf{x}_{\{k+1\}}$ – object's state in the $k + 1$ step, which depends on state $\mathbf{x}_{\{k\}}$, and control $\mathbf{u}_{\{k\}}$ in the $k$-th step. The state of a controlled object in any step of the iterative process may be determined based on the knowledge of state $\mathbf{x}_{\{k\}}$ and the sequence of selected controls $\mathbf{u}_{\{k\}}, \ldots, \mathbf{u}_{\{n-1\}}$, thus being the Markov chain [7].

From Bellman's principle of optimality it follows

$$V_{\{k\}}^*\left(\mathbf{x}_{\{k\}}^*, \mathbf{u}_{\{k\}}\right) = \min_{\mathbf{u}_{\{k\}}, \ldots, \mathbf{u}_{\{n-1\}}} \left\{ L_{C\{k\}}\left(\mathbf{x}_{\{k\}}^*, \mathbf{u}_{\{k\}}\right) \right.$$
$$\left. + V_{\{k+1\}}^*\left(\mathbf{x}_{\{k+1\}}^*, \mathbf{u}_{\{k+1\}}, \ldots, \mathbf{u}_{\{n-1\}}\right) \right\}, \tag{4.7}$$

where $V_{\{k\}}^*\left(\mathbf{x}_{\{k\}}^*, \mathbf{u}_{\{k\}}\right)$ – the optimal value function. The control law $\mathbf{u}_{\{k\}}, \ldots, \mathbf{u}_{\{n-1\}}$ for which relation (4.7) is satisfied, is the optimal control law $\mathbf{u}_{\{k\}}^*, \ldots, \mathbf{u}_{\{n-1\}}^*$, due to the assumed performance index. The value function $V_{\{k\}}^*$ of the optimal trajectory in the $k$-th step of the iterative process is independent from the object's states in the preceding stages, and an optimal control policy has the property that whatever the initial state and initial decision are, the remaining decisions must constitute an optimal policy with regard to the state resulting from the first decision [1], thus

$$V^*_{\{k\}}\left(\mathbf{x}^*_{\{k\}}, \mathbf{u}_{\{k\}}\right) = \min_{\mathbf{u}_{\{k\}}} \left\{ L_{C\{k\}}\left(\mathbf{x}^*_{\{k\}}, \mathbf{u}_{\{k\}}\right) + V^*_{\{k+1\}}\left(\mathbf{x}^*_{\{k+1\}}, \mathbf{u}^*_{\{k+1\}}, \dots, \mathbf{u}^*_{\{n-1\}}\right) \right\} \ .$$

(4.8)

To simplify and reduce the notation the value of optimal value function $V^*_{\{k+1\}}$ $\left(\mathbf{x}^*_{\{k+1\}}, \mathbf{u}^*_{\{k+1\}}, \dots, \mathbf{u}^*_{\{n-1\}}\right)$ was written as $V^*_{\{k+1\}}\left(\mathbf{x}^*_{\{k+1\}}, \mathbf{u}^*_{\{k+1\}}\right)$.

Equation (4.8) is a recursive, mathematical form of Bellman's principle of optimality with regard to discrete systems. Despite its concise mathematical form, DP is difficult to implement in practice.

The determination of the optimal trajectory of state $\mathbf{x}^*_{\{k\}}$ and the corresponding optimal control $\mathbf{u}^*_{\{k\}}$ starts from the last step of the iterative process $k = n$ and continues backwards to the first step $k = 0$, in a discretized state-space. Let us assume that in the last $n$-th step of the process, the system's target state is to be $m$, in step $k = n - 1$ the system may take any physical positions in states $g, h, j, k$, from where it can be transitioned to state $m$ by setting control actions, accordingly $u_{gm\{n-1\}}$, $u_{hm\{n-1\}}, u_{jm\{n-1\}}, u_{km\{n-1\}}$, whose local cost of respective transition is $L_{Cgm\{n-1\}}$, $L_{Chm\{n-1\}}, L_{Cjm\{n-1\}}, L_{Ckm\{n-1\}}$, and $V_{gm\{n-1\}} = L_{Cgm\{n-1\}} + \Phi_{m\{n\}}$, where $\Phi_{m\{n\}}$ – the local cost in the last $n$-th step of the iterative process. Similarly $V_{hm\{n-1\}} = L_{Chm\{n-1\}} + \Phi_{m\{n\}}, V_{jm\{n-1\}} = L_{Cjm\{n-1\}} + \Phi_{m\{n\}}, V_{km\{n-1\}} = L_{Ckm\{n-1\}} + \Phi_{m\{n\}}$. Then, in the $k = n - 2$ step, the system may be found in any physically allowable state $a, \dots, q$, from where it can be transitioned to state $g, h, j$, or $k$ by setting appropriate control actions, which incurs a certain local cost $L_{C\{k\}}$ resulting from the transition. For instance, the local cost of transition from state $a$ to state $g$ is $L_{Cag\{n-2\}}$, given a set control of $u_{ag\{n-2\}}$. The value of the value function $V_{agm\{n-2\}}$ is $V_{agm\{n-2\}} = L_{Cag\{n-2\}} + L_{Cgm\{n-1\}} + \Phi_{m\{n\}}$. Similarly, cost function values, sequences of state vector values and controls of any physically feasible trajectory of state must be determined. The number of computations related to the determination of control value $u_{\{k\}}$ and local cost $L_{C\{k\}}$ is rising with the number of states attainable by the system and the number of iteration steps. This phenomenon is referred to as the "curse of dimensionality" [1]. Once the computations are performed for all stages of the process, we obtain certain trajectories of the system state $\mathbf{x}_{\{k=0,\dots,n\}}$ and the corresponding control sequences $\mathbf{u}_{\{k=0,\dots,n-1\}}$, that allow to reach the pre-defined final position, with a certain transition cost being incurred. The optimal trajectory $\mathbf{x}^*_{\{k=0,\dots,n\}}$ is the trajectory, for which value of function $V_{\{k\}}$ is minimal and the control executing such a trajectory is optimal $\mathbf{u}^*_{\{k=0,\dots,n-1\}}$. The process of state trajectory determination in Bellman's DP is shown in Fig. 4.1. The DP method for determination of optimal state trajectory $\mathbf{x}^*_{\{k=0,\dots,n\}}$ executed under optimal control law $\mathbf{u}^*_{\{k=0,\dots,n-1\}}$, starting from the last step of a discrete process $k = n$ and going backwards to the first step $k = 0$, does not allow the optimal control law to be applied on-line, in real-time.

**Fig. 4.1** Schematic diagram
of DP state trajectory
determination

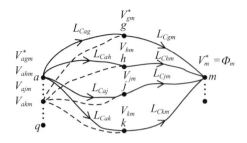

Bellman's DP limitations:

– the model of a controlled object or the process must be known,
– control values and cost function values are computed for all allowable object states
  in each step of the iterative process,
– high computational complexity,
– computations are performed from the last step of a discrete process to the first step
  of the process,
– optimal state trajectory may not be determined on-line.

   For the abovementioned reasons, DP was primarily limited to simple problems
with short time horizons. The rapid development of microprocessor technology in
the last 20 years has allowed for the expansion of potential DP applications in the
definition of optimal control law for complex systems. Nevertheless, the computation
process is still time-consuming. Imposing constraints on the object's state vector and
the control vector reduces computational complexity and allows for the determination
of optimal control law by means of DP.

### *Bellman's dynamic programming in optimal control of a linear dynamic system*

Discrete dynamic system is described by equation

$$x_{\{k+1\}} = Fx_{\{k\}} + Gu_{\{k\}} , \qquad (4.9)$$

where $F = 1$, $G = 1$. Determine optimal control law $u^*_{DP\{k\}}$, allowing for system
transition from the initial state $x_{\{0\}} = 8.9$ to state $x_{\{n\}}$, for $n = 9$, along the optimal
state trajectory $x^*_{\{k\}}$, that minimizes value function $V_{\{k\}}$, given the cost function

$$L_{C\{k\}}\left(x_{\{k\}}, u_{\{k\}}\right) = \frac{1}{2}\left(Rx^2_{\{k\}} + Qu^2_{\{k\}}\right) , \qquad (4.10)$$

where $R = 1$, $Q = 3$. To limit computational complexity, while solving the
problem with Bellman's DP method, the allowable system's state space $x_{\{k\}} \in$

**Table 4.1** Values of optimal state trajectory $x_{\{k\}}^*$, control $u_{DP\{k\}}^*$, and value function $V_{\{k\}}^*$

| $k$ | $x_{\{k\}}^*$ [−] | $u_{DP\{k\}}^*$ [−] | $V_{\{k\}}^*$ [−] |
|---|---|---|---|
| 0 | 8.9 | −3.9 | 91.22 |
| 1 | 5 | −2.2 | 28.8 |
| 2 | 2.8 | −1.2 | 9.04 |
| 3 | 1.6 | −0.7 | 2.96 |
| 4 | 0.9 | −0.4 | 0.945 |
| 5 | 0.5 | −0.2 | 0.3 |
| 6 | 0.3 | −0.1 | 0.115 |
| 7 | 0.2 | −0.1 | 0.055 |
| 8 | 0.1 | −0.1 | 0.02 |
| 9 | 0 | 0 | 0 |

$\{0, 0.1, \ldots, 9.9, 10\}$ and control space $u_{\{k\}} \in \{-5, -4.9, \ldots, -0.1, 0\}$ were discretized.

The optimal control law $u_{DP\{k\}}^*$, obtained with Bellman's DP, executes system transition from the initial state $x_{\{0\}}$ to state $x_{\{n\}}$, for $n = 9$, along the optimal state trajectory $x_{\{k\}}^*$, minimizing the value function $V_{\{k\}}$, given the number of steps in a discrete-time process. The values of individual variables in ten subsequent iteration steps $k = 0, \ldots, 9$, are shown in Table 4.1.

The optimal state trajectory $x_{\{k\}}^*$, optimal control law $u_{DP\{k\}}^*$, and optimal value function $V_{\{k\}}^*$, obtained by solving an example problem of linear object optimal control with Bellman's DP is shown in Fig. 4.2.

## 4.2 Linear-Quadratic Regulator

The synthesis of linear-quadratic regulator (LQR) [4, 7] was carried out in a discrete-time domain with finite-horizon.

Let us consider a discrete, linear, stationary dynamic system, described by equation

$$\mathbf{x}_{\{k+1\}} = \mathbf{F}\mathbf{x}_{\{k\}} + \mathbf{G}\mathbf{u}_{\{k\}} \,, \tag{4.11}$$

where system state $\mathbf{x}_{\{k\}}$ and control $\mathbf{u}_{\{k\}}$ are not subjected to any constraints, $k = 0, \ldots, n$. Find optimal control law, dependent on system state $\mathbf{u}_{\{k\}}^*\left(\mathbf{x}_{\{k\}}\right)$, which minimizes values of the performance index defined by (4.2), given the cost function (4.3) and (4.4), assuming $\gamma = 1$. The value function is given by

$$V_{\{k\}}\left(\mathbf{x}_{\{k\}}, \mathbf{u}_{\{k\}}\right) = \frac{1}{2}\left(\mathbf{x}_{\{n\}}^T \mathbf{R}_E \mathbf{x}_{\{n\}}\right) + \frac{1}{2}\sum_{k=0}^{n-1}\left(\mathbf{x}_{\{k\}}^T \mathbf{R}\mathbf{x}_{\{k\}} + \mathbf{u}_{\{k\}}^T \mathbf{Q}\mathbf{u}_{\{k\}}\right) \,, \tag{4.12}$$

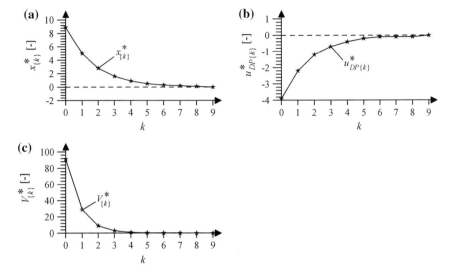

**Fig. 4.2**  **a** Optimal state trajectory $x^*_{\{k\}}$, $k = 0, \ldots, 9$, **b** optimal control $u^*_{DP\{k\}}$, **c** optimal value function $V^*_{\{k\}}$

where $\mathbf{R}$, $\mathbf{R}_E$, $\mathbf{Q}$ – diagonal design matrices with positive coefficients. $\mathbf{R}$ and $\mathbf{R}_E$ matrices coefficients may differ, as more emphasis is placed on the reduction of the state vector value in the final $n$-th step of the process than in other iteration stages; in a specific case $\mathbf{R} = \mathbf{R}_E$.

Defining the value function $V_{\{k\}} \left( \mathbf{x}_{\{n\}} \right)$ in the $k = n$ step by relation

$$V_{\{n\}} \left( \mathbf{x}_{\{n\}} \right) = \frac{1}{2} \left( \mathbf{x}^T_{\{n\}} \mathbf{R}_E \mathbf{x}_{\{n\}} \right) \equiv \frac{1}{2} \left( \mathbf{x}^T_{\{n\}} \mathbf{P}_{\{n\}} \mathbf{x}_{\{n\}} \right) , \tag{4.13}$$

where $\mathbf{R}_E \equiv \mathbf{P}_{\{n\}}$. The optimal value of the value function is $V^*_{\{n\}} \left( \mathbf{x}^*_{\{n\}} \right)$, while the optimal value of the system state vector $\mathbf{x}^*_{\{n\}}$. The value function in the $k = n - 1$ step is expressed by relation

$$V_{\{n-1\}} \left( \mathbf{x}_{\{n-1\}}, \mathbf{u}_{\{n-1\}} \right) = \frac{1}{2} \left( \mathbf{x}^T_{\{n-1\}} \mathbf{R} \mathbf{x}_{\{n-1\}} + \mathbf{u}^T_{\{n-1\}} \mathbf{Q} \mathbf{u}_{\{n-1\}} \right) + \frac{1}{2} \left( \mathbf{x}^{\dot{T}}_{\{n\}} \mathbf{P}_{\{n\}} \mathbf{x}_{\{n\}} \right) . \tag{4.14}$$

The optimal value of function $V^*_{\{n-1\}} \left( \mathbf{x}^*_{\{n-1\}}, \mathbf{u}^*_{\{n-1\}} \right)$ in the $n - 1$ step, given the Bellman's principle of optimality [1] and incorporating (4.11) into (4.14), is

$$V^*_{\{n-1\}} \left( \mathbf{x}^*_{\{n-1\}}, \mathbf{u}^*_{\{n-1\}} \right) = \min_{\mathbf{u}_{\{n-1\}}} \left\{ \frac{1}{2} \left( \mathbf{x}^{*T}_{\{n-1\}} \mathbf{R} \mathbf{x}^*_{\{n-1\}} + \mathbf{u}^T_{\{n-1\}} \mathbf{Q} \mathbf{u}_{\{n-1\}} \right) \right.$$

$$\left. + \frac{1}{2} \left[ \mathbf{F} \mathbf{x}^*_{\{n-1\}} + \mathbf{G} \mathbf{u}_{\{n-1\}} \right]^T \mathbf{P}_{\{n\}} \left[ \mathbf{F} \mathbf{x}^*_{\{n-1\}} + \mathbf{G} \mathbf{u}_{\{n-1\}} \right] \right\} , \tag{4.15}$$

where $\mathbf{x}^*_{\{n-1\}}$ – value of the optimal trajectory vector of state. System transition from state $\mathbf{x}^*_{\{n-1\}}$ to $\mathbf{x}^*_{\{n\}}$, along the optimal trajectory, requires a definition of optimal control $\mathbf{u}^*_{\{n-1\}}$, that satisfies the following relation

$$\frac{\partial V_{\{n-1\}}\left(\mathbf{x}^*_{\{n-1\}},\mathbf{u}_{\{n-1\}}\right)}{\partial \mathbf{u}_{\{n-1\}}} = 0 , \tag{4.16}$$

which leads to equation

$$\mathbf{Q}\mathbf{u}_{\{n-1\}} + \mathbf{G}^T\mathbf{P}_{\{n\}}\left[\mathbf{F}\mathbf{x}^*_{\{n-1\}} + \mathbf{G}\mathbf{u}_{\{n-1\}}\right] = 0 . \tag{4.17}$$

Control values $\mathbf{u}_{\{n-1\}}$ being the solution of Eq. (4.17) may correspond to the minimum or the maximum of the value function (4.15). It is therefore necessary to test the determinant of the matrix of second-order partial derivatives of the value function with respect to the control variables

$$\frac{\partial^2 V_{\{n-1\}}\left(\mathbf{x}^*_{\{n-1\}},\mathbf{u}_{\{n-1\}}\right)}{\partial \mathbf{u}^2_{\{n-1\}}} = \mathbf{Q} + \mathbf{G}^T\mathbf{P}_{\{n\}}\mathbf{G} . \tag{4.18}$$

It is assumed that matrices $\mathbf{G}$, $\mathbf{Q}$ and $\mathbf{R}_E$ are positive-definite, thus matrices $\mathbf{P}_{\{n\}} = \mathbf{R}_E$ and $\mathbf{G}^T\mathbf{P}_{\{n\}}\mathbf{G}$ are positive-definite as well. The sum of positive-definite matrices $\mathbf{Q} + \mathbf{G}^T\mathbf{P}_{\{n\}}\mathbf{G}$ is a positive-define matrix. Since the value function (4.15) is a quadratic function with respect to the control $\mathbf{u}_{\{n-1\}}$, and matrix $\mathbf{G}^T\mathbf{P}_{\{n\}}\mathbf{G}$ is positive-definite, the control $\mathbf{u}^*_{\{n-1\}}$ that satisfies Eq. (4.17), converges to a global minimum of the value function $V^*\left(\mathbf{x}^*_{\{n-1\}},\mathbf{u}^*_{\{n-1\}}\right)$. The solution of Eq. (4.17) with respect to $\mathbf{u}_{\{n-1\}}$ constitutes the optimal control law given by

$$\mathbf{u}^*_{\{n-1\}} = -\left[\mathbf{Q} + \mathbf{G}^T\mathbf{P}_{\{n\}}\mathbf{G}\right]^{-1}\mathbf{G}^T\mathbf{P}_{\{n\}}\mathbf{F}\mathbf{x}^*_{\{n-1\}}$$
$$\equiv -\mathbf{K}_{LQ\{n-1\}}\mathbf{x}^*_{\{n-1\}}. \tag{4.19}$$

where $\mathbf{K}_{LQ\{n-1\}}$ – gain matrix of the linear-quadratic regulator. Since matrix $\mathbf{Q} + \mathbf{G}^T\mathbf{P}_{\{n\}}\mathbf{G}$ is positive-definite, it has an inverse matrix. Substituting the optimal control law (4.17)–(4.15), a relation is obtained, which allows for the calculation of the function optimal value

$$V^*_{\{n-1\}}\left(\mathbf{x}^*_{\{n-1\}},\mathbf{u}^*_{\{n-1\}}\right) = \frac{1}{2}\mathbf{x}^{*T}_{\{n-1\}}\left\{\mathbf{R} + \mathbf{K}^T_{LQ\{n-1\}}\mathbf{Q}\mathbf{K}_{LQ\{n-1\}}\right.$$
$$\left. + \left[\mathbf{F} - \mathbf{G}\mathbf{K}_{LQ\{n-1\}}\right]^T\mathbf{P}_{\{n\}}\left[\mathbf{F} - \mathbf{G}\mathbf{K}_{LQ\{n-1\}}\right]\right\}\mathbf{x}^*_{\{n-1\}}$$
$$\equiv \frac{1}{2}\mathbf{x}^{*T}_{\{n-1\}}\mathbf{P}_{\{n-1\}}\mathbf{x}^*_{\{n-1\}}. \tag{4.20}$$

It can be noticed that the optimal value of function $V^*_{\{n\}}\left(\mathbf{x}^*_{\{n\}}\right)$ calculated from (4.13) and $V^*_{\{n-1\}}\left(\mathbf{x}^*_{\{n-1\}},\mathbf{u}^*_{\{n-1\}}\right)$ obtained from Eq. (4.20), have an identical form.

By continuing calculations for the $k = n - 2$ step we get the following expressions

$$\mathbf{u}^*_{\{n-2\}} = -\left[\mathbf{Q} + \mathbf{G}^T\mathbf{P}_{\{n-1\}}\mathbf{G}\right]^{-1}\mathbf{G}^T\mathbf{P}_{\{n-1\}}\mathbf{F}\mathbf{x}^*_{\{n-2\}}$$
$$\equiv -\mathbf{K}_{LQ\{n-2\}}\mathbf{x}^*_{\{n-2\}} , \qquad (4.21)$$

and

$$V^*_{\{n-2\}}\left(\mathbf{x}^*_{\{n-2\}}, \mathbf{u}^*_{\{n-2\}}\right) = \frac{1}{2}\mathbf{x}^{*T}_{\{n-2\}}\left\{\mathbf{R} + \mathbf{K}^T_{LQ\{n-2\}}\mathbf{Q}\mathbf{K}_{LQ\{n-2\}}\right.$$
$$\left. + \left[\mathbf{F} - \mathbf{G}\mathbf{K}_{LQ\{n-2\}}\right]^T\mathbf{P}_{\{n-1\}}\left[\mathbf{F} - \mathbf{G}\mathbf{K}_{LQ\{n-2\}}\right]\right\}\mathbf{x}^*_{\{n-2\}}$$
$$\equiv \frac{1}{2}\mathbf{x}^{*T}_{\{n-2\}}\mathbf{P}_{\{n-2\}}\mathbf{x}^*_{\{n-2\}} . \qquad (4.22)$$

The above-derived equations can be generalized by introducing the index $j = 0, \ldots, n$, thus the optimal control law $\mathbf{u}^*_{\{k\}}$ and the optimal value function $V^*_{\{k\}}\left(\mathbf{x}^*_{\{k\}}, \mathbf{u}^*_{\{k\}}\right)$ in any $k$-th iteration step, $k = n - j$, can be calculated according to the relations

$$\mathbf{u}^*_{\{k\}} = -\left[\mathbf{Q} + \mathbf{G}^T\mathbf{P}_{\{k+1\}}\mathbf{G}\right]^{-1}\mathbf{G}^T\mathbf{P}_{\{k+1\}}\mathbf{F}\mathbf{x}^*_{\{k\}}$$
$$\equiv -\mathbf{K}_{LQ\{k\}}\mathbf{x}^*_{\{k\}} , \qquad (4.23)$$

and

$$V^*_{\{k\}}\left(\mathbf{x}^*_{\{k\}}, \mathbf{u}^*_{\{k\}}\right) = \frac{1}{2}\mathbf{x}^{*T}_{\{k\}}\left\{\mathbf{R} + \mathbf{K}^T_{LQ\{k\}}\mathbf{Q}\mathbf{K}_{LQ\{k\}}\right.$$
$$\left. + \left[\mathbf{F} - \mathbf{G}\mathbf{K}_{LQ\{k\}}\right]^T\mathbf{P}_{\{k+1\}}\left[\mathbf{F} - \mathbf{G}\mathbf{K}_{LQ\{k\}}\right]\right\}\mathbf{x}^*_{\{k\}}$$
$$\equiv \frac{1}{2}\mathbf{x}^{*T}_{\{k\}}\mathbf{P}_{\{k\}}\mathbf{x}^*_{\{k\}} , \qquad (4.24)$$

where $\mathbf{P}_{\{k\}}$ is derived from the Riccati equation

$$\mathbf{P}_{\{k\}} = \mathbf{R} + \mathbf{K}_{LQ\{k\}}\mathbf{Q}\mathbf{K}^T_{LQ\{k\}} + \left[\mathbf{F} - \mathbf{G}\mathbf{K}_{LQ\{k\}}\right]^T\mathbf{P}_{\{k+1\}}\left[\mathbf{F} - \mathbf{G}\mathbf{K}_{LQ\{k\}}\right] . \quad (4.25)$$

An important conclusion follows from Eqs. (4.23) and (4.24) – the optimal control law $\mathbf{u}^*_{\{k\}}$ in each iteration step is a linear combination of the controlled object's optimal state $\mathbf{x}^*_{\{k\}}$ and the linear-quadratic regulator's gain matrix $\mathbf{K}_{LQ\{k\}}$. A major inconvenience in the application of the LQR in the control process is the necessity to calculate the values of regulator's matrix $\mathbf{P}_{\{k\}}$ and gain matrix $\mathbf{K}_{LQ\{k\}}$ starting from the last step of the discrete process $k = n$ and continuing backwards to the first step $k = 0$. The values of these matrices are first stored and then used for calculation of the optimal control law $\mathbf{u}^*_{\{k\}}$ and the optimal trajectory of the system's state $\mathbf{x}^*_{\{k\}}$.

It should be noted that although matrices $F$, $G$, $Q$ and $R$ are constant, the gain $K_{LQ\{k\}}$ of the LQR is dependent on the $k$ step of the iterative process.

### Linear-quadratic regulator in optimal control of a linear dynamic system

The problem from Sect. 4.1 was solved with a linear-quadratic regulator by generating signals in order to control linear dynamic objects described by Eq. (4.9).

The values of individual parameters are adopted as defined in Sect. 4.1, object parameters $F = 1$, $G = 1$, cost function parameters $R = R_F = 1$, $Q = 3$, initial state of the system $x_{\{0\}} = 8.9$, free final state in the $n$-th iteration step. Presentation of results is limited to $n = 9$ steps of the process, as the yielded values can be compared to solutions obtained with other methods.

Applying LQR defined by (4.23), the optimal control law $u^*_{LQ\{k\}}$ was determined

$$u^*_{LQ\{k\}} = -\left[Q + G^2 P_{\{k+1\}}\right]^{-1} G P_{\{k+1\}} F x^*_{\{k\}} = -K_{LQ\{k\}} x^*_{\{k\}}, \qquad (4.26)$$

that ensures the system's transition from the initial state $x_{\{0\}}$ to the final state $x_{\{n\}}$, when $n = 9$, along the optimal state trajectory $x^*_{\{k\}}$, minimizing values of the function $V_{\{k\}}$, with a defined number of steps in the discrete process. It is noted that in this illustrative example the system state $x_{\{k\}}$, control $u^*_{LQ\{k\}}$ and coefficients $F, G, Q, R$ are scalar quantities, including the LQR gain $K_{LQ\{k\}}$. The values of variable $P_{\{k\}}$ are calculated based on the Riccati equation

$$P_{\{k\}} = R + K_{LQ\{k\}} Q K_{LQ\{k\}} + \left[F - G K_{LQ\{k\}}\right]^2 P_{\{k+1\}}. \qquad (4.27)$$

Table 4.2 presents values of the optimal state trajectory $x^*_{\{k\}}$, optimal control law $u^*_{LQ\{k\}}$, optimal value function $V^*_{\{k\}}$ and coefficient $K_{LQ\{k\}}$ of the LQR algorithm in ten subsequent iteration steps $k = 0, \ldots, 9$.

Values of the optimal state trajectory $x^*_{\{k\}}$, optimal control law $u^*_{LQ\{k\}}$, optimal value function $V^*_{\{k\}}$, and gain $K_{LQ\{k\}}$ of the LQR obtained by solving an example problem with the application of LQR are shown in Fig. 4.3.

Optimal control (4.26) transits the dynamic system from the defined initial state $x_{\{0\}}$ to the free final state $x^*_{\{n\}}$, along the optimal trajectory, where the minimum value of function $V^*_{\{k\}}$, is conserved. The optimal control law is determined from the last step of the process $k = 9$ to the first step $k = 0$, by solving Riccati equation (4.27) and yielding values of the gain coefficient $K_{LQ\{k\}}$.

Analyzing the process of the generation of the LQR gain coefficient $K_{LQ\{k\}}$, from the last step $k = 9$ to $k = 0$ it can be noticed that the gain value increases from $K_{LQ\{8\}} = 0.25$ to $K_{LQ\{5\}} = 0.43$, and ultimately becomes stable. It can be concluded that assuming the optimal control law from state $u_{P\{k\}} = K_P x_{\{k\}}$, with a fixed gain coefficient value of $K_P = 0.43$, a suboptimal system response can be

**Table 4.2** Values of the optimal state trajectory $x^*_{\{k\}}$, optimal control law $u^*_{LQ\{k\}}$, optimal value function $V^*_{\{k\}}$ and coefficient $K_{LQ\{k\}}$ of the LQR algorithm

| $k$ | $x^*_{\{k\}}$ [−] | $u^*_{LQ\{k\}}$ [−] | $V^*_{\{k\}}$ [−] | $K_{LQ\{k\}}$ [−] |
|---|---|---|---|---|
| 0 | 8.9 | −3.86 | 91.20 | 0.43 |
| 1 | 5.04 | −2.19 | 29.19 | 0.43 |
| 2 | 2.85 | −1.24 | 9.34 | 0.43 |
| 3 | 1.61 | −0.7 | 2.99 | 0.43 |
| 4 | 0.91 | −0.39 | 0.96 | 0.43 |
| 5 | 0.52 | −0.22 | 0.31 | 0.43 |
| 6 | 0.3 | −0.12 | 0.1 | 0.41 |
| 7 | 0.18 | −0.06 | 0.03 | 0.37 |
| 8 | 0.12 | −0.03 | 0.01 | 0.25 |
| 9 | 0.09 | − | 0 | − |

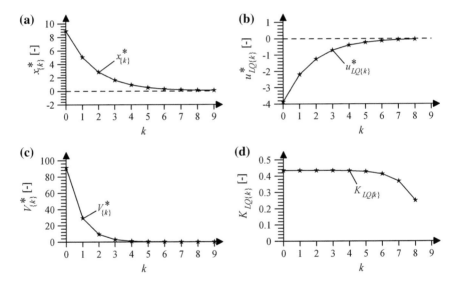

**Fig. 4.3**  **a** Optimal state trajectory $x^*_{\{k\}}$, $k = 0, \ldots, 9$, **b** optimal control $u^*_{LQ\{k\}}$, **c** optimal value function $V^*_{\{k\}}$, **d** gain coefficient $K_{LQ\{k\}}$

obtained, that in the first steps of the process $k = 0, \ldots, 6$ shall not diverge from the optimal solution.

## 4.3 Pontryagin's Maximum Principle

Pointryagin's maximum principle [4, 5, 7] provides the necessary conditions for control optimality, however they are not sufficient for nonlinear systems. In the case of linear systems these conditions are deemed necessary and sufficient. That is why it is not easy to determine the optimal control law.

For a controlled object of the $n$-th order, described in a continuous-time domain $t$ by a set of differential equations, and with a given value function, the $(M + 1)$ - dimensional vector equation is given by

$$\frac{d\mathbf{X}}{dt} = \mathbf{C}(\mathbf{x}, \mathbf{u}) , \tag{4.28}$$

where the first state variable $x_0$ of the augmented vector of state $\mathbf{X}$ is the performance index $V$, called the value function,

$$V(\mathbf{x}, \mathbf{u}) = \int_0^{t_1} c_0(\mathbf{x}, \mathbf{u}) \, dt , \tag{4.29}$$

while the remaining state variables $x_1, \ldots, x_n$ are the components of the state vector $\mathbf{x}$ of the controlled nonlinear object, $\mathbf{u}$ is the $r$-dimensional control vector, which may be subjected to constraints $\mathbf{u} \in \mathbf{U}$. It should be noted that $\mathbf{x}, \mathbf{u}$ and all their functions depend on time $t$ ($\mathbf{x}(t), \mathbf{u}(t)$), yet by reducing the notation this shall not apply.

Now, let us find the optimal control $\mathbf{u}^*$, to transfer an object from the initial state $\mathbf{x}(0)$ to the desired final state $\mathbf{x}(t_1)$, with a minimum value of the performance index (4.29).

We shall introduce $(M + 1)$-dimensional conjugate vector $\mathbf{\Psi}$, sometimes called the adjoint vector, defined by homogeneous linear relation

$$\frac{d\mathbf{\Psi}}{dt} = - \begin{bmatrix} \dfrac{\partial c_0}{\partial x_0} & \dfrac{\partial c_1}{\partial x_0} & \cdots & \dfrac{\partial c_M}{\partial x_0} \\ \vdots & & & \vdots \\ \dfrac{\partial c_0}{\partial x_M} & \dfrac{\partial c_1}{\partial x_M} & \cdots & \dfrac{\partial c_M}{\partial x_M} \end{bmatrix} \mathbf{\Psi} . \tag{4.30}$$

Scalar function in the form of Hamiltonian $H(\mathbf{\Psi}, \mathbf{x}, \mathbf{u})$ is defined based on relation (4.28) and the conjugate vector $\mathbf{\Psi}$ as follows

$$H(\mathbf{\Psi}, \mathbf{x}, \mathbf{u}) = \mathbf{\Psi}^T \mathbf{C} = \mathbf{\Psi}^T \frac{d\mathbf{X}}{dt} = \sum_{i=0}^{M} \psi_i c_i(\mathbf{x}, \mathbf{u}) . \tag{4.31}$$

Hamiltonian constitutes function $\mathbf{u} \in \mathbf{U}$, whose maximum value, with respect to control vector $\mathbf{u}$, is denoted by $M_H(\mathbf{\Psi}, \mathbf{x})$, where

$$M_H\left(\boldsymbol{\Psi}, \mathbf{x}\right) = \max_{\mathbf{u} \in \mathbf{U}} H\left(\boldsymbol{\Psi}, \mathbf{x}, \mathbf{u}\right) \ . \tag{4.32}$$

Pontryagin's maximum theory provides[1]:

Let $\mathbf{u}\,(t)$, $0 \geq t \geq t_1$ be admissible control that transfers the system from a given initial state $\mathbf{x}\,(0)$ to the desired final (terminal) state $\mathbf{x}\,(t_1)$ over time $t_1$. In order for $\mathbf{u}^*$ and the resulting trajectory $\mathbf{X}^*\,(t)$ in $(M + 1)$-dimensional state space to be optimal, it is necessary that there exists a non-zero continuous vector $\boldsymbol{\Psi}^*\,(t)$ corresponding to the vector $\mathbf{u}^*$ and the trajectory $\mathbf{X}^*$, such that everywhere in the interval $t\,(0 \geq t \geq t_1)$

1. function $H\,(\boldsymbol{\Psi}, \mathbf{x}, \mathbf{u})$ of variable $\mathbf{u} \in \mathbf{U}$ attains its maximum at the point $\mathbf{u} = \mathbf{u}^*$

$$H\left(\boldsymbol{\Psi}^*, \mathbf{x}^*, \mathbf{u}^*\right) = M_H\left(\boldsymbol{\Psi}^*, \mathbf{x}^*\right) \ , \tag{4.33}$$

2. terminal value $M_H\,(\boldsymbol{\Psi}^*, \mathbf{x}^*)$ equals zero, while terminal value $\psi_0$ is non-positive.

Terminal value $M_H\,(\boldsymbol{\Psi}^*, \mathbf{x}^*)$ is zero, for a stationary problem and free terminal time. It can be established that $M_H\,(\boldsymbol{\Psi}^*, \mathbf{x}^*)$ and $\psi_0$ are constant within adopted time interval $(0 \geq t \geq t_1)$, since $M_H\,(\boldsymbol{\Psi}^*, \mathbf{x}^*) = 0$ and relation (4.30) is homogeneous, it can be assumed that $\psi_0 = -1$ within adopted time interval.

Equations (4.28) and (4.30) written with respect to $H\,(\boldsymbol{\Psi}, \mathbf{x}, \mathbf{u})$ are as follows

$$\frac{dx_i}{dt} = \frac{dH}{d\psi_i}, \qquad \frac{d\psi_i}{dt} = -\frac{dH}{dx_i} \ , \tag{4.34}$$

for $i = 0, \ldots, M$.

The maximum principle is also applied to discrete systems, but with certain modifications. In the general case a discrete, nonlinear, stationary, $M$-dimensional dynamic system is described by relation

$$\mathbf{x}_{\{k+1\}} = \mathbf{f}\left(\mathbf{x}_{\{k\}}, \mathbf{u}_{\{k\}}\right), \tag{4.35}$$

where $\mathbf{f}\left(\mathbf{x}_{\{k\}}, \mathbf{u}_{\{k\}}\right)$ – column matrix of nonlinear functions describing the object's dynamics, with elements $f_{1\{k\}}, \ldots, f_{M\{k\}}$. Assuming $\gamma = 1$, the value of performance index $V_{\{k\}}\left(\mathbf{x}_{\{k+1\}}, \mathbf{u}_{\{k\}}\right)$ of a complete $n$ step iterative process, is given by

$$V_{\{k\}}\left(\mathbf{x}_{\{k+1\}}, \mathbf{u}_{\{k\}}\right) = \sum_{k=0}^{n-1} L_{C\{k\}}\left(\mathbf{x}_{\{k+1\}}, \mathbf{u}_{\{k\}}\right) \ . \tag{4.36}$$

Denotations of individual quantities in relation (4.36) were thoroughly discussed in subchapter 4.1. Similarly as in the continuous case, the zero state variable of the

---

[1]Takahashi, Y., Rabins, M.J., Auslander, D.M.: Control and Dynamical Systems. (in Polish) WNT, Warsaw (1976), p. 568.

augmented state vector $\mathbf{X}_{\{k\}}$ is assigned to the performance index (4.36) according to the expression

$$x_{0\{k+1\}} = x_{0\{k\}} + L_{C\{k\}}\left(\mathbf{f}\left(\mathbf{x}_{\{k\}}, \mathbf{u}_{\{k\}}\right), \mathbf{u}_{\{k\}}\right) \equiv d_0\left(\mathbf{x}_{\{k\}}, \mathbf{u}_{\{k\}}\right), \qquad (4.37)$$

where $d_0\left(\mathbf{x}_{\{k\}}, \mathbf{u}_{\{k\}}\right)$ – the first element of the column matrix $\mathbf{D}\left(\mathbf{x}_{\{k\}}, \mathbf{u}_{\{k\}}\right)$, containing the performance index and description of the controlled object's nonlinearity in the form of the right-hand side of Eq. (4.35). The augmented state vector $\mathbf{X}_{\{k\}}$ with dimension $(M + 1)$, consists of $x_{0\{k\}}$ and state vector $\mathbf{x}_{\{k\}}$ of the controlled object. Equation of motion for the variations of vector $\delta\mathbf{X}_{\{k\}}$ of the controlled object of the second order is given by

$$\delta\mathbf{X}_{\{k+1\}} = \begin{bmatrix} 1 & \dfrac{\partial d_0\left(\mathbf{x}_{\{k\}}, \mathbf{u}_{\{k\}}\right)}{\partial x_{1\{k\}}} & \dfrac{\partial d_0\left(\mathbf{x}_{\{k\}}, \mathbf{u}_{\{k\}}\right)}{\partial x_{2\{k\}}} \\ 0 & \dfrac{\partial d_1\left(\mathbf{x}_{\{k\}}, \mathbf{u}_{\{k\}}\right)}{\partial x_{1\{k\}}} & \dfrac{\partial d_1\left(\mathbf{x}_{\{k\}}, \mathbf{u}_{\{k\}}\right)}{\partial x_{2\{k\}}} \\ 0 & \dfrac{\partial d_2\left(\mathbf{x}_{\{k\}}, \mathbf{u}_{\{k\}}\right)}{\partial x_{1\{k\}}} & \dfrac{\partial d_2\left(\mathbf{x}_{\{k\}}, \mathbf{u}_{\{k\}}\right)}{\partial x_{2\{k\}}} \end{bmatrix} \delta\mathbf{X}_{\{k\}} \equiv \mathbf{M}_J \delta\mathbf{X}_{\{k\}}. \quad (4.38)$$

The first element of the Jacobian matrix equals one, which results from Eq. (4.37), while the remaining elements in the first column equal zero, as expressions $d_1\left(\mathbf{x}_{\{k\}}, \mathbf{u}_{\{k\}}\right)$ and $d_2\left(\mathbf{x}_{\{k\}}, \mathbf{u}_{\{k\}}\right)$ are not functions of $x_{0\{k\}}$.

Conjugate vector $\mathbf{\Psi}_{\{k\}}$ is defined by the Jacobian transpose

$$\begin{bmatrix} \psi_{0\{k\}} \\ \psi_{1\{k\}} \\ \psi_{2\{k\}} \end{bmatrix} = \begin{bmatrix} 1 & 0 & 0 \\ \dfrac{\partial d_0\left(\mathbf{x}_{\{k\}}, \mathbf{u}_{\{k\}}\right)}{\partial x_{1\{k\}}} & \dfrac{\partial d_1\left(\mathbf{x}_{\{k\}}, \mathbf{u}_{\{k\}}\right)}{\partial x_{1\{k\}}} & \dfrac{\partial d_2\left(\mathbf{x}_{\{k\}}, \mathbf{u}_{\{k\}}\right)}{\partial x_{1\{k\}}} \\ \dfrac{\partial d_0\left(\mathbf{x}_{\{k\}}, \mathbf{u}_{\{k\}}\right)}{\partial x_{2\{k\}}} & \dfrac{\partial d_1\left(\mathbf{x}_{\{k\}}, \mathbf{u}_{\{k\}}\right)}{\partial x_{2\{k\}}} & \dfrac{\partial d_2\left(\mathbf{x}_{\{k\}}, \mathbf{u}_{\{k\}}\right)}{\partial x_{2\{k\}}} \end{bmatrix} \begin{bmatrix} \psi_{0\{k+1\}} \\ \psi_{1\{k+1\}} \\ \psi_{2\{k+1\}} \end{bmatrix} \equiv \mathbf{M}_J^T \mathbf{\Psi}_{\{k+1\}},$$
$$(4.39)$$

where transfer from $k$ to $k + 1$ is inverse. From Eqs. (4.38) and (4.39) it follows

$$\mathbf{\Psi}_{\{k\}}^T \delta\mathbf{X}_{\{k\}} = \left(\mathbf{M}_J^T \mathbf{\Psi}_{\{k+1\}}\right)^T \delta\mathbf{X}_{\{k\}} = \mathbf{\Psi}_{\{k+1\}}^T \left(\mathbf{M}_J \delta\mathbf{X}_{\{k\}}\right) = \mathbf{\Psi}_{\{k+1\}}^T \delta\mathbf{X}_{\{k+1\}} = \text{const}.$$
$$(4.40)$$

From relation (4.39) it follows that $\psi_{0\{k\}}$ is a constant quantity, thus, as with continuous system, it may be assumed $\psi_{0\{k\}} = -1$ when $k = 0, \ldots, n$.

Hamiltonian takes the form

$$H = \mathbf{\Psi}_{\{k+1\}}^T \mathbf{D}\left(\mathbf{x}_{\{k\}}, \mathbf{u}_{\{k\}}\right) = -d_0\left(\mathbf{x}_{\{k\}}, \mathbf{u}_{\{k\}}\right) + \mathbf{\psi}_{\{k+1\}}^T \mathbf{f}\left(\mathbf{x}_k, \mathbf{u}_k\right), \qquad (4.41)$$

where $\mathbf{\psi}_{\{k\}} = \left[\psi_{1\{k\}}, \ldots, \psi_{M\{k\}}\right]$ – $M$-dimensional column matrix, obtained from conjugate vector $\mathbf{\Psi}_{\{k\}}$.

As with continuous systems, a discrete maximum principle is derived, which implies that Hamiltonian (4.41) takes the maximum value when control $\mathbf{u}_{\{k\}}$ is optimal.

A discrete principle is not applied to systems with constraints imposed upon the vector of state $\mathbf{x}_{\{k\}}$, or control $\mathbf{u}_{\{k\}}$, and the Hamiltonian $H$ maximum does not take constant or zero value.

Application of discrete form of Pontryagin's maximum principle to the optimal control of discrete linear object allows of the derivation of control law that transfers an object from the initial state $\mathbf{x}_{\{k=0\}}$ to state $\mathbf{x}_{\{k=n\}}$ with the minimum value of the pre-defined performance index $V_{\{k\}}\left(\mathbf{x}_{\{k+1\}}, \mathbf{u}_{\{k\}}\right)$. In the general case a discrete, linear, stationary, $M$-dimensional dynamic system is described with relation

$$\mathbf{x}_{\{k+1\}} = \mathbf{F}\mathbf{x}_{\{k\}} + \mathbf{G}\mathbf{u}_{\{k\}} , \tag{4.42}$$

where $\mathbf{F}$ – system's state matrix, $\mathbf{G}$ – control matrix. The performance index (4.36) was assumed in a modified form (4.12), that allows for the assignment of a different final state weight $\mathbf{R}_F$ and which includes term $\mathbf{x}_{\{0\}}^T \mathbf{R}\mathbf{x}_{\{0\}}$ that yields the initial value of the performance index. The following form of relation is assumed

$$V_{\{k\}}\left(\mathbf{x}_{\{k\}}, \mathbf{u}_{\{k\}}\right) = \frac{1}{2}\mathbf{x}_{\{n\}}^T \mathbf{R}_F \mathbf{x}_{\{n\}} + \frac{1}{2}\sum_{k=0}^{n-1}\left(\mathbf{x}_{\{k\}}^T \mathbf{R}\mathbf{x}_{\{k\}} + \mathbf{u}_{\{k\}}^T \mathbf{Q}\mathbf{u}_{\{k\}}\right) , \tag{4.43}$$

where $\mathbf{R}, \mathbf{R}_F, \mathbf{Q}$ – appropriate symmetric weight matrices. Equation (4.39) is written as

$$\boldsymbol{\psi}_{\{k\}} = -\frac{\partial d_0\left(\mathbf{x}_{\{k\}}, \mathbf{u}_{\{k\}}\right)}{\partial \mathbf{x}_{\{k\}}} + \mathbf{F}^T \boldsymbol{\psi}_{\{k+1\}} , \tag{4.44}$$

while Hamiltonian (4.41) is given by

$$H = \boldsymbol{\Psi}_{\{k+1\}}^T \mathbf{D}\left(\mathbf{x}_{\{k\}}, \mathbf{u}_{\{k\}}\right) = -d_0\left(\mathbf{x}_{\{k\}}, \mathbf{u}_{\{k\}}\right) + \boldsymbol{\psi}_{\{k+1\}}^T\left(\mathbf{F}\mathbf{x}_{\{k\}} + \mathbf{G}\mathbf{u}_{\{k\}}\right). \tag{4.45}$$

Given that matrix $\mathbf{R}$ is constant, except for the final state in step $n$, where it has the value $\mathbf{R}_F$, the problem can be solved as a stationary one, defining cost increment

$$\begin{aligned}
d_0\left(\mathbf{x}_{\{k\}}, \mathbf{u}_{\{k\}}\right) &= \frac{1}{2}\left(\mathbf{x}_{\{k+1\}}^T \mathbf{R}_F \mathbf{x}_{\{k+1\}} - \mathbf{x}_{\{k\}}^T \mathbf{R}_F \mathbf{x}_{\{k\}}\right) + \frac{1}{2}\mathbf{x}_{\{k\}}^T \mathbf{R}\mathbf{x}_{\{k\}} + \frac{1}{2}\mathbf{u}_{\{k\}}^T \mathbf{Q}\mathbf{u}_{\{k\}} \\
&= \frac{1}{2}\left(\mathbf{x}_{\{k\}}^T \mathbf{F}^T + \mathbf{u}_{\{k\}}^T \mathbf{G}^T\right)\mathbf{R}_F\left(\mathbf{F}\mathbf{x}_{\{k\}} + \mathbf{G}\mathbf{u}_{\{k\}}\right) - \frac{1}{2}\mathbf{x}_{\{k\}}^T \mathbf{R}_F \mathbf{x}_{\{k\}} \\
&\quad + \frac{1}{2}\mathbf{x}_{\{k\}}^T \mathbf{R}\mathbf{x}_{\{k\}} + \frac{1}{2}\mathbf{u}_{\{k\}}^T \mathbf{Q}\mathbf{u}_{\{k\}} .
\end{aligned} \tag{4.46}$$

Equation (4.44) is written as

$$\boldsymbol{\psi}_{\{k\}} = -\mathbf{F}^T\left[\mathbf{R}_F \mathbf{x}_{\{k+1\}} - \boldsymbol{\psi}_{\{k+1\}}\right] + [\mathbf{R}_F - \mathbf{R}]\mathbf{x}_{\{k\}} . \tag{4.47}$$

According to the discrete version of Pontryagin's maximum principle, the optimal control satisfies the condition $\partial H/\partial \mathbf{u}_{\{k\}}^* = 0$, where

$$-\frac{\partial d_0\left(\mathbf{x}_{\{k\}}, \mathbf{u}_{\{k\}}\right)}{\partial \mathbf{u}_{\{k\}}} + \mathbf{G}^T \boldsymbol{\psi}_{\{k+1\}} = 0 , \tag{4.48}$$

thus

$$\mathbf{G}^T \left[\boldsymbol{\psi}_{\{k+1\}} - \mathbf{R}_F \mathbf{x}_{\{k+1\}}\right] - \mathbf{Q}\mathbf{u}^*_{\{k\}} = 0 . \tag{4.49}$$

By solving Eq. (4.49) with respect to $\mathbf{u}^*_{\{k\}}$ the following optimal control is attained

$$\mathbf{u}^*_{M\{k\}} = \mathbf{Q}^{-1}\mathbf{G}^T \left[\boldsymbol{\psi}_{\{k+1\}} - \mathbf{R}_F \mathbf{x}_{\{k+1\}}\right], \tag{4.50}$$

where matrix $\mathbf{Q}$ is positive-definite, thus it has an inverse matrix.

### Pontryagin's maximum principle in optimal control of a linear dynamic system

The problem from Sect. 4.1 was solved with the control law resulting from the Pontryagin's maximum principle that was applied to control of a linear dynamic objet described by Eq. (4.9).

The values of individual parameters are assumed as defined in Sect. 4.1, object parameters $F = 1$, $G = 1$, cost function parameters $R = R_F = 1$, $Q = 3$, initial state of the system $x_{\{0\}} = 8.9$, free final state with $n$ iteration step. Presentation of results is limited to $n = 9$ steps of the process, as the yielded values can be easily compared to solutions obtained with other methods.

In the considered case the Hamiltonian is given by

$$H = -\frac{1}{2}\left[\left(R_F F^2 - R_F + R\right) x_{\{k\}}^2 + \left(R_F G^2 + Q\right) u_{\{k\}}^2 + 2R_F F G x_{\{k\}} u_{\{k\}}\right]$$
$$+ \psi_{1\{k+1\}}\left(F x_{\{k\}} + G u_{\{k\}}\right), \tag{4.51}$$

and the optimal control law

$$u^*_{M\{k\}} = Q^{-1}G\left[\psi_{1\{k+1\}} - R_F x_{\{k+1\}}\right], \tag{4.52}$$

Substituting the optimal control law $u^*_{M\{k\}}$ to object Eq. (4.42) and (4.47), we obtain

$$\begin{bmatrix} x_{\{k+1\}} \\ \psi_{1\{k\}} \end{bmatrix} = \begin{bmatrix} \frac{3}{4} & \frac{1}{4} \\ -\frac{3}{4} & \frac{3}{4} \end{bmatrix} \begin{bmatrix} x_{\{k\}} \\ \psi_{1\{k+1\}} \end{bmatrix}, \tag{4.53}$$

where $k = 0, \ldots, 9$, $x_{\{0\}} = 8.9$. Conjugate vector component $\psi_{1\{9\}} = 0$, as we do know the initial state and number of steps in the discrete process (time of the process in the case of a continuous system), whereas the final state is unknown.

**Table 4.3** Values of the optimal trajectory of state $x^*_{\{k\}}$, optimal control $u^*_{M\{k\}}$, optimal value function $V^*_{\{k\}}$, and function $\psi_{1\{k\}}$, algorithm MP

| $k$ | $x^*_{\{k\}}$ [−] | $u^*_{M\{k\}}$ [−] | $V^*_{\{k\}}$ [−] | $\psi_{1\{k\}}$ [−] |
|-----|------------------|-------------------|------------------|-------------------|
| 0 | 8.9 | −3.86 | 91.20 | −23.19 |
| 1 | 5.04 | −2.19 | 29.19 | −13.12 |
| 2 | 2.85 | −1.24 | 9.34 | −7.42 |
| 3 | 1.61 | −0.7 | 2.99 | −4.19 |
| 4 | 0.91 | −0.39 | 0.96 | −2.37 |
| 5 | 0.52 | −0.22 | 0.31 | −1.33 |
| 6 | 0.3 | −0.12 | 0.1 | −0.74 |
| 7 | 0.18 | −0.06 | 0.03 | −0.39 |
| 8 | 0.12 | −0.03 | 0.01 | −0.17 |
| 9 | 0.09 | − | 0 | 0 |

To determine a sequence of optimal object state values $x^*_{\{k\}}$ and optimal control $u^*_{M\{k\}}$, a two-boundary problem must be solved, namely, a set of $2(n-1)$ equations must be solved.

Table 4.3 includes values of the optimal trajectory of object state $x^*_{\{k\}}$, with a given initial state, values of optimal control law derived from a discrete version of the Pontryagin's maximum principle (MP) $u^*_{M\{k\}}$ according to relation (4.52), values of optimal function $V^*_{\{k\}}$ and function $\psi_{1\{k\}}$.

Figure 4.4 shows the values of the optimal trajectory of state $x^*_{\{k\}}$, optimal control law $u^*_{M\{k\}}$, optimal value function $V^*_{\{k\}}$, and values of function $\psi_{1\{k\}}$ obtained by solving the problem of optimal control of a linear dynamic object with the application of the Pontryagin's maximum principle.

Application of the optimal control law $u^*_{M\{k\}}$ (4.52) results in transition of the system from the given initial state $x^*_{\{0\}}$ to the free final state $x^*_{\{n\}}$ in $n$ iteration steps, while value of given performance index $V^*_{\{k\}}$ is minimal. According to the discrete Pontryagin's maximum principle theory, conjugate vector components $\psi_{\{k\}}$ take negative values for $k = 0, \ldots, n-1$, whereas $k = n$, $\psi_{\{k\}} = 0$. Performed computations confirm, that obtained values $\psi_{1\{k\}}$ satisfy these assumptions.

The best effects in terms of dynamic objects control are achieved by employing (vector) state feedback. Such control can be derived to a linear system described by (4.42) with performance index (4.43), applying the Pontryagin's maximum principle. Let us introduce a symmetric, non-stationary matrix $\mathbf{H}_{\{k+1\}}$, defined as

$$\mathbf{R}_F \mathbf{x}_{\{k+1\}} - \boldsymbol{\psi}_{\{k+1\}} = \mathbf{H}_{\{k+1\}} \mathbf{x}_{\{k+1\}} . \tag{4.54}$$

Substituting Eq. (4.54) to relation (4.50) we get

$$\mathbf{u}^*_{MS\{k\}} = -\mathbf{Q}^{-1}\mathbf{G}^T \mathbf{H}_{\{k+1\}} \mathbf{x}_{\{k+1\}} , \tag{4.55}$$

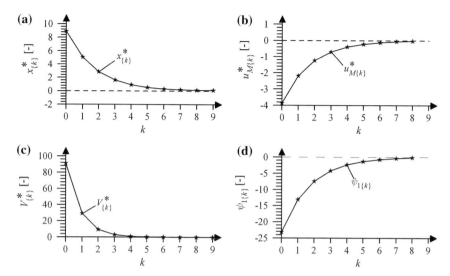

**Fig. 4.4   a** Values of optimal state trajectory $x^*_{\{k\}}$, $k = 0, \ldots, 9$, **b** values of optimal control law $u^*_{M\{k\}}$, **c** graph of optimal value function $V^*_{\{k\}}$, **d** values of function $\psi_{1\{k\}}$

thus, given the equation of model (4.42), the optimal control, dependent on state, is written as

$$\mathbf{u}^*_{MS\{k\}} = -\left[\mathbf{Q} + \mathbf{G}^T \mathbf{H}_{\{k+1\}} \mathbf{G}\right]^{-1} \mathbf{G}^T \mathbf{H}_{\{k+1\}} \mathbf{F} \mathbf{x}_{\{k\}}$$
$$\equiv -\mathbf{K}_{MS\{k\}} \mathbf{x}_{\{k\}} , \tag{4.56}$$

where $\mathbf{K}_{MS\{k\}}$ – gain matrix of optimal state control, derived according to the maximum principle. To find gain coefficients' values of matrix $\mathbf{K}_{MS\{k\}}$, the values of $\mathbf{H}_{\{k+1\}}$ must be known, which can be determined by comparing $\boldsymbol{\psi}_{\{k\}}$ from relations (4.47) and (4.54) in step $k$

$$\boldsymbol{\psi}_{\{k\}} = -\mathbf{H}_{\{k\}} \mathbf{x}_{\{k\}} + \mathbf{R}_F \mathbf{x}_{\{k\}} ,$$
$$\boldsymbol{\psi}_{\{k\}} = -\mathbf{F}^T \mathbf{H}_{\{k+1\}} \mathbf{x}_{\{k+1\}} + [\mathbf{R}_F - \mathbf{R}] \mathbf{x}_{\{k\}}. \tag{4.57}$$

By comparing the right-hand sides of (4.57) we obtain an equation that must be satisfied for every value $\mathbf{x}_{\{k\}}$. Replacing $\mathbf{x}_{\{k+1\}}$ with the right-hand side of Eq. (4.42), where control $\mathbf{u}^*_{\{k\}}$ is expressed by (4.56), the Riccati equation for $\mathbf{H}_{\{k\}}$ has the form

$$\mathbf{H}_{\{k\}} = \mathbf{F}^T \mathbf{H}_{\{k+1\}} \left(\mathbf{F} - \mathbf{G}\left[\mathbf{Q} + \mathbf{G}^T \mathbf{H}_{\{k+1\}} \mathbf{G}\right]^{-1} \mathbf{G}^T \mathbf{H}_{\{k+1\}} \mathbf{F}\right) + \mathbf{R}. \tag{4.58}$$

The considered case deals with the generation of optimal control in a process, where the initial state and number of steps are known, and where the final state is free, thus from the transversality condition it follows

$$\boldsymbol{\psi}_{\{n\}} = \mathbf{0}, \tag{4.59}$$

then Eq. (4.58) leads to expression

$$\mathbf{H}_{\{n\}} = \mathbf{R}_F. \tag{4.60}$$

The computation of the optimal control law is based on backward induction. Starting from the initial condition (4.60) in Riccati equation (4.58) the values of matrix $\mathbf{H}_{\{k\}}$ and gain matrix $\mathbf{K}_{MS\{k\}}$, are calculated in subsequent steps $k = n - 1, n - 2, \ldots, 1, 0$. Then, based on the initial condition $\mathbf{x}_{\{0\}}$ the optimal control law $\mathbf{u}^*_{MS\{k\}}$ is determined (4.56), that transfers an object along the optimal trajectory $\mathbf{x}^*_{\{k\}}$, in $n$ steps of the process, to the free final state with a minimum value of the pre-defined performance index (4.43).

### Pontryagin's maximum principle in optimal (state-dependent) control of linear dynamic system

The problem from Sect. 4.1 was solved with state-dependent control law resulting from the Pontryagin's maximum principle that was applied to control of a linear dynamic objet described by Eq. (4.9).

The values of individual parameters are assumed as defined in Sect. 4.1, object parameters $F = 1$, $G = 1$, cost function parameters $R = R_F = 1$, $Q = 3$, initial state of the system $x_{\{0\}} = 8.9$, free final state in the $n$-th iteration step. Presentation of results is limited to $n = 9$ steps of the process, as the yielded values can be easily compared to solutions obtained with other methods.

Riccati equation (4.58) has the form

$$H_{\{k\}} = F H_{\{k+1\}} \left( F - G \left[ Q + G^2 H_{\{k+1\}} \right]^{-1} G H_{\{k+1\}} F \right) + R, \tag{4.61}$$

thus, given the initial condition $H_{\{n\}} = R_F$, a gain sequence of the control system $K_{MS\{k\}}$ is determined, according to relation

$$K_{MS\{k\}} = \left[ Q + G^2 H_{\{k+1\}} \right]^{-1} G H_{\{k+1\}} F. \tag{4.62}$$

Then, by means of relation

$$u^*_{MS\{k\}} = -K_{MS\{k\}} x^*_{\{k\}}, \tag{4.63}$$

and the initial condition $x_{\{0\}} = 8.9$, the optimal control law $u^*_{MS\{k\}}$ and optimal state trajectory $x^*_{\{k\}}$ were determined.

**Table 4.4**  Values of the optimal state trajectory $x^*_{\{k\}}$, optimal control $u^*_{MS\{k\}}$, optimal value function $V^*_{\{k\}}$, and coefficient $K_{MS\{k\}}$, algorithm MPS

| $k$ | $x^*_{\{k\}}$ [−] | $u^*_{MS\{k\}}$ [−] | $V^*_{\{k\}}$ [−] | $K_{MS\{k\}}$ |
|---|---|---|---|---|
| 0 | 8.9 | −3.86 | 91.20 | 0.43 |
| 1 | 5.04 | −2.19 | 29.19 | 0.43 |
| 2 | 2.85 | −1.24 | 9.34 | 0.43 |
| 3 | 1.61 | −0.7 | 2.99 | 0.43 |
| 4 | 0.91 | −0.39 | 0.96 | 0.43 |
| 5 | 0.52 | −0.22 | 0.31 | 0.43 |
| 6 | 0.3 | −0.12 | 0.1 | 0.41 |
| 7 | 0.18 | −0.06 | 0.03 | 0.37 |
| 8 | 0.12 | −0.03 | 0.01 | 0.25 |
| 9 | 0.09 | − | 0 | − |

Table 4.4 presents values of the optimal trajectory of object's state $x^*_{\{k\}}$, with the pre-defined initial state, values of the state-dependent optimal control law, derived from a discrete version of the Pontryagin's maximum principle (MPS) $u^*_{MS\{k\}}$ according to relation (4.63), values of the optimal function $V^*_{\{k\}}$ and coefficient $K_{MS\{k\}}$ calculated according to relation (4.62). The calculated values are identical to those obtained from the solution of LQR problem in Sect. 4.2.

Figure 4.5 presents the optimal state trajectory $x^*_{\{k\}}$, state-dependent optimal control law $u^*_{MS\{k\}}$, optimal value function $V^*_{\{k\}}$ and values of gain coefficient $K_{MS\{k\}}$ obtained by solving the problem of optimal control of a linear dynamic object with the application of the Pontryagin's maximum principle.

The optimal control law (4.63) is a linear combination of the gain coefficient $K_{MS\{k\}}$ and the object's state $x^*_{\{k\}}$. It transfers a dynamic system from the pre-defined initial state $x_{\{0\}}$ to the free final state $x^*_{\{n\}}$ along the optimal trajectory, with the minimum value of function $V^*_{\{k\}}$. The calculation of the optimal control law is carried out from the last step of the process $k = 9$ to the first step $k = 0$ by solving the discrete Riccati equation (4.61) and yielding values of the gain coefficient $K_{MS\{k\}}$.

Analyzing the process of the generation of the regulator's gain coefficient $K_{MS\{k\}}$, from the last step $k = 9$ to $k = 0$ it can be noticed that the gain value increases from $K_{MS\{8\}} = 0.25$ to $K_{MS\{5\}} = 0.43$, and ultimately becomes stable. It can be concluded that assuming the linear control law from state $u_{P\{k\}} = K_P x_{\{k\}}$, with a fixed value of proportionality coefficient $K_P = 0.43$, a suboptimal system response can be obtained, that in the first steps of the process $k = 0, \ldots, 6$ will not diverge from the optimal solution.

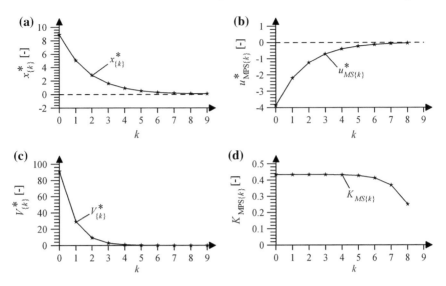

**Fig. 4.5  a** Values of the optimal state trajectory $x^*_{\{k\}}$, $k = 0, \ldots, 9$, **b** values of the optimal control $u^*_{MS\{k\}}$, **c** values of the optimal value function $V^*_{\{k\}}$, **d** values of the gain coefficient $K_{MS\{k\}}$

## 4.4  Summary

The summary is a comparison of computation results related to the optimal control of a stationary linear object of the first order with the application of presented methods. The results obtained when solving the optimal control law generation problem with the use of Bellman's DP (Sect. 4.1), LQR (Sect. 4.2) and the Pontryagin's maximum principle (Sect. 4.3) were compared to those obtained with the use of a control system derived from the Pontryagin's maximum principle, with the control in the state vector function (Sect. 4.3).

Table 4.5 provides a comparison of values related to the optimal control law for a stationary linear object of the first order, that was derived by applying Bellman's DP $u^*_{DP\{k\}}$, obtained on the basis of analytical calculations in the form of LQR $u^*_{LQ\{k\}}$, and determined with the use of the discrete Pontryagin's maximum principle $u^*_{M\{k\}}$, including the state-dependent version $u^*_{MS\{k\}}$.

The optimal control generated with the use of LQR, MP and MPS has the same values, whereas the optimal control generated by applying DP slightly differs from the other three, which is due to the adopted discretization of the DP admissible control interval.

Table 4.6 includes the values of the object's optimal state trajectories $x^*_{\{k\}}$ obtained under control determined with the application of DP, LQR and MP, and the control with a structure that was derived from the discrete maximum principle, in the state-dependent version, MPS.

**Table 4.5** Values of the optimal control law for a linear object obtained by the application of Bellman's DP $\left(u^*_{DP\{k\}}\right)$, LQR $\left(u^*_{LQ\{k\}}\right)$, Pontryagin's maximum principle $\left(u^*_{M\{k\}}\right)$, including the state-dependent version $\left(u^*_{MS\{k\}}\right)$

| $k$ | $u^*_{DP\{k\}}$ [−] | $u^*_{LQ\{k\}}$ [−] | $u^*_{M\{k\}}$ [−] | $u^*_{MS\{k\}}$ [−] |
|---|---|---|---|---|
| 0 | −3.9 | −3.86 | −3.86 | −3.86 |
| 1 | −2.2 | −2.19 | −2.19 | −2.19 |
| 2 | −1.2 | −1.24 | −1.24 | −1.24 |
| 3 | −0.7 | −0.7 | −0.7 | −0.7 |
| 4 | −0.4 | −0.39 | −0.39 | −0.39 |
| 5 | −0.2 | −0.22 | −0.22 | −0.22 |
| 6 | −0.1 | −0.12 | −0.12 | −0.12 |
| 7 | −0.1 | −0.06 | −0.06 | −0.06 |
| 8 | −0.1 | −0.03 | −0.03 | −0.03 |
| 9 | 0 | − | − | − |

**Table 4.6** Values of the optimal trajectory $x^*_{\{k\}}$, obtained from the solution of an optimal control problem with the application of Bellman's DP, LQR, MP and MPS

| $k$ | $x^*_{\{k\}}$ [−] DP | $x^*_{\{k\}}$ [−] LQR | $x^*_{\{k\}}$ [−] MP | $x^*_{\{k\}}$ [−] MPS |
|---|---|---|---|---|
| 0 | 8.9 | 8.9 | 8.9 | 8.9 |
| 1 | 5 | 5.04 | 5.04 | 5.04 |
| 2 | 2.8 | 2.85 | 2.85 | 2.85 |
| 3 | 1.6 | 1.61 | 1.61 | 1.61 |
| 4 | 0.9 | 0.91 | 0.91 | 0.91 |
| 5 | 0.5 | 0.52 | 0.52 | 0.52 |
| 6 | 0.3 | 0.3 | 0.3 | 0.3 |
| 7 | 0.2 | 0.18 | 0.18 | 0.18 |
| 8 | 0.1 | 0.12 | 0.12 | 0.12 |
| 9 | 0 | 0.09 | 0.09 | 0.09 |

Comparison of the object's state values, obtained with the application of control algorithms derived with individual methods, indicates that the results for the LQR and for the control based on the Pontryagin's maximum principle (MP and MPS) are identical. These three slightly differ from the state trajectory values yielded with the Bellman's DP. It is noted that in the case of Bellman's DP the values of system's state $x_{\{k\}}$ and control $u_{\{k\}}$ were constrained to discretized sets, whereas in the case of LQR, MP and MPS the system's state and control could take any values.

Table 4.7 presents the values of the optimal function $V^*_{\{k\}}$ obtained from the solution of an optimal control problem for a linear object with the application of Bellman's DP, LQR and control systems derived with the Pontryagin's maximum principle, including the state-dependent version.

**Table 4.7** Values of the optimal function $V_{\{k\}}^*$ obtained from the solution of an optimal control problem with the application of Bellman's DP, LQR, MP and MPS

| k | $V_{\{k\}}^*$ [−] DP | $V_{\{k\}}^*$ [−] LQR | $V_{\{k\}}^*$ [−] MP | $V_{\{k\}}^*$ [−] MPS |
|---|---|---|---|---|
| 0 | 91.22 | 91.20 | 91.20 | 91.20 |
| 1 | 28.8 | 29.19 | 29.19 | 29.19 |
| 2 | 9.04 | 9.34 | 9.34 | 9.34 |
| 3 | 2.96 | 2.99 | 2.99 | 2.99 |
| 4 | 0.945 | 0.96 | 0.96 | 0.96 |
| 5 | 0.3 | 0.31 | 0.31 | 0.31 |
| 6 | 0.115 | 0.1 | 0.1 | 0.1 |
| 7 | 0.055 | 0.03 | 0.03 | 0.03 |
| 8 | 0.02 | 0.01 | 0.01 | 0.01 |
| 9 | 0 | 0 | 0 | 0 |

**Table 4.8** Values of the gain coefficient $K_{LQ\{k\}}$ of the LQR algorithm and $K_{MS\{k\}}$ of the MPS algorithm

| k | $K_{LQ\{k\}}$ [−] | $K_{MS\{k\}}$ [−] |
|---|---|---|
| 0 | 0.43 | 0.43 |
| 1 | 0.43 | 0.43 |
| 2 | 0.43 | 0.43 |
| 3 | 0.43 | 0.43 |
| 4 | 0.43 | 0.43 |
| 5 | 0.43 | 0.43 |
| 6 | 0.41 | 0.41 |
| 7 | 0.37 | 0.37 |
| 8 | 0.25 | 0.25 |
| 9 | − | − |

The values of the optimal function $V_{\{k\}}^*$ depend on the object's state and the control applied. In the case of the LQR, MP and MPS, the functions take identical values, whereas DP related values are slightly different from those calculated for the other three algorithms, which is due to the adopted discretization method for the object's state and control.

The controlled object (4.9) is of the first order, thus the LQR gain matrix and matrix of the control system, whose structure was derived with the discrete maximum principle, in the state-dependent version, reduce to scalar gain coefficients $K_{LQ\{k\}}$ of the LQR algorithm and $K_{MS\{k\}}$ of the MPS algorithm, whose values in individual steps are shown in Table 4.8.

The optimal LQRs and MPSs gain coefficients of the pre-defined quadratic performance index and stationary linear object take identical values in each iteration of the process. When comparing relations (4.23) in the case of the LQR and (4.56) in the case of the algorithm derived with the MP in the state-dependent version, one may notice similarities in the structure of both control systems. In both cases, the optimal control law is a linear combination of the gain coefficient $\mathbf{K}_{LQ\{k\}}$ or $\mathbf{K}_{MS\{k\}}$ and the object's state $\mathbf{x}^*_{\{k\}}$. The gain coefficients are generated from step $k = 9$ to $k = 0$ which entails the need to solve Riccati equation (4.25) in the case of LQR and (4.58) in the case of MPS. It should be noted that for a stationary object and constant values of matrices $\mathbf{R}$, $\mathbf{R}_F$ and $\mathbf{Q}$, the gain coefficients $\mathbf{K}_{LQ\{k\}}$ or $\mathbf{K}_{MS\{k\}}$ do not take constant values. Considering the process of their generation, based on the calculations performed in Sect. 4.2 and 4.3, from the last step $k = 9$ to $k = 0$, it may be noticed that initially small gain values $K_{LQ\{8\}} = K_{MS\{8\}} = 0.25$ increase to $K_{LQ\{5\}} = K_{MS\{5\}} = 0.43$, and ultimately remain stable. It can be concluded that introduction of fixed values of proportional gain coefficients $K_P = 0.43$ into the state-dependent control $u_{P\{k\}} = K_P x_{\{k\}}$ shall result in a suboptimal state trajectory that in steps $k = 0, \ldots, 6$, shall not diverge from the optimal solution.

The analysis of results indicates strong convergence of solutions achieved by the application of the Bellman's DP method and the other control systems (LQR, MP and MPS). It is also noted, that the Bellman's DP algorithm was applied to a discretized set of admissible controls and states of the controlled object, which accounts for the difference in the obtained results. LQR, MP and MPS algorithms produce identical optimal control signal, despite their structural differences and differing methods for derivation of control law.

All the aforementioned methods for optimal control of a linear object have the same drawback, which is that in order to be applied the mathematical model of a controlled object must be known. Furthermore, the calculation of optimal control law from the last step of the process $k = n$ to the first step $k = 0$ for the DP, LQR and MPS methods, and in the case of the MP the necessity to solve a set of $2(n - 1)$ equations in a two-boundary problem, prevent their application in an online control process.

# References

1. Bellman, R.: Dynamic Programming. Princeton University Press, New York (1957)
2. Bertsekas, D.P.: Dynamic Programming and Optimal Control, vol. I. Athena Scientific, Belmont, II (2005)
3. Ferrari, S., Stengel, R.F.: Model-based adaptive critic designs in learning and approximate dynamic programming. In: Si, J., Barto, A., Powell, W., Wunsch, D.J. (eds.) Handbook of Learning and Approximate Dynamic Programming, pp. 64–94. Wiley & Sons, New York (2004)
4. Grecki, H.: Optimization and Control of Dynamic Systems. (in Polish) AGH University of Science and Technology Press, Krakow (2006)
5. Kaczorek, T.: Control Theory. (in Polish) vol. II, PWN, Warsaw (1981)
6. Kirk, D.E.: Optimal Control Theory. Dover Publications, New York (2004)
7. Takahashi, Y., Rabins, M.J., Auslander, D.M.: Control and Dynamic Systems. (in Polish) WNT, Warsaw (1976)

# Chapter 5
# Learning Methods for Intelligent Systems

This chapter provides a description of learning methods for artificial intelligence algorithms. It presents selected information on supervised learning algorithms that constitute a group of the most preferred NN learning methods that are applied to problems where the values of approximate representations are known. These algorithms are characterized by a fast learning rate, accurate mapping of desired signals and a simple notation as opposed to other learning methods. This chapter is an introduction to reinforcement learning methods, the application of which allows for solving the policy selection problem in the absence of sufficient information that could serve as training examples for supervised learning. The chapter also describes selected methods of unsupervised learning that is applied to neural networks with simple structures that produce representations (maps) of certain features and are used for data clustering, recognition and compression.

## 5.1 Supervised Learning

Supervised learning, also known as learning with a teacher, is one of the neural network learning concepts [6–9]. In supervised learning data is a set of training examples consisting of vector pairs i.e. network's input vectors and the corresponding output values. An NN is trained off-line by adjusting values of networks's adaptive parameters, e.g. weights $\mathbf{W}_{\{k\}}$, according to the learning algorithm. The process shall be carried out in such a way that the value of the $g$th output from the network $\hat{f}_{g\{k\}}$ for the given input vector $\mathbf{x}_j$ corresponds to the desired value $d_{g,l}$ of the $j$th example from the training set, thereby minimizing values of the objective function

$$E\left(\mathbf{W}_{\{k\}}\right) = \frac{1}{2} \sum_{j=1}^{p} \sum_{g=1}^{M} \left(\hat{f}_{g\{k\}} - d_{j,g}\right)^2 , \tag{5.1}$$

© Springer International Publishing AG 2018
M. Szuster and Z. Hendzel, *Intelligent Optimal Adaptive Control
for Mechatronic Systems*, Studies in Systems, Decision and Control 120,
https://doi.org/10.1007/978-3-319-68826-8_5

where $k = 1, \ldots, n$ – index of subsequent steps of the learning process, $n$ – number of learning steps, $j = 1, \ldots, p$ – index of training samples, $p$ – number of training samples, $g = 1, \ldots, M$ – index of network outputs, $M$ – number of NN outputs.

In the simplest case, with one training sample, the objective function takes the form of a mean square error. This occurs in the online process of NN weight training, where in each $k$th step of the iterative process only current training sample is available $\mathbf{d}_{\{k\}}$ that corresponds to the network input vector $\mathbf{x}_{\{k\}}$. This type of neural network learning is known as adaptation of weights, relation (5.1) can be written as

$$E\left(\mathbf{W}_{\{k\}}\right) = \frac{1}{2} \sum_{g=1}^{M} \left(\hat{f}_{g\{k\}} - d_{g\{k\}}\right)^2 . \tag{5.2}$$

Given the continuity of the objective function based on the mapping error between the desired value and the network output value, the most effective network learning methods are gradient-based optimization methods, where the weight vector is updated according to the formula

$$\mathbf{W}_{\{k+1\}} = \mathbf{W}_{\{k\}} + \mathbf{\Gamma}_N \, \mathbf{p}\left(\mathbf{W}_{\{k\}}\right) , \tag{5.3}$$

where $\mathbf{\Gamma}_N$ – diagonal matrix of positive reinforcement coefficients, $\mathbf{p}\left(\mathbf{W}_{\{k\}}\right)$ – direction in the multidimensional space $\mathbf{W}_{\{k\}}$. Multi-layer NN learning based on gradient methods requires the determination of a gradient vector with respect to weights of all network layers. This is not a difficult task in the case of output layer weights, however the weights of the remaining layers require the application of the backpropagation algorithm.

The most commonly applied supervised learning methods include e.g. the steepest descent algorithm, variable metric algorithm, Levenberg–Marquardt algorithm and the conjugate gradient method.

### 5.1.1   Steepest Descent Algorithm

The steepest descent algorithm [7] is based on a linear approximation of the objective function $E\left(\mathbf{W}_{\{k\}}\right)$ expanded into a Taylor series, in a small neighborhood of the known solution $\mathbf{W}_{\{k\}}$ with accuracy (truncation error) of $O\left(h^2\right)$

$$E\left(\mathbf{W}_{\{k\}} + \mathbf{p}_{\{k\}}\right) = E\left(\mathbf{W}_{\{k\}}\right) + \left[\nabla E_{\{k\}}\right]^T \mathbf{p}_{\{k\}} + O\left(h^2\right) , \tag{5.4}$$

where $\mathbf{p}\left(\mathbf{W}_{\{k\}}\right) = \mathbf{p}_{\{k\}}$, $\nabla E_{\{k\}} = \left[\frac{\partial E\left(\mathbf{W}_{\{k\}}\right)}{\partial W_{1\{k\}}}, \frac{\partial E\left(\mathbf{W}_{\{k\}}\right)}{\partial W_{2\{k\}}}, \ldots, \frac{\partial E\left(\mathbf{W}_{\{k\}}\right)}{\partial W_{m\{k\}}}\right]^T$ – gradient vector, $m$ – number of network neurons. The NN weight learning algorithm shall be selected in such a way as to minimize the objective function, $E\left(\mathbf{W}_{\{k+1\}}\right) < E\left(\mathbf{W}_{\{k\}}\right)$, thus $\left[\nabla E_{\{k\}}\right]^T \mathbf{p}_{\{k\}} < 0$. Assuming

$$\mathbf{p}_{\{k\}} = -\nabla \mathbf{E}_{\{k\}} \,, \tag{5.5}$$

this condition is satisfied, Eq. (5.3) can be written in the form of relation

$$\mathbf{W}_{\{k+1\}} = \mathbf{W}_{\{k\}} - \boldsymbol{\Gamma}_N \, \nabla \mathbf{E}_{\{k\}} \,. \tag{5.6}$$

The steepest descent algorithm does not use the information included in the Hessian, its convergence is linear, and the learning process slows down in the neighborhood of the optimal point, where the gradient has low values. Despite the aforementioned disadvantages and low effectiveness, the method is commonly applied to multilayer NN learning, due to simplicity and low computational cost. The steepest algorithm performance can be improved by implementing the so-called momentum [7, 9].

### 5.1.2 Variable Metric Algorithm

In the variable metric algorithm [9] the expansion into a Taylor series includes quadratic approximation to the objective function $E\left(\mathbf{W}_{\{k\}}\right)$ in the neighborhood of the closest solution $\mathbf{W}_{\{k\}}$ with accuracy (truncation error) $O\left(h^3\right)$

$$E\left(\mathbf{W}_{\{k\}} + \mathbf{p}_{\{k\}}\right) = E\left(\mathbf{W}_{\{k\}}\right) + \left[\nabla \mathbf{E}_{\{k\}}\right]^T \mathbf{p}_{\{k\}} + \frac{1}{2}\mathbf{p}_{\{k\}}^T \mathbf{H}\left(\mathbf{W}_{\{k\}}\right)\mathbf{p}_{\{k\}} + O\left(h^3\right) \,, \tag{5.7}$$

where $\mathbf{H}\left(\mathbf{W}_{\{k\}}\right)$ – square symmetric matrix of second-order derivatives, commonly called a Hessian,

$$\mathbf{H}\left(\mathbf{W}_{\{k\}}\right) = \begin{bmatrix} \dfrac{\partial E\left(\mathbf{W}_{\{k\}}\right)}{\partial W_{1\{k\}}\partial W_{1\{k\}}} & \cdots & \dfrac{\partial E\left(\mathbf{W}_{\{k\}}\right)}{\partial W_{m\{k\}}\partial W_{1\{k\}}} \\ \vdots & & \vdots \\ \dfrac{\partial E\left(\mathbf{W}_{\{k\}}\right)}{\partial W_{1\{k\}}\partial W_{m\{k\}}} & \cdots & \dfrac{\partial E\left(\mathbf{W}_{\{k\}}\right)}{\partial W_{m\{k\}}\partial W_{m\{k\}}} \end{bmatrix}. \tag{5.8}$$

The condition for a minimum at the considered point is that the function has zero first derivative in the direction $\mathbf{p}_{\{k\}}$ and positive value of second derivative. Accordingly, by differentiating and transforming (5.7) we get

$$\mathbf{p}_{\{k\}} = -\left[\mathbf{H}\left(\mathbf{W}_{\{k\}}\right)\right]^{-1} \nabla \mathbf{E}_{\{k\}} \,, \tag{5.9}$$

where $\mathbf{p}_{\{k\}}$ – descent direction of the objective function. Determining the direction, which is the substance of the Newton's method for optimization, requires the calculation of the gradient vector $\nabla \mathbf{E}_{\{k\}}$ and the Hessian $\mathbf{H}\left(\mathbf{W}_{\{k\}}\right)$, in each step of the process.

The Hessian shall be positive definite which, in a general case, is difficult to satisfy, thus its approximation $\mathbf{G}\left(\mathbf{W}_{\{k\}}\right)$ is used for practical applications.

The most common methods for recursive definition of the inverse Hessian approximation matrix are: the Davidon–Fletcher–Powell formula and the Broyden–Goldfarb–Fletcher–Shanno formula [7, 9, 10]. The variable metric method outperforms the steepest descent algorithm, as it is characterized by quadratic convergence. Despite its computational complexity that requires determination of $m^2$ elements of the Hessian in each step, it is considered one of the best methods for optimization of multivariable functions, provided the applied NN does not have a great number of neurons $m$.

## 5.1.3  Levenberg–Marquardt Algorithm

The Levenberg–Marquardt algorithm also uses the Newton's method for optimization in the NN weight training process [7, 9]. In the Levenberg–Marquardt algorithm the Hessian matrix $\mathbf{H}\left(\mathbf{W}_{\{k\}}\right)$ from relation (5.9) is replaced with its approximation $\mathbf{G}\left(\mathbf{W}_{\{k\}}\right)$ that is defined based on the gradient vector and given the regularization factor.

Assuming the error measure in the form of Eq. (5.2), relevant literature provides the following formulas defining the gradient vector

$$\nabla \mathbf{E}_{\{k\}} = \left[\mathbf{J}\left(\mathbf{W}_{\{k\}}\right)\right]^{T}\mathbf{e}\left(\mathbf{W}_{\{k\}}\right) , \tag{5.10}$$

and approximated Hessian matrix

$$\mathbf{G}\left(\mathbf{W}_{\{k\}}\right) = \left[\mathbf{J}\left(\mathbf{W}_{\{k\}}\right)\right]^{T}\mathbf{J}\left(\mathbf{W}_{\{k\}}\right) + \mathbf{R}\left(\mathbf{W}_{\{k\}}\right) , \tag{5.11}$$

where $\mathbf{R}\left(\mathbf{W}_{\{k\}}\right)$ – term of the Hessian expansion that includes higher derivatives with respect to vector $\mathbf{W}_{\{k\}}$,

$$\mathbf{e}\left(\mathbf{W}_{\{k\}}\right) = \left[e_1\left(\mathbf{W}_{\{k\}}\right), \ldots, e_M\left(\mathbf{W}_{\{k\}}\right)\right]^{T} , \tag{5.12}$$

is the error vector of individual network outputs, while

$$\mathbf{J}\left(\mathbf{W}_{\{k\}}\right) = \begin{bmatrix} \dfrac{\partial e_1\left(\mathbf{W}_{\{k\}}\right)}{\partial W_{1\{k\}}} & \cdots & \dfrac{\partial e_1\left(\mathbf{W}_{\{k\}}\right)}{\partial W_{m\{k\}}} \\ \vdots & & \vdots \\ \dfrac{\partial e_M\left(\mathbf{W}_{\{k\}}\right)}{\partial W_{1\{k\}}} & \cdots & \dfrac{\partial e_M\left(\mathbf{W}_{\{k\}}\right)}{\partial W_{m\{k\}}} \end{bmatrix} . \tag{5.13}$$

Values $\mathbf{R}\left(\mathbf{W}_{\{k\}}\right)$ are difficult to define, hence this term is approximated with the regularization factor $\mathbf{R}\left(\mathbf{W}_{\{k\}}\right) = v_L\mathbf{I}$, where $v_L$ is a scalar Levenberg–Marquardt parameter, whose value changes during the optimization process. The descent direction $\mathbf{p}_{\{k\}}$ may be written as

$$\mathbf{p}_{\{k\}} = -\left[\left[\mathbf{J}\left(\mathbf{W}_{\{k\}}\right)\right]^T \mathbf{J}\left(\mathbf{W}_{\{k\}}\right) + v_L\mathbf{I}\right]^{-1}\left[\mathbf{J}\left(\mathbf{W}_{\{k\}}\right)\right]^T \mathbf{e}\left(\mathbf{W}_{\{k\}}\right) . \tag{5.14}$$

Parameter $v_L$ has large value at the beginning of the network learning process that tends to zero while approach optimal solution. Despite its high convergence rate the Levenberg–Marquardt algorithm is complex in computational terms as it involves inversion of an $m^2$ matrix. A modified method, free from this drawback, is proposed in [12]. The algorithm may be used for online weight adaptation [4], provided the applied network's structure is not a complex one.

### 5.1.4 Conjugate Gradient Method

The conjugate gradient method [7] does not use the information included in the Hessian and the search direction $\mathbf{p}_{\{k\}}$ is built in such a way as to ensure orthogonality and conjugation to the set of previous directions $\mathbf{p}_{\{0\}}, \mathbf{p}_{\{1\}}, \ldots, \mathbf{p}_{\{k-1\}}$. The above condition is satisfied with respect to matrix $\mathbf{G}$, when

$$\mathbf{p}_{\{i\}}^T\mathbf{G}\mathbf{p}_{\{j\}} = 0, \qquad i \neq j , \tag{5.15}$$

where $i = 0, \ldots, k$, $j = 0, \ldots, k$, vector $\mathbf{p}_{\{k\}}$ has the form

$$\mathbf{p}_{\{k\}} = -\nabla\mathbf{E}_{\{k\}} + \sum_{i=0}^{k-1} \beta_{Si\{k\}}\mathbf{p}_{\{i\}} , \tag{5.16}$$

where $\beta_{Si\{k\}}$ – conjugation coefficient. The summation in the second part of the formula concerns all previous directions. Relation (5.16) can be simplified based on the orthogonality condition and given the conjugation between vectors, then

$$\mathbf{p}_{\{k\}} = -\nabla\mathbf{E}_{\{k\}} + \beta_{S\{k-1\}}\mathbf{p}_{\{k-1\}} . \tag{5.17}$$

The cost function descent direction depends on the gradient value at the solution point, the product of the previous search direction $\mathbf{p}_{\{k-1\}}$ and the conjugation coefficient $\beta_{S\{k-1\}}$ that stores information about the previous search directions. The relevant literature provides many formulas for the determination of the coefficient, the most known being [7]

$$\beta_{S\{k-1\}} = \frac{\nabla \mathbf{E}_{\{k\}}^{T} \left( \nabla \mathbf{E}_{\{k\}} - \nabla \mathbf{E}_{\{k-1\}} \right)}{\nabla \mathbf{E}_{\{k\}}^{T} \nabla \mathbf{E}_{\{k-1\}}} \,. \qquad (5.18)$$

Practically, in the conjugate gradient method the property of orthogonality between the descent direction vectors is lost, due to the accumulation of round-off errors in subsequent computational cycles, this is why the method has to be restarted after every few steps. It demonstrates higher convergence than the steepest descent method and a lower computational complexity, which is why it is considered the only effective optimization algorithm applied to problems with several thousand of variables or more.

## 5.2  Learning with a Critic

Reinforcement learning consists in dynamic interaction with the environment where actions are taken and evaluated in the form of numerical rewards that correspond to a particular goal-oriented behavior [1, 3]. The idea of the learning process is to establish an action selection policy that would maximize the values of rewards, also known as reinforcement values.

The origins of reinforcement learning are in the observation of nature and mainly research into animal learning. The term of reinforcement learning relates to the adaptation of artificial intelligence algorithms and terms from experimental psychology. The aforementioned observations show that if an action is followed by a satisfactory state or an improvement in the current state the tendency to produce that action is strengthened (reinforced). In the animal learning, food or an electric shock may serve as such reinforcers. The above analogy implies that the reinforcement value, often referred to as the reward, may in certain situations be considered as a punishment. These terms, however, shall not be taken literally with respect to the adaptive process of artificial intelligence algorithms, as there is no fixed threshold value that identifies a given reinforcement as a reward or punishment. It is also a relative term, for example, in system's states where high rewards are received a lower reinforcement value might be considered a punishment, whereas in other states even much lower reinforcement value may be recognized as a reward. In these considerations the sign of the reinforcement value is of secondary importance. It does not determine whether we are dealing with a reward or a punishment. Therefore, for the purpose of this work, equivalent terms shall apply regardless of the values they take i.e. reward, reinforcement or the cost function.

Reinforcement learning is classified as a learning without a teacher method. It uses training information that evaluates agent's actions rather than instructs by giving types of actions to be undertaken. The source of such knowledge may be construed as coming from a critic rather than a teacher. Thus, reinforcement learning methods are often referred to as learning with a critic methods and further in this work they are used interchangeably. The learning process is carried out in an environment that is unknown to the agent and often non-deterministic and non-stationary. In simple terms reinforcement learning may be defined as trial-and-error learning where agent is supposed to act in a certain way within the surrounding environment. Having no prior knowledge on evaluation of decisions made agent's actions are error-prone, however evaluative feedback indicates how good the action taken is and allows for improvement of the control policy so as to maximize rewards. The idea of reinforcement learning is to modify action-selection policy based on the observed states, history of actions taken and rewards received.

Adaptive algorithms that utilize reinforcement learning methods usually operate in the discrete-time domain. These are decision-making processes where based on the current state of an object or environment $\mathbf{x}_{\{k\}}$ in the $k$th iteration step, and action taken $\mathbf{u}_{\{k\}}$, state $\mathbf{x}_{\{k+1\}}$ is determined. The reinforcement learning process in the $k$th step requires the knowledge of object's state $\mathbf{x}_{\{k\}}$, action taken under the current control policy $\mathbf{u}_{\{k\}}$, and the reinforcement value $L_{C\{k\}}$, construed as value of taking action $\mathbf{u}_{\{k\}}$ in state $\mathbf{x}_{\{k\}}$, and state achieved $\mathbf{x}_{\{k+1\}}$ via the action. Based on thus obtained set of information the reinforcement learning algorithm makes improvements to the current control policy.

The learning environment may not be known to the agent and involves uncertainty. This has a significant impact on the difficulty in task performance and the practical applicability of learning with a critic methods. The environmental uncertainty means that the mechanisms for reinforcement generation and transition of states under actions taken may be stochastic, and a multiple performance of the very same action $\mathbf{u}_{\{k\}}$ in a given state $\mathbf{x}_{\{k\}}$ may lead to other states $\mathbf{x}_{\{k+1\}}$ and reinforcement values $L_{C\{k\}}$. The knowledge about the environment model is not a necessary condition for the application of reinforcement learning methods, however additional information on the agent-environment can be used by some algorithms to speed up the learning process.

The goal of reinforcement learning is to find a control policy that maximizes the performance criterion that is defined based on the rewards received by the agent. The criterion selection determines the type of reinforcement learning, and the most common one encountered in the relevant literature is the performance index, whereby the received rewards are maximized over the long time horizon. In such a case the applied policy might not be optimal in the short-term, however it shall maximize the long-term reward. This type of reinforcement learning requires taking into account delayed consequences of actions and is defined as learning from delayed rewards, where applied algorithms solve the problem of temporal reward assignment. It consists in the evaluation of currently undertaken actions against the long-term control goal, where the reinforcement value depends not only on the immediate effect of the action, but also on its predicted impact on the task realisation in the future. On the contrary, in learning with instantaneous reinforcement the agent maximizes the performance criterion being a reward function related to the current action, ignoring future effects of these actions.

In conclusion, in reinforcement learning methods the agent is not presented with a training sample, but it uses information that evaluates actions taken under the adopted objective function. Learning has an interactive character and is carried out by trial-

and-error, where the agent arbitrarily selects a policy that leads to the achievement
of the goal. There is no distinction between learning and algorithm testing phase,
both these processes take place simultaneously.

The most common algorithms that use reinforcement learning methods include
e.g. the neural dynamic programming, Q-learning, R-learning and SARSA. The Q-
learning algorithm is presented below, while a detailed description of neural dynamic
programming algorithms can be found in Chap. 6.

## 5.2.1   Q-Learning Algorithm

Q-learning is one of the algorithms, whose adaptation is based on the idea of reinforcement learning. The primary version of the algorithm worked in the discrete-
time domain [11]. Due to its qualities it is commonly applied [1–3, 5, 13] and
has undergone numerous modifications. In its simplest form, the so-called 1-step Q-
learning, the algorithm execution involves the estimation of control and cost function
$Q_{L\{k\}}\left(\mathbf{x}_{\{k\}}, \mathbf{u}_{\{k\}}\right)$

$$Q_{L\{k\}}\left(\mathbf{x}_{\{k\}}, \mathbf{u}_{\{k\}}\right) = \sum_{k=0}^{n} \gamma^k L_{C\{k\}}\left(\mathbf{x}_{\{k\}}, \mathbf{u}_{\{k\}}\right) , \qquad (5.19)$$

where $n$ – number of steps in the iterative process, $\gamma$ – the so-called discount factor for
future rewards/punishments, selected from the interval $0 \leq \gamma \leq 1$, $L_{C\{k\}}\left(\mathbf{x}_{\{k\}}, \mathbf{u}_{\{k\}}\right)$
– cost function, also known as the local cost or reinforcement coefficient, in the $k$th
step, $\mathbf{x}_{\{k\}}$ – state vector, $\mathbf{u}_{\{k\}}$ – control signals vector resulting from action taken in
the $k$th step. In the Q-learning algorithm the cost function always depends on the
object's current state-value and control signals values.

The goal of function $Q_{L\{k\}}\left(\mathbf{x}_{\{k\}}, \mathbf{u}_{\{k\}}\right)$ estimation is to minimize the temporal
difference error $e_{TD\{k\}}$ that is given by

$$e_{TD\{k\}} = L_{C\{k\}}\left(\mathbf{x}_{\{k\}}, \mathbf{u}_{\{k\}}\right) + \gamma \min_{\mathbf{u}_{\{k\}}} Q_{L\{k+1\}}\left(\mathbf{x}_{\{k+1\}}, \mathbf{u}_{\{k\}}\right) - Q_{L\{k\}}\left(\mathbf{x}_{\{k\}}, \mathbf{u}_{\{k\}}\right) .$$

$$(5.20)$$

In the simplest case the estimation of function $Q_{L\{k\}}$ is carried out by entering into
a table the values of function $Q_{L\{k\}}\left(\mathbf{x}_{\{k\}}, \mathbf{u}_{\{k\}}\right)$ for individual state $\mathbf{x}_{\{k\}}$ – action $\mathbf{u}_{\{k\}}$
pairs. This approach is only suitable for discrete systems and cannot be applied to
object control in the continuous-time domain. There is a great number of publications
that address the problem of Q-learning in the continuous-time domain [2, 5, 13].
Based on the tabular representation of function $Q_{L\{k\}}$, in each $k$th step of the process,
the control $\mathbf{u}_{\{k\}}$ is selected that minimizes the value of function $Q_{L\{k\}}$, which leads to
the so-called greedy policy. It has been proven in [11] that control generated by the
Q-learning algorithm converges to the optimal value, when $k \to \infty$, on condition
that each of the states $\mathbf{x}_{\{k\}}$ is visited adequately often.

The major advantage of Q-learning, compared to the actor-critic algorithms, is the intensity of the exploration process. In Q-learning the adaptive process may be carried out regardless of the currently applied control policy. The drawback of the considered version is the discretization of the input space and generation of discrete control law, as opposed to smooth control signals of neural dynamic programming algorithms. Q-learning algorithms that operate in the continuous-time domain have no such disadvantages. To execute the Q-learning algorithm only one adaptive structure is needed (e.g. in a tabular form) that maps the values of function $Q_{L\{k\}}$ into the state and control space. The actor-critic algorithms require two adaptive structures.

## 5.3 Learning Without a Teacher

In learning without a teacher [7, 9], also referred to as the unsupervised learning, the learning sequence consists only of network input values, with no desired output signal. These methods are applied to training of self-organizing networks. They are usually single-layer NNs with uncomplicated structures that are used for e.g. pattern recognition, data clustering and compression. In competitive learning, NN's neurons compete with each other. The goal is to arrange competitive neurons by selecting their weight values so as to minimize the performance index construed as an error between the set of input vectors $\mathbf{x}_j$ and the weight values of the winning-neuron, where $j$ – index, $j = 1, \ldots, p$, $p$ – number of all training examples. If the training set contains $p$ training samples in the form of network input vectors, then by applying Euclidean norm the performance index can be expressed by relation

$$E\left(\mathbf{W}_{\{k\}}\right) = \frac{1}{p} \sum_{j=1}^{p} \left\| \mathbf{x}_j - \mathbf{W}_{i\{k\}} \right\|, \qquad (5.21)$$

where $\mathbf{W}_{i\{k\}}$ – weight vector of the $i$th neuron that is deemed the winner during the presentation of vector $\mathbf{x}_j$ in the $k$th cycle of the network learning process, $k = 0, \ldots, n$, $\mathbf{W}_{\{k\}} = \left[\mathbf{W}_{1\{k\}}, \ldots, \mathbf{W}_{m\{k\}}\right]^T$, $m$ – number of neurons.

Self-organizing networks can be divided into two groups, assuming the criterion of classification is the selection of neuron/group of neurons, whose weights undergo the learning process. The first group comprises NNs trained by the WTA algorithm (Winner Takes All), the second being the networks trained by means of the WTM algorithm (Winner Takes Most).

### 5.3.1 Winner-Take-All Networks

In self-organizing networks the $j$th vector of NN input signals $\mathbf{x}_j = \left[x_{j,1}, \ldots, x_{j,N}\right]^T$ is fed into the input of all $m$ neurons, where the distance between the input signal $\mathbf{x}_j$ and all the weight vectors $\mathbf{W}_{i\{k\}}$, is measured, mostly by using the Euclidean metric

$$d_i\left(\mathbf{x}_j, \mathbf{W}_{i\{k\}}\right) = \left\|\mathbf{x}_j - \mathbf{W}_{i\{k\}}\right\| = \sqrt{\sum_{h=1}^{N}\left(x_{j,h} - W_{i,h\{k\}}\right)^2}. \qquad (5.22)$$

The neuron with the shortest distance calculated according to relation (5.22) yields the output value of 1 and is thus called the winner. The outputs of the remaining neurons yield the value of 0. The weights of the winner-neuron are adjusted according to the relation

$$\mathbf{W}_{i\{k+1\}} = \mathbf{W}_{i\{k\}} + \mathbf{\Gamma}_{N\{k\}}\left[\mathbf{x}_j - \mathbf{W}_{i\{k\}}\right], \qquad (5.23)$$

where $\mathbf{\Gamma}_{N\{k\}}$ – the diagonal matrix of NN positive weight reinforcement coefficients with values dependent on the $k$ step of the learning process. The initial values of coefficients shall be high and later decrease. In self-organizing networks the key issue is the normalization of the input signals into the NN.

Operation of a self-organizing network consists in a division of $p$ training samples into $m$ classes represented by network's neurons. Once the learning process is complete the weight vectors $\mathbf{W}_{i\{n\}}$ become the centers of classes identified by the NN.

### 5.3.2  Winner-Take-Most Networks

In learning without a teacher with the WTA method only one neuron's weights are adjusted that have the closest values to the input vector, given the adopted distance. In the case of the WTM learning algorithms not only are the winner weights adjusted but also the weights of neurons in the winner's neighborhood. The Kohonen algorithm is an example of a self-organizing network, where weight training is carried out by the WTM method. In the Kohonen network the arrangement of neighbouring neurons into a grid is permanent. In the learning process a winning neuron has to be found whose weight vector lies closest to the NN input vector according to the adopted distance norm. The best results are achieved when applying the Gaussian-type neighborhood function

$$G\left(i, \mathbf{x}_j\right) = \exp\left(-\frac{d_i^2\left(\mathbf{x}_j, \mathbf{W}_{i\{k\}}\right)}{2l_K^2}\right), \qquad (5.24)$$

where $l_K$ – distance known as the radius of the neighborhood. Depending on the value of the radius of the neighborhood and distance (5.22), the weight values of neurons in the winner's neighborhood change to a varying degree, according to the relation

$$\mathbf{W}_{i\{k+1\}} = \mathbf{W}_{i\{k\}} + \Gamma_{Ni,i\{k\}} G\left(i, \mathbf{x}_j\right)\left[\mathbf{x}_j - \mathbf{W}_{i\{k\}}\right], \qquad (5.25)$$

where values of the diagonal matrix of reinforcement coefficients $\boldsymbol{\Gamma}_{N\{k\}}$ are initially high and decrease as the learning process continues.

Another example of the WTM application to the arrangement of neurons in a network trained without a teacher is the Neural Gas algorithm, where distance between weight vectors of all neurons and the input vector is calculated according to the adopted distance norm. Neurons are arranged by increasing distance order (5.22), and have a number assigned $h_{NGi}$, where the winner-neuron $h_{NGi} = 0$, and the neuron with the most distant weight vector $h_{NGi} = m - 1$. The neighborhood function defines the degree of the $i$th neuron weight adjustment

$$G\left(i, \mathbf{x}_j\right) = \exp^{\left(-\dfrac{h_{NGi}}{l_{NG}}\right)}, \qquad (5.26)$$

where $l_{NG}$ – distance known as the radius of the neighborhood, which decreases as the NN learning process continues. In the case when $l_{NG} = 0$, the algorithm in fact operates under the WTA method. Neurons' weights are trained according to relation (5.25).

Neural gas is considered effective algorithm for organizing neurons and with a proper selection of process parameters it performs better than the Kohonen algorithm.

# References

1. Barto, A., Sutton, R.: Reinforcement Learning: An Introduction. MIT Press, Cambridge (1998)
2. Barto, A., Mahadevan, S.: Recent advances in hierarchical reinforcement learning. Discret. Event Dyn. Syst. **13**, 343–379 (2003)
3. Cichosz, P.: Learning Systems (in Polish). WNT, Warsaw (2000)
4. Dias, F.M., et al.: Using the Levenberg–Marquardt for on–line training of a variant system. Lecture Notes in Computer Science, vol. 3697, pp. 359–364 (2005)
5. Hagen, S., Krose, B.: Neural Q-learning. Neural Comput. Appl. **12**, 81–88 (2003)
6. Jamshidi, M., Zilouchian, A.: Intelligent Control Systems Using Soft Computing Methodologies. CRC Press, London (2001)
7. Osowski, S.: Neural Networks (in Polish). Warsaw University of Technology Publishing House, Warsaw (1996)
8. Osowski, S.: Neural Networks - An Algorithmic Approach (in Polish). WNT, Warsaw (1996)
9. Rutkowski, L.: Computational Intelligence - Methods and Techniques (in Polish). Polish Scientific Publishers PWN, Warsaw (2005)
10. Stadnicki, J.: Theory and Practice of Optimization Task Solving (in Polish). WNT, Warsaw (2006)
11. Watkins, C.: Learning from delayed rewards. Ph.D. thesis, Cambridge University, Cambridge, England (1989)
12. Wilamowski, B., Kaynak, O.: An algorithm for fast convergence in training neural networks. In: Proceedings of IJCNN, vol. 3, pp. 1178–1782 (2001)
13. Zelinsky, A., Gaskett, C., Wettergreen, D.: Q-learning in continuous state and action spaces. In: Proceedings of the Australian Joint Conference on Artificial Intelligence, pp. 417–428. Springer, Berlin (1999)

# Chapter 6
# Adaptive Dynamic Programming - Discrete Version

This chapter presents the application of adaptive structures to the Bellman's DP method to approximate the value function. Such action resulted in the creation of a family of neural dynamic programming algorithms that can be used for on-line control of a dynamic objects. The chapter also looks at the main features of the aforementioned family of algorithms and provides a descripion of selected actor-critic learning methods such as heuristic dynamic programming, dual-heuristic dynamic programming and global dual-heuristic dynamic programming which assume availability of a mathematical model, as well as model-free methods i.e. action-dependent heuristic dynamic programming algorithm.

## 6.1 Neural Dynamic Programming

Neural dynamic programming algorithms (NDP) are usually based on the actor-critic structure also defined as adaptive critic designs (ACD) [1, 6, 9, 14, 30, 41–43]. Actor and critic can be implemented in the form of any type of an adaptive structure e.g. NN. The NDP algorithms described in this chapter operate in discrete-time domain.

In control based on NDP algorithms the determination of suboptimal control law and the corresponding suboptimal state trajectory is carried out from the first to the last step of the iterative process. Determination of the state trajectory in NDP is shown in the schematic diagram in Fig. 6.1. In the first step of the iterative process $k = 0$, the system's state is $c$, from where it is to be transferred to state $h$. Due to the application of adaptive structures to implement the actor and critic into an NDP algorithm, in the first trial ($\xi = 1$), construed as an online performance of the sequence of all discrete steps $k$ of the process, $k = 0, \ldots, n$, where $n$ is the number of iteration steps, the determined suboptimal state trajectory 1 may diverge from the optimal trajectory 5. This is due to the selection of initial values of parameters that are adapted e.g. NN

© Springer International Publishing AG 2018
M. Szuster and Z. Hendzel, *Intelligent Optimal Adaptive Control
for Mechatronic Systems*, Studies in Systems, Decision and Control 120,
https://doi.org/10.1007/978-3-319-68826-8_6

**Fig. 6.1** Schematic diagram
- Determination of
suboptimal state trajectory
with NDP algorithm

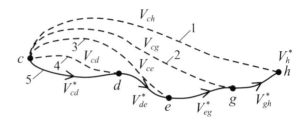

weights. The value function $V_{ch}$ in the first sequence of the adaptive process might
not be optimal.

Retaining the knowledge base within the adaptive structures e.g. in the form of
NN stored weights values, allows for the determination of suboptimal trajectory 2 in
the second trial ($\xi = 2$) that is "closer" to the desired optimal trajectory compared
to the one obtained in the previous trial. The value function in the second sequence
is $V_{cg} + V_{gh}^*$, its value is lower than the one yielded in the first sequence $V_{cg} +
V_{gh}^* < V_{ch}$. Similarly, in the third trial ($\xi = 3$), the value function is $V_{ce} + V_{eg}^* + V_{gh}^*$.
Continuation of actor-critic parametric adaptation in an online process allows for the
determination of the suboptimal trajectory and the suboptimal control law that are
near-optimal.

NDP algorithms are free from Bellman's DP limitations and have the following
advantages:

- lower computational complexity due to the application of NNs,
- control values and the value function are computed only for the current object's
  state in a given step,
- computations are performed from the first to the last step of an iterative process,
- suboptimal control may be generated in online processes.

The goal of the NDP structure is to generate the control law $\mathbf{u}_{A\{k\}}$, that minimizes
the value function $V_{\{k\}}$, which defines the cost of transition from the current state of
the process $k = 0$, to the final state $k = n$ and is expressed by relation

$$V_{\{k\}}\left(\mathbf{x}_{\{k\}}, \mathbf{u}_{A\{k\}}\right) = \sum_{k=0}^{n} \gamma^k L_{C\{k\}}\left(\mathbf{x}_{\{k\}}, \mathbf{u}_{A\{k\}}\right) \, , \qquad (6.1)$$

where $n$ – number of steps in the iterative process, $\gamma$ – the so-called discount factor for
future rewards/punishments selected from the interval $0 \leq \gamma \leq 1$, $L_{C\{k\}}\left(\mathbf{x}_{\{k\}}, \mathbf{u}_{A\{k\}}\right)$
– cost function, referred to as the local cost in the $k$th step, $\mathbf{x}_{\{k\}}$ – object's state vector,
$\mathbf{u}_{A\{k\}}$ – vector of control signals generated by the actor's adaptive structure (where
control algorithm includes no other elements generating a control signal e.g. PD
controller or robust control term).

The cost function is usually assumed as a quadratic form of the object's state and/or
control:

$$L_{C\{k\}}\left(\mathbf{x}_{\{k\}}, \mathbf{u}_{A\{k\}}\right) = \frac{1}{2}\left(\mathbf{x}_{\{k\}}^T \mathbf{R}\mathbf{x}_{\{k\}} + \mathbf{u}_{A\{k\}}^T \mathbf{Q}\mathbf{u}_{A\{k\}}\right) , \qquad (6.2)$$

where $\mathbf{R}, \mathbf{Q}$ – constant, positive-definite design matrices.

In the general case the NDP structure consists of:

- **Predictive model**, whose task is to predict the process state $\hat{\mathbf{x}}_{\{k+1\}}$ in step $k + 1$. In the absence of analytical description of the controlled process an adaptive model can be used that includes e.g. an NN. The object model is not required for some NDP algorithms.
- **Actor**, whose task is to generate the suboptimal control law $\mathbf{u}_{A\{k\}}$. In the case where the actor structure is implemented in the form of an NN, the actor weights are adapted based on the signal generated by the critic NN.
- **Critic**, whose task is to approximate the value function $V_{\{k\}}$ or its derivative with respect to the state $\boldsymbol{\lambda}_{\{k\}} = \partial V_{\{k\}}\left(\mathbf{x}_{\{k\}}, \mathbf{u}_{A\{k\}}\right)/\partial \mathbf{x}_{\{k\}}$ (depending on the type of the applied algorithm).

Considering the function executed by the critic structure, the algorithm used for the actor-critic parametric adaptation, and requirements to the model of a controlled object, the following NDP algorithms have been identified [12–14, 41, 42, 44, 45]:

- **Heuristic Dynamic Programming** (HDP) is the basic NDP algorithm, where the critic adaptive structure approximates the value function $V_{\{k\}}$, while the actor adaptive structure generates the suboptimal control law $\mathbf{u}_{A\{k\}}$.
- **Dual-Heuristic Dynamic Programming** (DHP) differs from the HDP algorithm in that the DHP critic structure approximates the value function derivative with respect to the controlled object's state $\boldsymbol{\lambda}_{\{k\}} = \partial V_{\{k\}}\left(\mathbf{x}_{\{k\}}, \mathbf{u}_{A\{k\}}\right)/\partial \mathbf{x}_{\{k\}}$, which complicates the algorithms for the actor-critic NN weights adaptation, but speeds up the process of NN parametric adaptation and improves performance of the generated control.
- **Globalized Dual-Heuristic Dynamic Programming** (GDHP) where actor and critic are constructed in the same manner as in the HDP algorithm. The adaptive process of the critic structure is more complex than that described above as it is composed of the adaptive algorithms for the critic structures characteristic for the HDP and DHP algorithms.
- **Action-Dependent Heuristic Dynamic Programming** (ADHDP) differs from the HDP algorithm in that the control signal generated by the actor is fed into the input of the critic NN which simplifies the actor's parameter-adaptation algorithm.
- **Action Dependent Dual-Heuristic Dynamic Programming** (ADDHP), differs from the DHP algorithm in that the critic NN input vector includes the generated control signal thereby simplifying the actor's parameter-adaptation algorithm.
- **Action-Dependent Globalized Dual-Heuristic Dynamic Programming** (ADGDHP) differs from the GDHP in that the critic NN input vector includes a control signal thereby simplifying the actor's parameter-adaptation algorithm.

The family of NDP algorithms is shown in Fig. 6.2.

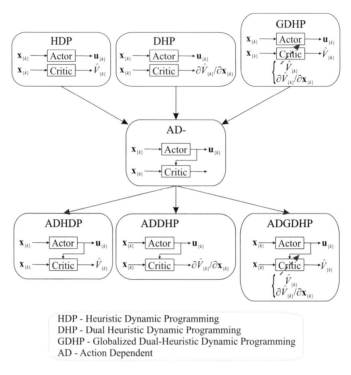

**Fig. 6.2**  Schematic diagram - Family of NDP algorithms

NDP algorithms are frequently used for nonlinear object control such as the WMR [57] as their properties outperform those observed in classical Bellman's DP. One of the first publications on the implementation of adaptive critic designs into the control of a nonlinear system with a tendency to understeer is [6]. A comprehensive description of NDP structures can be found in [4, 14, 31, 32, 42, 44, 62, 65]. NDP algorithms were primarily formulated in a discrete version [6, 42, 44] and further extended into continuous versions [2, 18, 53, 54, 58].

The growing interest in reinforcement learning algorithms is evident in the number of publications on their potential applications in a variety of fields. The use of reinforcement learning algorithms has contributed to the solution of difficult problems related to autonomous control of aerial robots such as unmanned helicopters [39], or underwater robots such as autonomous submarines [7]. The worldwide scientific publications provide descriptions of numerous applications of algorithms trained by reinforcement learning method such as trajectory planning for a mobile robot [28], traffic light control [3], power system control [10]. However, most of them are implementations of the Q-learning algorithm [4, 5, 8, 16, 67].

In recent years only a few publications have been released that focus on the application of NDP algorithms e.g. the use of an ADHDP algorithm for a static compensator connected to a multimachine power system [35] or the use of HDP and DHP algorithms for target recognition [27]. The application of NDP algorithms to mobile robot control is presented in [51], and trajectory generation in [34]. Publications [59, 68] describe the application of an HDP algorithm to nonlinear system control by presenting results of simulation studies. In work [29] a DHP algorithm was used in numerical tests for the control of a four-wheeled vehicle. In [52] an adaptive critic design was applied to tracking control of a two-wheeled mobile robot, supplemented with numerical test results. An actor-critic algorithm was successfully applied to sub-optimal missile control under constrained conditions [17], where the authors prove other methods ineffective. Publication [55] compared the quality of turbogenerator control with off-line trained HDP and DHP algorithms to a conventional PID controller, and found the NDP algorithms to produce better results. In [64] an adaptive critic design was used for automatic selection of parameters for a PD controller in tracking control of a selected WMR point. Interesting results are presented in [63], where based on HDP and DHP structures new kernel versions of algorithms are proposed that are more efficient compared to their standard versions. Implementation of GDHP algorithms in the control of a linear object is presented in publication [61], the control of a nonlinear object in publications [11, 33, 60], the control of a turbogenerator in [56]. Work [40] presents a new type of architecture for reinforcement learning algorithms - the natural actor-critic (NAC), and provides numerical tests for the presented algorithm in a cart-pole balancing problem. Recent publications present results of research into application of NDP algorithms to dynamic objects control, whose models are unknown [53, 54, 58]. Another direction of research into NDP algorithms involves the extension of the actor-critic structure with a third element, the so-called goal network. The value function is formulated based on the goal network output values which in consequence enhances the actor-critic structure adaptive process and improves control performance [36–38, 69]. It is noted that the worldwide literature mainly includes theoretical considerations and numerical test results concerned with the application of NDP structures to control systems.

The remainder of this chapter provides a detailed description of NDP structures in the configuration of HDP [21, 23], DHP [15, 19, 20, 23–25, 48], GDHP [26, 47, 49, 50] and ADHDP [22] that have been used for the synthesis of the WMR's and robotic manipulator's tracking control algorithms. The following considerations are limited to the aforementioned NDP algorithms due to the specific character of the problem addressed in this book. The implementation of designed tracking control algorithms for a selected WMR point and a robotic manipulator with NDP structures, in real-time and by means of rapid prototyping of control systems methods, requires that the proposed algorithms have simple forms and easy-to-algorithmize procedures for parametric adaptation of actor-critic structures.

## 6.2 Model-Based Learning Methods

A mathematical model of a controlled object is required by five of six algorithms of the NDP family. Depending on the function approximated by the critic structure (value function $V_{\{k\}}$ or its derivative with respect to object's state $\lambda_{\{k\}} =$

**Table 6.1** Requirement for a model in NDP structure-adaptation procedures

| Algorithm | Actor | Critic |
|-----------|-------|--------|
| HDP       | +     | −      |
| DHP       | +     | +      |
| GDHP      | +     | +      |
| ADHDP     | −     | −      |
| ADDHP     | −     | +      |
| ADGDHP    | −     | +      |

$\partial V_{\{k\}} \left( \mathbf{x}_{\{k\}}, \mathbf{u}_{A\{k\}} \right) / \partial \mathbf{x}_{\{k\}}$), and the extension of the critic NN input vector by a generated control signal, the knowledge of a mathematical model may be required by the procedure of actor's NN weights adaptation, the procedure of critic's NN weights adaptation, by both procedures of NN parameter-adaptation, or may not be required at all. Table 6.1 indicates which NDP actor-critic parameter-adaptation procedures require a mathematical model. Sign "+" means that in the synthesis of a parameter-adaptation algorithm for any given structure a model is required, whereas sign "-" indicates that the knowledge of such a model is not necessary.

This subchapter discusses three NDP structures that require a mathematical description of a controlled object for the actor and critic adaptive processes i.e. HDP, DHP and GDHP.

## 6.2.1   Heuristic Dynamic Programming

HDP algorithm is the basic algorithm in the family of NDP algorithms. It consists of:

- **Predictive model** - whose task is to predict the controlled object's state $\mathbf{x}_{\{k+1\}}$ in step $k + 1$.
- **Actor** - that may be implemented in the form of an adaptive structure that can map any nonlinear function e.g. NN. The actor generates suboptimal control law $\mathbf{u}_{A\{k\}}$ in the $k$- th step, according to the following relation

$$\mathbf{u}_{A\{k\}} \left( \mathbf{x}_{A\{k\}}, \mathbf{W}_{A\{k,l\}} \right) = \mathbf{W}_{A\{k,l\}}^{T} \mathbf{S} \left( \mathbf{x}_{A\{k\}} \right) \ , \tag{6.3}$$

where $\mathbf{W}_{A\{k,l\}}$ – matrix of actor's NN output layer weights in the $l$th step of inner calculation loop weights adaptation in the $k$th step of the process, $\mathbf{x}_{A\{k\}}$ – actor's NN inputs vector, whose construction depends on the specific character of the controlled process, $\mathbf{S} \left( \mathbf{x}_{A\{k\}} \right)$ – neuron activation function vector. In the general case $\mathbf{x}_{A\{k\}} = \left[ \kappa_A \mathbf{x}_{\{k\}} \right]$, where $\kappa_A$ is a positive-definite diagonal matrix of scaling coefficients for actor's NN inputs values.

The procedure of actor's NN weights adaptation in HDP algorithm is to minimize the performance index

$$
\mathbf{e}_{A\{k\}} = \frac{\partial L_{C\{k\}}\left(\mathbf{x}_{\{k\}}, \mathbf{u}_{A\{k\}}\right)}{\partial \mathbf{u}_{A\{k\}}} + \gamma \left[\frac{\partial \mathbf{x}_{\{k+1\}}}{\partial \mathbf{u}_{A\{k\}}}\right]^{T} \frac{\partial \hat{V}_{\{k+1\}}\left(\mathbf{x}_{C\{k+1\}}, \mathbf{W}_{C\{k,l\}}\right)}{\partial \mathbf{x}_{\{k+1\}}} , \quad (6.4)
$$

by applying any weights adaptation method e.g. according to relation

$$
\mathbf{W}_{A\{k,l+1\}} = \mathbf{W}_{A\{k,l\}} - \mathbf{e}_{A\{k\}}\mathbf{\Gamma}_{A}\frac{\mathbf{u}_{A\{k\}}\left(\mathbf{x}_{A\{k\}}, \mathbf{W}_{A\{k,l\}}\right)}{\partial \mathbf{W}_{A\{k,l\}}} , \quad (6.5)
$$

where $\mathbf{\Gamma}_{A}$ – constant diagonal matrix of positive reinforcement coefficients for actor's NN weights learning, $\Gamma_{Ai,i} \in \langle 0, 1\rangle$, $i$ – index.
- **Critic** - that may be implemented in the form of an adaptive structure that can map any nonlinear function e.g. NN. The critic estimates the value function $V_{\{k\}}\left(\mathbf{x}_{\{k\}}, \mathbf{u}_{A\{k\}}\right)$. The critic's NN output in the $k$th step is written as

$$
\hat{V}_{\{k\}}\left(\mathbf{x}_{C\{k\}}, \mathbf{W}_{C\{k,l\}}\right) = \mathbf{W}_{C\{k,l\}}^{T}\mathbf{S}\left(\mathbf{x}_{C\{k\}}\right) , \quad (6.6)
$$

where $\mathbf{W}_{C\{k,l\}}$ – critic's NN output layer weights vector in the $l$th step of inner calculation loop weights adaptation in the $k$th step of the process, $\mathbf{x}_{C\{k\}}$ – critic's NN inputs vector, whose construction is based on the adopted value function, $\mathbf{S}\left(\mathbf{x}_{C\{k\}}\right)$ – neuron activation function vector. In the general case, when the local cost $L_{C\{k\}}\left(\mathbf{x}_{\{k\}}, \mathbf{u}_{A\{k\}}\right)$ is the function of object's state $\mathbf{x}_{\{k\}}$ and control $\mathbf{u}_{A\{k\}}$, $\mathbf{x}_{C\{k\}} = \left[\kappa_{C}\left[\mathbf{x}_{\{k\}}^{T}, \mathbf{u}_{A\{k\}}^{T}\right]^{T}\right]$ where $\kappa_{C}$ is a positive-definite diagonal matrix of scaling coefficients for critic's NN inputs values.

Critic's NN output layer weights $\mathbf{W}_{C\{k,l\}}$ are adapted by minimizing the temporal difference error $e_{C\{k\}}$ described by relation

$$
e_{C\{k\}} = L_{C\{k\}}\left(\mathbf{x}_{\{k\}}, \mathbf{u}_{A\{k\}}\right) + \gamma \hat{V}_{\{k+1\}}\left(\mathbf{x}_{C\{k+1\}}, \mathbf{W}_{C\{k,l\}}\right) - \hat{V}_{\{k\}}\left(\mathbf{x}_{C\{k\}}, \mathbf{W}_{C\{k,l\}}\right) ,
$$
$$(6.7)$$

where $\mathbf{x}_{C\{k+1\}}$ – input vector into the critic's NN in step $k + 1$, obtained from the predictive model, $\mathbf{u}_{A\{k\}} = \mathbf{u}_{A\{k\}}\left(\mathbf{x}_{A\{k\}}, \mathbf{W}_{A\{k,l+1\}}\right)$.
Critic's NN output layer weights $\mathbf{W}_{C\{k,l\}}$ are adapted according to the following relation

$$
\mathbf{W}_{C\{k,l+1\}} = \mathbf{W}_{C\{k,l\}} - e_{C\{k\}}\mathbf{\Gamma}_{C}\frac{\hat{V}_{\{k\}}\left(\mathbf{x}_{C\{k\}}, \mathbf{W}_{C\{k,l\}}\right)}{\partial \mathbf{W}_{C\{k,l\}}} \quad (6.8)
$$

where $\mathbf{\Gamma}_{C}$ – constant diagonal matrix of reinforcement coefficients for critic's NN weights learning, $\Gamma_{Ci,i} \in \langle 0, 1\rangle$.

**Fig. 6.3** Schematic diagram
of NDP structure in HDP
configuration

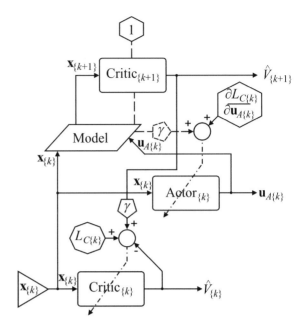

A general schematic diagram of NDP structure in HDP configuration is shown in Fig. 6.3.

HDP algorithm adaptive procedure [14] consists of two adaptive processes that occur cyclically in the $k$th step.

- **Policy-Improvement Routine** consists in cyclical adaptation of control law in subsequent inner loop iteration steps $l = 0, 1, \ldots, l_{AC}$ in the $k$th step of the process. The goal is to adjust actor's NN weights $\mathbf{W}_{A\{k,l\}}$, in such a way that the generated control signals $\mathbf{u}_{A\{k\}} \left( \mathbf{x}_{A\{k\}}, \mathbf{W}_{A\{k,l\}} \right)$ converge to the optimal values $\mathbf{u}^{*}_{A\{k\}}$, when $l = 0, 1, \ldots, l_{AC}$, where $l_{AC}$ – number of inner loop iteration steps of the $k$th step of the process. Parameter $l_{AC}$ may be constant or variable in each $k$ step of the discrete process.

The policy improvement routine for the HDP algorithm is carried out in the following manner:

- The following values are known $\mathbf{x}_{\{k\}}$, $\mathbf{u}_{A\{k\}} \left( \mathbf{x}_{A\{k\}}, \mathbf{W}_{A\{k,l\}} \right)$, and $\hat{V}_{\{k\}} \left( \mathbf{x}_{C\{k\}}, \mathbf{W}_{C\{k,l\}} \right)$, where $l = 0$, $\mathbf{W}_{A\{k,l=0\}} = \mathbf{W}_{A\{k-1,l_{AC}\}}$, $\mathbf{W}_{C\{k,l=0\}} = \mathbf{W}_{C\{k-1,l_{AC}\}}$.
- In the inner loop of the $k$th step of the process ($l = 0, 1, \ldots, l_{AC}$) the following calculations are performed:

  – Calculation of the local cost difference value $L_{C\{k\}} \left( \mathbf{x}_{\{k\}}, \mathbf{u}_{A\{k\}} \right)$ with respect to the generated control law $\mathbf{u}_{A\{k\}}$

$$\frac{\partial L_{C\{k\}} \left( \mathbf{x}_{\{k\}}, \mathbf{u}_{A\{k\}} \right)}{\partial \mathbf{u}_{A\{k\}}}. \tag{6.9}$$

– Determination of the object's state value $\mathbf{x}_{\{k+1\}}$ in step $k + 1$ based on the predictive model

$$\mathbf{x}_{\{k+1\}} = \mathbf{f}\left(\mathbf{x}_{\{k\}}, \mathbf{u}_{A\{k\}}\right) . \tag{6.10}$$

– Calculation of the derivative value for the value function estimate $\hat{V}_{\{k+1\}}$ $\left(\mathbf{x}_{C\{k+1\}}, \mathbf{W}_{C\{k,l\}}\right)$, with respect to state $\mathbf{x}_{\{k+1\}}$, according to the following relation

$$\frac{\partial \hat{V}_{\{k+1\}}\left(\mathbf{x}_{C\{k+1\}}, \mathbf{W}_{C\{k,l\}}\right)}{\mathbf{x}_{\{k+1\}}} . \tag{6.11}$$

– Determination of object's state difference value in step $k + 1$ with respect to the control signal

$$\frac{\partial \mathbf{x}_{\{k+1\}}}{\partial \mathbf{u}_{A\{k\}}} . \tag{6.12}$$

– Calculation of the error value $\mathbf{e}_{A\{k\}}$ according to relation (6.4).
– Adaptation of actor's NN output layer weights by means of a gradient method, according to relation (6.5).

• The implementation of policy improvement routine produces new values of actor's NN weights $\mathbf{W}_{A\{k,l+1\}}$, and $\mathbf{u}_{A\{k\}}\left(\mathbf{x}_{A\{k\}}, \mathbf{W}_{A\{k,l+1\}}\right)$.

- **Value Determination Operation** consists in a cyclical adaptation of the value function estimate in subsequent inner loop iteration steps $l = 1, \ldots, l_{AC}$ in the $k$th step of the controlled process. The goal is to adjust the critic's NN weights $\mathbf{W}_{C\{k,l\}}$, in such a way that the generated signal of the value function estimate $\hat{V}_{\{k\}}\left(\mathbf{x}_{C\{k\}}, \mathbf{W}_{C\{k,l\}}\right)$ converges to the optimal value $V^*_{\{k\}}$, when $l = 0, 1, \ldots, l_{AC}$.
The algorithm for the determination of HDP structure's value function is executed as follows:

• The following values are known $\mathbf{x}_{\{k\}}$, $\mathbf{u}_{A\{k\}}\left(\mathbf{x}_{A\{k\}}, \mathbf{W}_{A\{k,l+1\}}\right)$, and $\hat{V}_{\{k\}}$ $\left(\mathbf{x}_{C\{k\}}, \mathbf{W}_{C\{k,l\}}\right)$.
• In the inner loop of the $k$th step of the process ($l = 0, 1, \ldots, l_{AC}$) the following calculations are performed:

– Calculation of the cost function value $L_{C\{k\}}\left(\mathbf{x}_{\{k\}}, \mathbf{u}_{A\{k\}}\right)$ based on the object's state $\mathbf{x}_{\{k\}}$ and the actor's NN control signal $\mathbf{u}_{A\{k\}}\left(\mathbf{x}_{A\{k\}}, \mathbf{W}_{A\{k,l+1\}}\right)$, obtained from the policy-improvement routine.
– Determination of the object's state value $\mathbf{x}_{\{k+1\}}$ in $k + 1$ step based on the predictive model and the adapted control law $\mathbf{u}_{A\{k\}}\left(\mathbf{x}_{A\{k\}}, \mathbf{W}_{A\{k,l+1\}}\right)$,

$$\mathbf{x}_{\{k+1\}} = \mathbf{f}\left(\mathbf{x}_{\{k\}}, \mathbf{u}_{A\{k\}}\right) . \tag{6.13}$$

**Fig. 6.4** Schematic diagram
of NN weights adaptation
process for HDP structure in
the inner loop

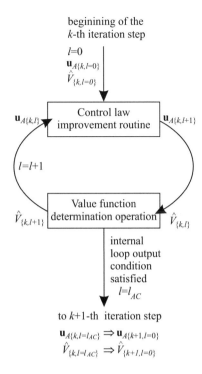

- Calculation of the value for the cost function estimate $\hat{V}_{\{k\}}\left(\hat{\mathbf{x}}_{C\{k+1\}}, \mathbf{W}_{C\{k,l\}}\right)$.
- Determination of the error $e_{C\{k\}}$ according to relation (6.7).
- Critic's NN output layer weights are adapted by means of a gradient method, according to relation (6.8).

- If $l = l_{AC}$ is not satisfied, or any other assumed condition for convergence, the inner loop calculations are repeated for the $k$th step of the process for the actor-critic NN weights adaptation procedure.

The actor-critic NNs weights adaptation process for NDP structure in HDP config-uration consists in a cyclical implementation of the control law improvement routine and the value function determination operation in the $k$th step of a discrete process. Figure 6.4 presents a schematic diagram of the process for $l_{AC}$ iterations of the inner calculation loop, where $\mathbf{u}_{A\{k,l\}} = \mathbf{u}_{A\{k\}}\left(\mathbf{x}_{A\{k\}}, \mathbf{W}_{A\{k,l\}}\right), \hat{V}_{\{k,l\}} = \hat{V}_{\{k\}}\left(\mathbf{x}_{C\{k\}}, \mathbf{W}_{C\{k,l\}}\right)$. A schematic diagram of NN weights adaptation process for NDP structure in HDP configuration is shown in Fig. 6.5.

### 6.2.2   Dual-Heuristic Dynamic Programming

DHP algorithm falls within the group of advanced ACD and is composed of:

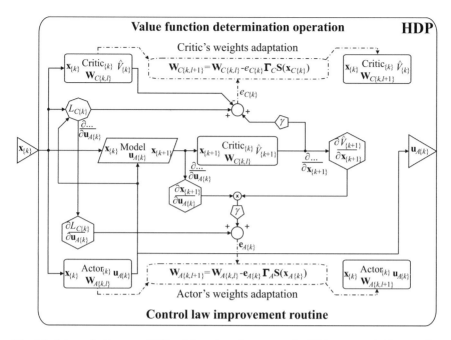

**Fig. 6.5** Schematic diagram of NN weights adaptation process for NDP structure in HDP configuration

- **Predictive model** - whose task is to predict the controlled object's state $\mathbf{x}_{\{k+1\}}$ in step $k + 1$.
- **Actor** - that may be implemented in the form of an adaptive structure that can map any nonlinear function e.g. NN. The actor generates suboptimal control law $\mathbf{u}_{A\{k\}}$ in the $k$th step, according to the following relation

$$\mathbf{u}_{A\{k\}}\left(\mathbf{x}_{A\{k\}}, \mathbf{W}_{A\{k,l\}}\right) = \mathbf{W}_{A\{k,l\}}^{T}\mathbf{S}\left(\mathbf{x}_{A\{k\}}\right) . \tag{6.14}$$

The procedure of the actor's NN weights adaptation in the DHP algorithm is to minimize the performance index

$$\mathbf{e}_{A\{k\}} = \frac{\partial L_{C\{k\}}\left(\mathbf{x}_{\{k\}}, \mathbf{u}_{A\{k\}}\right)}{\partial \mathbf{u}_{A\{k\}}} + \gamma \left[\frac{\partial \mathbf{x}_{\{k+1\}}}{\partial \mathbf{u}_{A\{k\}}}\right]^{T} \hat{\lambda}_{\{k+1\}}\left(\mathbf{x}_{C\{k+1\}}, \mathbf{W}_{C\{k,l\}}\right) , \tag{6.15}$$

according to relation

$$\mathbf{W}_{A\{k,l+1\}} = \mathbf{W}_{A\{k,l\}} - \mathbf{e}_{A\{k\}}\mathbf{\Gamma}_{A}\frac{\partial \mathbf{u}_{A\{k\}}\left(\mathbf{x}_{A\{k\}}, \mathbf{W}_{A\{k,l\}}\right)}{\partial \mathbf{W}_{A\{k,l\}}} , \tag{6.16}$$

where $\mathbf{\Gamma}_A$ – constant diagonal matrix of reinforcement coefficients for actor's NN weight learning, $\Gamma_{Ai,i} \in \langle 0, 1 \rangle$.

- **Critic** - that may be implemented in the form of an adaptive structure that can map any nonlinear function e.g. NN. The critic estimates the derivative of the value function with respect to the controlled object's state $\boldsymbol{\lambda}_{\{k\}} = \partial V_{\{k\}}\left(\mathbf{x}_{\{k\}}, \mathbf{u}_{A\{k\}}\right)/\partial \mathbf{x}_{\{k\}}$. The critic's structure NN output in the $k$th step is written as

$$\hat{\boldsymbol{\lambda}}_{\{k\}}\left(\mathbf{x}_{C\{k\}}, \mathbf{W}_{C\{k,l\}}\right) = \mathbf{W}_{C\{k,l\}}^T \mathbf{S}\left(\mathbf{x}_{C\{k\}}\right) \ , \tag{6.17}$$

where $\mathbf{W}_{C\{k,l\}}$ – critic's NN output layer weights vector, $\mathbf{x}_{C\{k\}}$ – critic's NN inputs vector, whose construction is based on the adopted value function.

Critic's NN output layer weights $\mathbf{W}_{C\{k,l\}}$ are adapted by minimizing the difference in temporal difference error $\mathbf{e}_{C\{k\}}$ described by relation

$$\mathbf{e}_{C\{k\}} = \frac{\partial L_{C\{k\}}\left(\mathbf{x}_{\{k\}}, \mathbf{u}_{A\{k\}}\right)}{\partial \mathbf{x}_{\{k\}}} + \left[\frac{\partial \mathbf{u}_{A\{k\}}}{\partial \mathbf{x}_{\{k\}}}\right]^T \frac{\partial L_{C\{k\}}\left(\mathbf{x}_{\{k\}}, \mathbf{u}_{A\{k\}}\right)}{\partial \mathbf{u}_{A\{k\}}} + \tag{6.18}$$

$$+\gamma\left[\frac{\partial \mathbf{x}_{\{k+1\}}}{\partial \mathbf{x}_{\{k\}}} + \left[\frac{\partial \mathbf{u}_{A\{k\}}}{\partial \mathbf{x}_{\{k\}}}\right]^T \frac{\partial \mathbf{x}_{\{k+1\}}}{\partial \mathbf{u}_{A\{k\}}}\right]^T \hat{\boldsymbol{\lambda}}_{\{k+1\}}\left(\mathbf{x}_{C\{k+1\}}, \mathbf{W}_{C\{k,l\}}\right) - \hat{\boldsymbol{\lambda}}_{\{k\}}\left(\mathbf{x}_{C\{k\}}, \mathbf{W}_{C\{k,l\}}\right),$$

where $\mathbf{x}_{C\{k+1\}}$ – the input vector into the critic's NN in step $k + 1$, obtained from the predictive model of the controlled object.

Critic's NN output layer weights $\mathbf{W}_{C\{k,l\}}$ are adapted according to the following relation

$$\mathbf{W}_{C\{k,l+1\}} = \mathbf{W}_{C\{k,l\}} - \mathbf{e}_{C\{k\}}\mathbf{\Gamma}_C \frac{\hat{\boldsymbol{\lambda}}_{\{k\}}\left(\mathbf{x}_{C\{k\}}, \mathbf{W}_{C\{k,l\}}\right)}{\partial \mathbf{W}_{C\{k,l\}}} \tag{6.19}$$

where $\mathbf{\Gamma}_C$ – constant diagonal matrix of positive reinforcement coefficients for critic's NN weight learning, $\Gamma_{Ci,i} \in \langle 0, 1 \rangle$, $i$ – index.

A general schematic diagram of NDP structure in DHP configuration is shown in Fig. 6.6.

The actor-critic NN weights adaptation procedure for DHP algorithm is analogous to that used for the HDP structure and described in Sect. 6.2.1. A schematic diagram of NN weights adaptation procedure for DHP structure is shown in Fig. 6.7.

### 6.2.2.1 Application of DHP Algorithm to Optimal Control of a Linear Dynamic System

The problem from Sect. 4.1 was solved using the NDP structure in the DHP configuration to generate a control signal for a linear dynamic object described by equation (4.9).

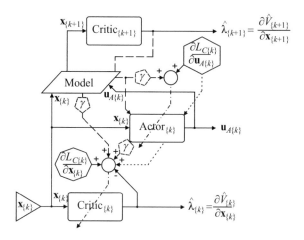

**Fig. 6.6** Schematic diagram of NDP structure in DHP configuration

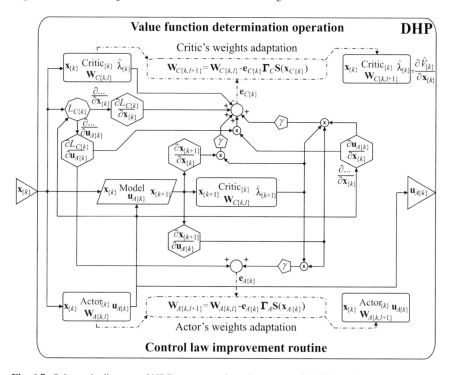

**Fig. 6.7** Schematic diagram of NDP structure adaptation process in DHP configuration

In DHP algorithm the actor and critic structures are implemented in the form of single-layer RVFL NNs that are linear with respect to the output layer weights, with bipolar sigmoid neuron activation functions described by the following relation

$$S_i \left( \mathbf{D}_A^T \mathbf{x}_{N\{k\}} \right) = \frac{2}{1 + \exp \left( -\beta_N \mathbf{D}_{A:,i}^T \mathbf{x}_{N\{k\}} \right)} - 1 . \qquad (6.20)$$

For the purpose of a numerical test, the following parametric values were adopted for DHP algorithm's adaptive structures:
- $m_A = 8$ – number of neurons in the actor's NN,
- $m_C = 8$ – number of neurons in the critic's NN,
- $\mathbf{\Gamma}_A = \text{diag} \{0.4\}$ – matrix of reinforcement coefficients for the actor's NN weights adaptation,
- $\mathbf{\Gamma}_C = \text{diag} \{0.6\}$ – matrix of reinforcement coefficients for the critic's NN weights adaptation,
- $\mathbf{W}_{A\{k=0\}} = \mathbf{0}, \mathbf{W}_{C\{k=0\}} = \mathbf{0}$ – zero initial values of NN output layer weights in the first stage of the adaptation process $\xi = 1$,
- $\mathbf{D}_A, \mathbf{D}_C$ – actor's and critic's NN input layer weights vectors, randomly selected in the initialization process from the interval $D_{Aj} \in \langle -0.5, 0.5 \rangle, D_{Cj} \in \langle -0.5, 0.5 \rangle,$ $j = 1, \ldots, m,$
- $\beta_N$ – slope coefficient of bipolar sigmoid functions at inflection point,
- $\kappa_u = 0.2$ – scaling coefficient of control signal $u_{A\{k\}}$ at the input to NNs,
- $\kappa_x = 0.1$ – scaling coefficient of object's state $x_{\{k\}}$ at the input to NNs,
- $l_{AC} = 3$ – number of inner loop iteration steps in the process of actor and critic structures adaptation,
- $\gamma = 1$ – discount factor for the local cost value in subsequent iteration steps.
It was assumed that $x_{\{k=0\}} = 8.9$ – initial condition for dynamic object's state, and $n = 9$ – number of steps in a discrete process.

**Trial 1** ($\xi = 1$)

In the first trial of the NNs weights adaptation process $\xi = 1$ zero initial values were assigned to the actor's and critic's NN output layer weights, which is the least favorable case, because the adaptive structures retain no information about the controlled object. This causes significant divergence between the suboptimal control signal $u_{A\{k\}}$ generated by the actor's NN and the optimal control $u^*_{DP\{k\}}$ obtained by using DP method in Sect. 4.1. Figure 6.8 compares the values $x_{\{k\}}, u_{A\{k\}}$ and $\hat{V}_{\{k\}}$, to the optimal ones obtained by solving a problem with Bellman's DP.

The state trajectory $x_{\{k\}}$, and control signal $u_{A\{k\}}$, generated by the actor's NN diverge from the optimal values and hence higher values of the function $\hat{V}_{\{k\}}$ compared to $V^*_{\{k\}}$. Figure 6.9 shows the values of the actor's and critic's NNs weights.

The actor's and critic's NNs zero initial weights change under the assumed weights adaptation algorithm. Their values stabilize around certain values.

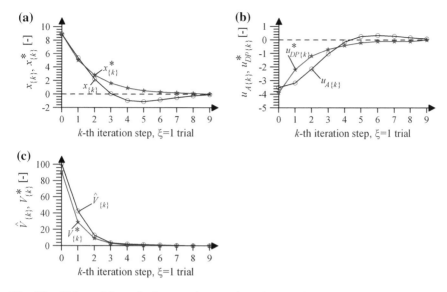

**Fig. 6.8** **a** Values of the optimal state trajectory $x^*_{\{k\}}$ and state trajectory $x_{\{k\}}$ in $\xi = 1$ trial, $k = 0, \ldots, 9$, **b** values of the optimal control signal $u^*_{DP\{k\}}$ and control signal $u_{A\{k\}}$, **c** optimal value function $V^*_{\{k\}}$ and value function $\hat{V}_{\{k\}}$

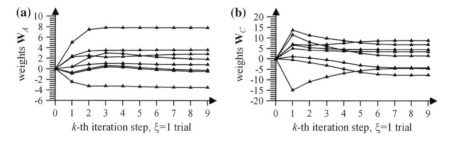

**Fig. 6.9** **a** Values of the actor's NN weights, $\xi = 1$, **b** values of the critic's NN weights

## Trial 2 ($\xi = 2$)

In the second trial of the NNs weights adaptation process, $\xi = 2$, the initial values of the actor's and critic's NN output layer weights were obtained in the last adaptation step of the first trial. This provides the adaptive structures with some information about the controlled object that is recorded in the form of output layer weights values which allows for the generation of a control law that is closer to the optimal control. The system's state trajectory $x_{\{k\}}$, shown in Fig. 6.10.a, is closer to the optimal one $x^*_{\{k\}}$, as the suboptimal control law (see Fig. 6.10.b) generated by the actor structure $u_{A\{k\}}$ is closer to the optimal control law $u^*_{DP\{k\}}$. The suboptimal value function $\hat{V}_{\{k\}}$ has similar values to those of $V^*_{\{k\}}$ (Fig. 6.10.c).

The actor's and critic's NNs output layers weights values are shown in Fig. 6.11.

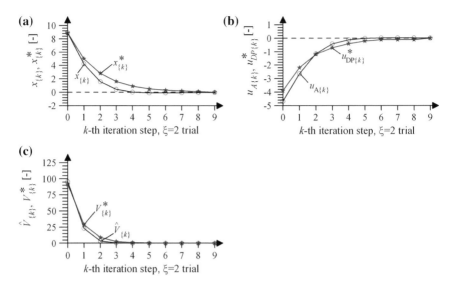

**Fig. 6.10  a** Values of the optimal state trajectory $x^*_{\{k\}}$ and state trajectory $x_{\{k\}}$ in $\xi = 2$ trial, $k = 0, \ldots, 9$, **b** values of the optimal control signal $u^*_{DP\{k\}}$ and control signal $u_{A\{k\}}$, **c** optimal value function $V^*_{\{k\}}$ and value function $\hat{V}_{\{k\}}$

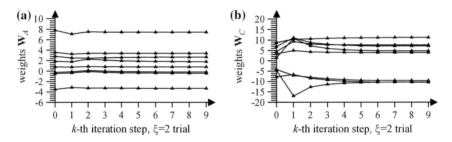

**Fig. 6.11  a** Values of the actor's NN weights, $\xi = 2$, **b** values of the critic's NN weights

The second weights adaptation trial produces only minor changes in the actor's and critic's NNs weights values in each step of the discrete process.

**Trial 5 ($\xi = 5$)**

In the fifth trial of the NNs weights adaptation process, $\xi = 5$, the initial values of the actor's and critic's NNs output layer weights were obtained in the last adaptation step of the preceding trial, $\xi = 4$. The system's state trajectory $x_{\{k\}}$, shown in Fig. 6.12.a, is close to the optimal trajectory $x^*_{\{k\}}$, as the suboptimal control law (see Fig. 6.12.b) generated by the actor structure $u_{A\{k\}}$ converges to the optimal control law $u^*_{DP\{k\}}$. The suboptimal value function $\hat{V}_{\{k\}}$ has similar values to those of $V^*_{\{k\}}$ (Fig. 6.12.c).

The actor's and critic's NNs output layer weights values are shown in Fig. 6.13. The weights values have stabilized.

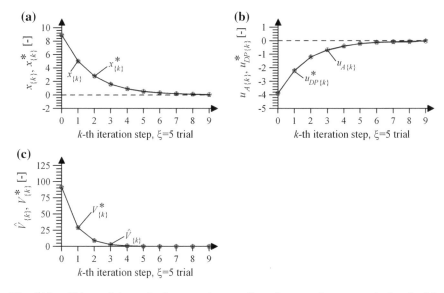

**Fig. 6.12 a** Values of the optimal state trajectory $x^*_{\{k\}}$ and state trajectory $x_{\{k\}}$ in $\xi = 5$ trial, $k = 0, \ldots, 9$, **b** values of the optimal control signal $u^*_{DP\{k\}}$ and control signal $u_{A\{k\}}$, **c** optimal value function $V^*_{\{k\}}$ and value function $\hat{V}_{\{k\}}$

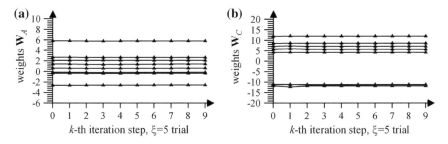

**Fig. 6.13 a** Values of the actor's NN weights, $\xi = 5$, b) values of the critic's NN weights

## Comparison of results, $\xi = 1, 2, 5$

Figure 6.14 provides a comparison of the suboptimal state trajectory $x_{\{k\}}$, suboptimal control law $u_{A\{k\}}$ generated by the actor's NN and the suboptimal value function $\hat{V}_{\{k\}}$ in trials $\xi = 1, 2, 5$, with regard to the optimal values.

The below graphs show how individual quantities converge to the optimal values $x^*_{\{k\}}$, $u^*_{DP\{k\}}$ and $V^*_{\{k\}}$ with each trial of the actor's and critic's NNs weights adaptation.

The results obtained in the $\xi = 5$ trial of the actor's and critic's NNs weights adaptation process were compared to the optimal values yielded in Sect. 4.1 where DP method was used and to the values from Sect. 4.3 where the Pontryagin's maximum principle (MP) was applied. Since the LQR and MPS control results are identical to the MP values, they are not included in the comparison. It should be noted that the

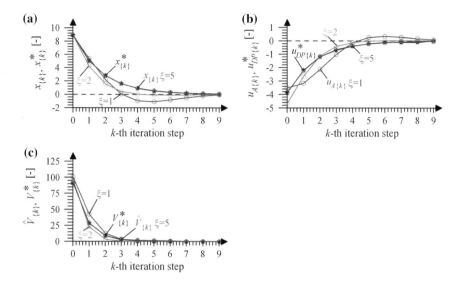

**Fig. 6.14** **a** Optimal state trajectory $x^*_{\{k\}}$, and state trajectory $x_{\{k\}}$ in $\xi = 1, 2, 5$ trial, $k = 0, \ldots, 9$, **b** optimal control signal $u^*_{DP\{k\}}$ and control signal $u_{A\{k\}}$, **c** optimal cost function $V^*_{\{k\}}$ and cost function $\hat{V}_{\{k\}}$

state trajectory and the control signal, obtained by solving the problem of linear object control with the application of DP, take on values from a defined, discretized interval under predefined conditions. However, when solving a problem with the application of MP or DHP algorithm, these values can be arbitrary, but in the considered case they were rounded to the nearest hundredth to facilitate interpretation of the results.

Table 6.2 includes the values of the optimal control law for a linear object that was determined by applying Bellman's DP $u^*_{DP\{k\}}$, MP $u^*_{M\{k\}}$ and the suboptimal control $u_{A\{k\}}$, generated by the actor's NN of the DHP algorithm, when $\xi = 5$.

The values of suboptimal control law $u_{A\{k\}}$, generated by the actor's NN, converge to the optimal control values $u^*_{DP\{k\}}$ and $u^*_{M\{k\}}$.

Tabel 6.3 presents the values of the optimal state trajectory $x^*_{\{k\}}$ obtained from the solution of a problem with the application of Bellman's DP and MP, and the suboptimal state trajectory $x_{\{k\}}$ after $\xi = 5$ trials in the actor's and critic's NNs weights adaptation process.

Imposing the suboptimal control law $u_{A\{k\}}$ on a stationary linear object causes its transition from the initial state $x_{\{0\}}$ to the final state $x_{\{n\}}$ along the suboptimal state trajectory, convergent to the optimal trajectory.

Tabel 6.4 presents the values of the optimal function $V^*_{\{k\}}$ obtained from the solution of the optimal control problem for a linear object with the use of Bellman's DP and MP, and the values of the suboptimal function $\hat{V}_{\{k\}}$ obtained from the solution of a linear object control problem with the application of the DHP algorithm, $\xi = 5$.

**Table 6.2** Values of the optimal control law for a linear object with the application of Bellman's DP $u^*_{DP\{k\}}$ and MP $u^*_{M\{k\}}$, and suboptimal control $u_{A\{k\}}$, generated by the actor's NN of the DHP algorithm, $\xi = 5$

| $k$ | $u^*_{DP\{k\}}$ [−] | $u^*_{M\{k\}}$ [−] | $u_{A\{k\}}$ [−] |
|---|---|---|---|
| 0 | −3.9 | −3.86 | −3.88 |
| 1 | −2.2 | −2.19 | −2.22 |
| 2 | −1.2 | −1.24 | −1.21 |
| 3 | −0.7 | −0.7 | −0.67 |
| 4 | −0.4 | −0.39 | −0.38 |
| 5 | −0.2 | −0.22 | −0.22 |
| 6 | −0.1 | −0.12 | −0.13 |
| 7 | −0.1 | −0.06 | −0.07 |
| 8 | −0.1 | −0.03 | −0.04 |
| 9 | 0 | − | −0.03 |

**Table 6.3** Values of the optimal state trajectory $x^*_{\{k\}}$ obtained from the solution of the optimal control problem with the application of Bellman's DP, MP, and values of the suboptimal state trajectory $x_{\{k\}}$ obtained from the solution of a problem with the application of the DHP algorithm, $\xi = 5$

| $k$ | $x^*_{\{k\}}$ [−] DP | $x^*_{\{k\}}$ [−] MP | $x_{\{k\}}$ [−] DHP |
|---|---|---|---|
| 0 | 8.9 | 8.9 | 8.9 |
| 1 | 5 | 5.04 | 5.02 |
| 2 | 2.8 | 2.85 | 2.8 |
| 3 | 1.6 | 1.61 | 1.59 |
| 4 | 0.9 | 0.91 | 0.92 |
| 5 | 0.5 | 0.52 | 0.54 |
| 6 | 0.3 | 0.3 | 0.32 |
| 7 | 0.2 | 0.18 | 0.19 |
| 8 | 0.1 | 0.12 | 0.12 |
| 9 | 0 | 0.09 | 0.08 |

Function $V_{\{k\}}$ values in individual iteration steps depend on the object's state and the control signal. The suboptimal cost function $\hat{V}_{\{k\}}$ is convergent to the optimal function $V^*_{\{k\}}$ determined by means of the MP algorithm.

Table 6.5 provides a comparison between the values of the value function $\hat{V}_{\{0\}}$ in individual trials of the NN weights adaptation ($\xi = 1, 2, 3, 4, 5$) and the values of the optimal value function $V^*_{\{0\}}$, obtained from the solution of the optimal control problem for a linear object with the use of Bellman's DP and MP. It can be noticed that the first three trials of the weights adaptation process $\xi = 1, 2, 3$ are the most crucial in terms of convergence to the optimal values (given the system parameters). It is

**Table 6.4** Values of the optimal function $V_{\{k\}}^*$, obtained from the solution of the optimal control problem with the use of Bellman's DP, MP, and values of the suboptimal function $\hat{V}_{\{k\}}$, obtained from the solution of a problem with the application of the DHP algorithm, $\xi = 5$

| $k$ | $V_{\{k\}}^*$ [−] DP | $V_{\{k\}}^*$ [−] MP | $\hat{V}_{\{k\}}$ [−] DHP |
|-----|---------------------|---------------------|--------------------------|
| 0 | 91.22 | 91.20 | 91.21 |
| 1 | 28.8 | 29.19 | 29.03 |
| 2 | 9.04 | 9.34 | 9.02 |
| 3 | 2.96 | 2.99 | 2.9 |
| 4 | 0.945 | 0.96 | 0.96 |
| 5 | 0.3 | 0.31 | 0.32 |
| 6 | 0.115 | 0.1 | 0.11 |
| 7 | 0.055 | 0.03 | 0.04 |
| 8 | 0.02 | 0.01 | 0.01 |
| 9 | 0 | 0 | 0 |

**Table 6.5** Values of the optimal value function $V_{\{0\}}^*$ and the suboptimal value functions $V_{\{0\}}$ in the $\xi$th trial, $k = 0$

| | | $\xi$ | $k = 0$ |
|---|---|-------|---------|
| $\hat{V}_{\{0\}}$ | DHP | 1 | 100.72 |
| | | 2 | 95.46 |
| | | 3 | 91.64 |
| | | 4 | 91.23 |
| | | 5 | 91.21 |
| $V_{\{0\}}^*$ | DP | − | 91.22 |
| $V_{\{0\}}^*$ | MP | − | 91.20 |

noted that in each $k$th step of the process, the NN weights values were updated with the application of control law improvement routine and value function determination operation in the inner calculation loop with $l_{AC}$ steps. The control space was searched and the near-optimal signal selected based on the current knowledge of the adaptive critic structure contained in the NN weights values.

The suboptimal control law $u_{A\{k\}}$ generated by the NDP structure in the DHP configuration is convergent to the optimal control law $u_{\{k\}}^*$ generated based on the Bellman's DP and MP. However, contrary to the optimal control calculated with the use of DP and MP, the suboptimal control may be generated in an online process, from step $k = 0$ to $k = n$.

### 6.2.3   Global Dual-Heuristic Dynamic Programming

GDHP algorithm, like DHP, falls within the group of advanced ACD and consists of:

- **Predictive model** – whose task is to predict the controlled object's state $\mathbf{x}_{\{k+1\}}$ in step $k + 1$.
- **Actor** – that may be implemented in the form of an adaptive structure e.g. NN. The actor generates suboptimal control law $\mathbf{u}_{A\{k\}}$ in the $k$th step, according to the following relation

$$\mathbf{u}_{A\{k\}}\left(\mathbf{x}_{A\{k\}}, \mathbf{W}_{A\{k,l\}}\right) = \mathbf{W}_{A\{k,l\}}^{T}\mathbf{S}\left(\mathbf{x}_{A\{k\}}\right) \ , \tag{6.21}$$

where $\mathbf{W}_{A\{k,l\}}$ – matrix of actor's NN output layer weights in the $l$th step of inner calculation loop weights adaptation in the $k$th step of the process, $\mathbf{x}_{A\{k\}}$ – actor's NN inputs vector, $\mathbf{S}\left(\mathbf{x}_{A\{k\}}\right)$ – neuron activation function vector.
The procedure of actor's NN weights adaptation in GDHP algorithm is to minimize the performance index

$$\mathbf{e}_{A\{k\}} = \frac{\partial L_{C\{k\}}\left(\mathbf{x}_{\{k\}}, \mathbf{u}_{A\{k\}}\right)}{\partial \mathbf{u}_{A\{k\}}} + \gamma \left[\frac{\partial \mathbf{x}_{\{k+1\}}}{\partial \mathbf{u}_{A\{k\}}}\right]^{T} \frac{\partial \hat{V}_{\{k+1\}}\left(\mathbf{x}_{C\{k+1\}}, \mathbf{W}_{C\{k,l\}}\right)}{\partial \mathbf{x}_{\{k+1\}}} \ , \tag{6.22}$$

by applying any weights adaptation method e.g. according to relation

$$\mathbf{W}_{A\{k,l+1\}} = \mathbf{W}_{A\{k,l\}} - \mathbf{e}_{A\{k\}}\mathbf{\Gamma}_{A}\mathbf{S}\left(\mathbf{x}_{A\{k\}}\right) \ , \tag{6.23}$$

where $\mathbf{\Gamma}_{A}$ – constant diagonal matrix of positive reinforcement coefficients for actor's NN weights adaptation, $\Gamma_{Ai,i} \in \langle 0, 1 \rangle$.
- **Critic** - that may be implemented in the form of an adaptive structure e.g. NN. The critic estimates the value function $V_{\{k\}}\left(\mathbf{x}_{\{k\}}, \mathbf{u}_{\{k\}}\right)$, hence it is constructed in the same manner as it is in the HDP structure Sect. 6.2.1. The critic's NN output in the $k$th step is written in the form of relation

$$\hat{V}_{\{k\}}\left(\mathbf{x}_{C\{k\}}, \mathbf{W}_{C\{k,l\}}\right) = \mathbf{W}_{C\{k,l\}}^{T}\mathbf{S}\left(\mathbf{x}_{C\{k\}}\right) \ , \tag{6.24}$$

where $\mathbf{W}_{C\{k,l\}}$ – critic's NN output layer weights vector in the $l$th step of inner calculation loop weights adaptation in the $k$th step of the process, $\mathbf{x}_{C\{k\}}$ – critic's NN inputs vector, whose construction is based on the adopted value function, $\mathbf{S}\left(\mathbf{x}_{C\{k\}}\right)$ – neuron activation function vector.
Critic's NN output layer weights $\mathbf{W}_{C\{k,l\}}$ are adapted by minimizing two performance indexes. The first one is the temporal difference error $e_{C1\{k\}}$, characteristic for the HDP algorithm's critic structure adaptation Sect. (6.2.2), described by relation

$$e_{C1\{k\}} = L_{C\{k\}}\left(\mathbf{x}_{\{k\}}, \mathbf{u}_{A\{k\}}\right) + \gamma \hat{V}_{\{k+1\}}\left(\mathbf{x}_{C\{k+1\}}, \mathbf{W}_{C\{k,l\}}\right) - \hat{V}_{\{k\}}\left(\mathbf{x}_{C\{k\}}, \mathbf{W}_{C\{k,l\}}\right),$$
$$(6.25)$$

where $\mathbf{x}_{C\{k+1\}}$ – input vector into the critic's NN in step $k + 1$, obtained from the predictive model, $\mathbf{u}_{A\{k\}} = \mathbf{u}_{A\{k\}}\left(\mathbf{x}_{A\{k\}}, \mathbf{W}_{A\{k,l+1\}}\right)$.

The second performance index is the derivative of the temporal difference error $e_{C2\{k\}}$ with respect to the controlled object's state vector $\mathbf{x}_{\{k\}}$. This index is characteristic for the DHP algorithm's critic structure adaptation (6.2.2), written as

$$e_{C2\{k\}} = \mathbf{I}_D^T \left\{ \frac{\partial L_{C\{k\}}\left(\mathbf{x}_{\{k\}}, \mathbf{u}_{A\{k\}}\right)}{\partial \mathbf{x}_{\{k\}}} + \left[\frac{\partial \mathbf{u}_{A\{k\}}}{\partial \mathbf{x}_{\{k\}}}\right]^T \frac{\partial L_{C\{k\}}\left(\mathbf{x}_{\{k\}}, \mathbf{u}_{A\{k\}}\right)}{\partial \mathbf{u}_{A\{k\}}} + \right. \qquad (6.26)$$

$$+ \gamma \left[\frac{\partial \mathbf{x}_{\{k+1\}}}{\partial \mathbf{x}_{\{k\}}} + \left[\frac{\partial \mathbf{u}_{A\{k\}}}{\partial \mathbf{x}_{\{k\}}}\right]^T \frac{\partial \mathbf{x}_{\{k+1\}}}{\partial \mathbf{u}_{A\{k\}}}\right]^T \frac{\partial \hat{V}_{\{k+1\}}\left(\mathbf{x}_{C\{k+1\}}, \mathbf{W}_{C\{k\}}\right)}{\partial \mathbf{x}_{\{k+1\}}} + $$

$$\left. - \frac{\partial \hat{V}_{\{k\}}\left(\mathbf{x}_{C\{k\}}, \mathbf{W}_{C\{k\}}\right)}{\partial \mathbf{x}_{\{k\}}} \right\},$$

where $\mathbf{I}_D$ – constant vector, $\mathbf{I}_D = [1, \ldots, 1]^T$.

Critic's NN output layer weights $\mathbf{W}_{C\{k,l\}}$ are adapted according to the following relation

$$\mathbf{W}_{C\{k,l+1\}} = \mathbf{W}_{C\{k,l\}} - \eta_1 e_{C1\{k,l\}} \mathbf{\Gamma}_C \frac{\partial \hat{V}_{\{k,l\}}\left(\mathbf{x}_{C\{k\}}, \mathbf{W}_{C\{k,l\}}\right)}{\partial \mathbf{W}_{C\{k,l\}}} + \qquad (6.27)$$

$$- \eta_2 e_{C2\{k,l\}} \mathbf{\Gamma}_C \frac{\partial^2 \hat{V}_{\{k,l\}}\left(\mathbf{x}_{C\{k\}}, \mathbf{W}_{C\{k,l\}}\right)}{\partial \mathbf{x}_{\{k\}} \partial \mathbf{W}_{C\{k,l\}}},$$

where $\mathbf{\Gamma}_C$ – constant diagonal matrix of positive reinforcement coefficients for the critic's NN weights adaptation, $\Gamma_{Ci,i} \in \langle 0, 1 \rangle$, $\eta_1, \eta_2$ – positive constants.

A general schematic diagram of NDP structure in GDHP configuration is shown in Fig. 6.15.

The actor-critic NN weights adaptation procedure for GDHP algorithm is analogous to that used for the HDP and DHP structures.

Computational complexity of individual algorithms of NDP family is varied and results mainly from the design and function of the critic structure linked to the size of the state vector of controlled object, and complexity of the weights adaptation law of the actor's and critic's NNs. In HDP and GDHP algorithms the critic is composed of one NN that approximates the value function. In the DHP algorithm, however, the critic approximates the derivative of the value function with respect to the state vector of the controlled object. In this case the critic's design depends on the dimensionality of the state vector. For instance, in the case of the WMR tracking control, where the state vector is adopted to be two-dimensional, the critic consists of $n = 2$ NNs, as the derivative of the value function with respect to the state vector is a

**Fig. 6.15** Schematic
diagram of NDP structure in
GDHP configuration

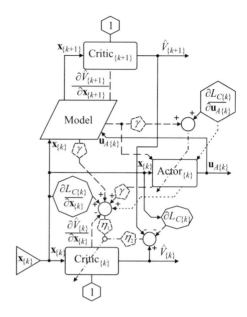

two-component vector. Similarly, the actor consists of $n = 2$ NNs, since the control signal is generated for two independent wheel drive units. In a comparable situation the HDP or GDHP algorithm shall consist of one critic's NN and two actor's NNs. The difference in the complexity of structure to the disadvantage of the DHP algorithm when compared to the HDP and GDHP algorithms, becomes more evident with the increase in dimensionality of the controlled object's state vector. In the tracking control of a robotic manipulator with six degrees of freedom and given $n = 6$ – component state vector, the DHP algorithm has $n = 6$ actor's NNs and $n = 6$ critic's NNs. In a similar situation the critic structure in the HDP and GDHP algorithms shall be composed of one NN, as the critic function in those algorithms is an approximation of the value function. The differences in the complexity of individual algorithms of the NDP family are shown in the schematic diagram in Fig. 6.16. A similar situation applies to the action dependent (AD) versions of the algorithms discussed above.

The worldwide scientific literature provides papers concerned with the implementation of GDHP algorithms with a more complex critic structure that approximates both the value function and its derivative with respect to the controlled object's state vector [42]. In this case the GDHP critic structure can be construed as a combination of HDP and DHP critic structures. However, based on the value function approximation and applying an NN with an input vector composed of the state vector, one can analytically determine the derivative of the approximated value function with respect to the state vector i.e. inputs into an NN. The above mentioned approach of additional NNs for the approximation of the value function is not developed in this work. This is due to the availability of the value function derivatives with respect to

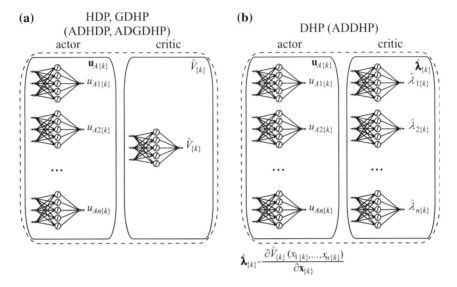

**Fig. 6.16** Schematic illustration of differences in the complexity of algorithms: **a** HDP (ADHDP) and GDHP (ADGDHP), and **b** DHP (ADDHP)

the state vector that were determined by analytical calculations, thereby meeting the requirement instituted by the actor's and critic's NN weights adaptation algorithm.

## 6.3  Model-Free Learning Methods

NDP algorithms include only one that does not require a mathematical model of the controlled object for parameter-adaptation procedures of the actor and critic structures. It is the ADHDP algorithm, in which a direct presentation of control signals to the critic structure allows for the simplification of the actor's adaptation law. Thus, the synthesis of the actor's adaptation law is possible without the necessity for a mathematical model of the controlled object. This unique property of the ADHDP algorithm makes it applicable to control actions in processes, where the model is unknown e.g. wheeled mobile robot trajectory generation in an unknown environment.

### 6.3.1  Action-Dependent Heuristic Dynamic Programming

ADHDP is the only algorithm of the NDP family that does not require a mathematical model of the controlled object in the actor-critic adaptation process [42, 44].

A characteristic feature of the algorithms in the action-dependent version is the necessity to present the actor generated control signal as an input to the critic's NN.

ADHDP algorithm consists of:

- **Actor** – that can be implemented in the form of an adaptive structure e.g. an NN. The actor generates the suboptimal control law $\mathbf{u}_{A\{k\}}$ in the $k$th step according to the following relation

$$\mathbf{u}_{A\{k\}}\left(\mathbf{x}_{A\{k\}}, \mathbf{W}_{A\{k,l\}}\right) = \mathbf{W}_{A\{k,l\}}^T \mathbf{S}\left(\mathbf{x}_{A\{k\}}\right) , \tag{6.28}$$

where $\mathbf{W}_{A\{k,l\}}$ – matrix of actor's NN output layer weights in the $l$th iteration of the inner loop in the $k$th step of the process, $\mathbf{S}\left(\mathbf{x}_{A\{k\}}\right)$ – neuron activation function vector, $\mathbf{x}_{A\{k\}}$ – actor's NN inputs vector.

The procedure of actor's NN weights adaptation in ADHDP algorithm is to minimize the performance index

$$\mathbf{e}_{A\{k\}} = \frac{\partial L_{C\{k\}}\left(\mathbf{x}_{\{k\}}, \mathbf{u}_{A\{k\}}\right)}{\partial \mathbf{u}_{A\{k\}}} + \gamma \frac{\partial \hat{V}_{\{k+1\}}\left(\mathbf{x}_{C\{k+1\}}, \mathbf{W}_{C\{k,l\}}\right)}{\mathbf{u}_{A\{k+1\}}} , \tag{6.29}$$

according to relation

$$\mathbf{W}_{A\{k,l+1\}} = \mathbf{W}_{A\{k,l\}} - \mathbf{e}_{A\{k\}} \mathbf{\Gamma}_A \frac{\mathbf{u}_{A\{k\}}\left(\mathbf{x}_{A\{k\}}, \mathbf{W}_{A\{k,l\}}\right)}{\partial \mathbf{W}_{A\{k,l\}}} , \tag{6.30}$$

where $\mathbf{\Gamma}_A$ – constant diagonal matrix of reinforcement coefficients for actor's NN weights learning, $\Gamma_{Ai,i} \in \langle 0, 1\rangle$. The control signal generated by the actor is a component of the input vector to the critic's NN, thus the derivative of $\hat{V}_{\{k+1\}}$ with respect to the control signal, from relation (6.29), can be calculated directly as backpropagation of the control signal by the critic's NN.

- **Critic** – that can be implemented in the form of an adaptive structure e.g. an NN. The critic approximates the value function $V_{\{k\}}\left(\mathbf{x}_{\{k\}}, \mathbf{u}_{A\{k\}}\right)$. The critic's NN output in the $k$th step is written as

$$\hat{V}_{\{k\}}\left(\mathbf{x}_{C\{k\}}, \mathbf{W}_{C\{k,l\}}\right) = \mathbf{W}_{C\{k,l\}}^T \mathbf{S}\left(\mathbf{x}_{C\{k\}}\right) , \tag{6.31}$$

where $\mathbf{W}_{C\{k,l\}}$ – critic's NN output layer weights vector in the $l$th iteration of the inner loop in the $k$-th step of the process, $\mathbf{S}\left(\mathbf{x}_{C\{k\}}\right)$ – neuron activation function vector, $\mathbf{x}_{C\{k\}}$ – critic's NN inputs vector.

Critic's NN output layer weights $\mathbf{W}_{C\{k,l\}}$ are adapted by minimizing the temporal difference error $e_{C\{k\}}$ described by relation

$$e_{C\{k\}} = L_{C\{k\}}\left(\mathbf{x}_{\{k\}}, \mathbf{u}_{A\{k\}}\right) + \gamma \hat{V}_{\{k+1\}}\left(\mathbf{x}_{C\{k+1\}}, \mathbf{W}_{C\{k,l\}}\right) - \hat{V}_{\{k\}}\left(\mathbf{x}_{C\{k\}}, \mathbf{W}_{C\{k,l\}}\right) , \tag{6.32}$$

where $\mathbf{x}_{C\{k+1\}}$ – critic's NN inputs vector in the $k + 1$ step, obtained by one discrete step delay of all input signals into the NN of ADHDP structure.

**Fig. 6.17** Schematic
diagram of NDP structure in
ADHPD configuration

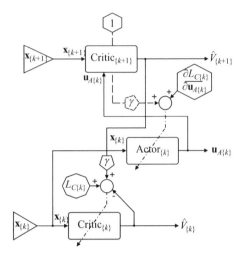

Critic's NN output layer weights $\mathbf{W}_{C\{k,l\}}$ are adapted according to the following relation

$$\mathbf{W}_{C\{k,l+1\}} = \mathbf{W}_{C\{k,l\}} - e_{C\{k\}}\mathbf{\Gamma}_C \frac{\hat{V}_{\{k\}}\left(\mathbf{x}_{C\{k\}}, \mathbf{W}_{C\{k,l\}}\right)}{\partial \mathbf{W}_{C\{k,l\}}} \tag{6.33}$$

where $\mathbf{\Gamma}_C$ – constant diagonal matrix of reinforcement coefficients for critic's NN weights adaptation, $\Gamma_{Ci,i} \in \langle 0, 1 \rangle$.

A schematic diagram of the ADHDP structure is shown in Fig. 6.17.
A schematic diagram of NN weights adaptation process for NDP structure in ADHDP configuration is shown in Fig. 6.18. ADHDP algorithm's NN weights adaptation, as in the case of the other NDP algorithms, is carried out in the inner loop as shown in Fig. 6.4. Iterations of the inner calculation loop are performed in each $k$th step of the process. The iterations are denoted by $l$ index and involve an alternate adaptation of the control law and value function approximation.

Compared to the HDP and DHP structures the ADHDP algorithm has a simplified design and uses less complex NN weights adaptation methods.

Further in the publication, the notation of quantities that change their values in the inner iterative loop of the NDP algorithms does not include the index $l$ ($l_{AC} = 1$), hence $\mathbf{u}_{A\{k\}}\left(\mathbf{x}_{A\{k\}}, \mathbf{W}_{A\{k\}}\right)$, $\hat{V}_{\{k\}}\left(\mathbf{x}_{C\{k\}}, \mathbf{W}_{C\{k\}}\right)$. This simplification shall be construed as execution of one step of NDP algorithm adaptation procedure consisting of the control law improvement routine and value function determination operation, in each $k$th step of the discrete process. This is to reduce the notation, however it should be noted that in all designed control algorithms with NDP structures the adaptation procedure of NDP algorithms was performed for $l_{AC} > 1$.

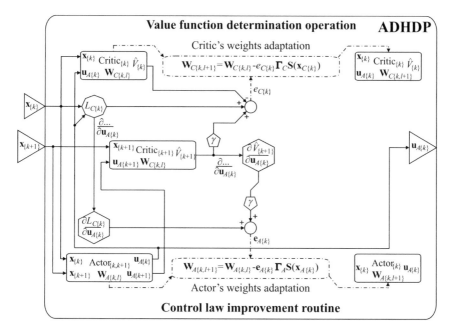

**Fig. 6.18** Schematic diagram of NDP structure adaptation in ADHPD configuration

# References

1. Astrom, K.J., Wittenmark, B.: Adaptive Control. Addison-Wesley, New York (1979)
2. Baird III, L.C.: Reinforcement learning in continuous time: advantage updating. In: Proceedings of the IEEE International Conference on Neural Networks, pp. 2448–2453 (1994)
3. Balaji, P.G., German, X., Srinivasan, D.: Urban traffic signal control using reinforcement learning agents. IET Intell. Transp. Sy. **4**, 177–188 (2010)
4. Barto, A., Sutton, R.: Reinforcement Learning: An Introduction. MIT Press, Cambridge (1998)
5. Barto, A., Mahadevan, S.: Recent advances in hierarchical reinforcement learning. Discrete Event Dyn. Syst. **13**, 343–379 (2003)
6. Barto, A., Sutton, R., Anderson, C.: Neuronlike adaptive elements that can solve difficult learning problems. EEE Trans. Syst., Man, Cybern., Syst. I **13**, 834–846 (1983)
7. Carreras, M., Yuh, J., Batlle, J., Ridao, P.: A behaviorbased scheme using reinforcement learning for autonomous underwater vehicles. IEEE J. Ocean. Eng. **30**, 416–427 (2005)
8. Cichosz, P.: Learning Systems. (in Polish). WNT, Warsaw (2000)
9. Doya, K.: Reinforcement learning in continuous time and space. Neural Comput. **12**, 219–245 (2000)
10. Ernst, D., Glavic M., Wehenkel, L.: Power systems stability control: reinforcement learning framework. IEEE Trans. Power Syst. **19**, 427–435 (2004)
11. Fairbank, M., Alonso, E., Prokhorov, D.: Simple and fast calculation of the second-order gradients for globalized dual heuristic dynamic programming in neural networks. IEEE Trans. Neural Netw. Learn. Syst. **23**, 1671–1676 (2012)
12. Ferrari, S.: Algebraic and Adaptive Learning in Neural Control Systems. Ph.D. Thesis, Princeton University, Princeton (2002)
13. Ferrari, S., Stengel, R.F.: An adaptive critic global controller. In: Proceedings of American Control Conference, vol. 4, pp. 2665–2670. Anchorage, Alaska (2002)

14. Ferrari, S., Stengel, R.F.: Model-based adaptive critic designs in learning and approximate dynamic programming. In: Si, J., Barto, A., Powell, W., Wunsch, D.J. (eds.) Handbook of Learning and Approximate Dynamic Programming, pp. 64–94. Wiley, New York (2004)

15. Gierlak, P., Szuster, M., ylski, W.: Discrete dual-heuristic programming in 3DOF manipulator control. Lect. Notes Artif. Int. **6114**, 256–263 (2010)

16. Hagen, S., Krose, B.: Neural Q-learning. Neural. Comput. Appl. **12**, 81–88 (2003)

17. Han, D., Balakrishnan, S.: Adaptive critic based neural networks for control-constrained agile missile control. Proc. Am. Control Conf. **4**, 2600–2605 (1999)

18. Hanselmann, T., Noakes, L., Zaknich, A.: Continuous-time adaptive critics. IEEE Trans. Neural Netw. **18**, 631–647 (2007)

19. Hendzel, Z., Burghardt, A., Szuster, M.: Reinforcement learning in discrete neural control of the underactuated system. Lect. Notes Artif. Int. **7894**, 64–75 (2013)

20. Hendzel, Z., Szuster, M.: Discrete model-based dual heuristic programming in wheeled mobile robot control. In: Awrejcewicz, J., Kamierczak, M., Olejnik, P., Mrozowski, J. (eds.) Dynamical Systems - Theory and Applications, pp. 745–752. Left Grupa, Lodz (2009)

21. Hendzel, Z., Szuster, M.: Heuristic dynamic programming in wheeled mobile robot control. In: Kaszyski, R., Pietrusewicz, K. (eds.) Methods and Models in Automation and Robotics, pp. 513–518. IFAC, Poland (2009)

22. Hendzel, Z., Szuster, M.: Discrete action dependant heuristic dynamic programming in wheeled mobile robot control. Solid State Phenom. **164**, 419–424 (2010)

23. Hendzel, Z., Szuster, M.: Discrete model-based adaptive critic designs in wheeled mobile robot control. Lect. Notes Artif. Int. **6114**, 264–271 (2010)

24. Hendzel, Z., Szuster, M.: Discrete neural dynamic programming in wheeled mobile robot control. Commun. Nonlinear. Sci. Numer. Simul. **16**, 2355–2362 (2011)

25. Hendzel, Z., Szuster, M.: Adaptive dynamic programming methods in control of wheeled mobile robot. Int. J. Appl. Mech. Eng. **17**, 837–851 (2012)

26. Hendzel, Z., Szuster, M.: Globalised dual heuristic dynamic programming in control of nonlinear dynamical system. In: Awrejcewicz, J., Kamierczak, M., Olejnik, P., Mrozowski, J. (eds.) Dynamical Systems: Applications, pp. 123–134. WPL, Lodz (2013)

27. Iftekharuddin, K.M.: Transformation invariant on-line target recognition. IEEE Trans. Neural Netw. **22**, 906–918 (2011)

28. Kareem Jaradat, M.A., Al-Rousan M., Quadan, L.: Reinforcement based mobile robot navigation in dynamic environment. Robot. Cim.-Int. Manuf. **27**, 135–149 (2011)

29. Lendaris, G., Schultz, L., Shannon, T.: Adaptive critic design for intelligent steering and speed control of a 2-axle vehicle. In: Proceedings of the IEEE INNS-ENNS International Joint Conference on Neural Networks, vol. 3, pp. 73–78 (2000)

30. Lendaris, G., Shannon, T.: Application considerations for the DHP methodology. In: Proceedings of the IEEE International Joint Conference on Neural Networks, vol. 2, pp. 1013–1018 (1998)

31. Lewis, F.L., Liu, D., Lendaris, G.G.: Guest editorial: special issue on adaptive dynamic programming and reinforcement learning in feedback control. IEEE Trans. Syst. Man Cybern. B Cybern. **38**, 896–897 (2008)

32. Lewis, F.L., Vrabie, D.: Reinforcement learning and adaptive dynamic programming for feedback control. IEEE Circuits Syst. Mag. **9**, 32–50 (2009)

33. Liu, D., Wang, D., Yang X.: An iterative adaptive dynamic programming algorithm for optimal control of unknown discrete-time nonlinear systems with constrained inputs. Inform. Sci. **220**, 331–342 (2013)

34. Millán, J.,del R.: Reinforcement learning of goal-directed obstacle-avoiding reaction strategies in an autonomous mobile robot. Robot. Auton. Syst. **15**, 275–299 (1995)

35. Mohagheghi, S., Venayagamoorthy, G.K., Harley, R.G.: Adaptive critic design based neuro-fuzzy controller for a static compensator in a multimachine power system. IEEE Trans. Power Syst. **21**, 1744–1754 (2006)

36. Ni, Z., He, H.: Heuristic dynamic programming with internal goal representation. Soft Comput. **17**, 2101–2108 (2013)

37. Ni, Z., He, H., Wen, J., Xu, X.: Goal representation heuristic dynamic programming on maze navigation. IEEE Trans. Neural Netw. Learn. Syst. **24**, 2038–2050 (2013)
38. Ni, Z., He, H., Zhao, D., Xu, X., Prokhorov, D.V.: Grdhp: A general utility function representation for dual heuristic dynamic programming. IEEE Trans. Neural Netw. Learn. Syst **26**, 614–627 (2015)
39. Ng, A.Y., Kim, H.J., Jordan, M.I., Sastry, S.: Autonomous helicopter flight via reinforcement learning. Adv. Neural Inf. Process. Syst. **16** (2004)
40. Peters, J., Schaal, S.: Natural actor-critic. Neurocomputing **71**, 1180–1190 (2008)
41. Powell, W.B.: Approximate Dynamic Programming: Solving the Curses of Dimensionality. Princeton, Willey-Interscience (2007)
42. Prokhorov, D., Wunch, D.: Adaptive critic designs. IEEE Trans. Neural Netw. **8**, 997–1007 (1997)
43. Rutkowski, L.: Computational Intelligence - Methods and Techniques (in Polish). Polish Scientific Publishers PWN, Warsaw (2005)
44. Si, J., Barto, A.G., Powell, W.B., Wunsch, D.: Handbook of Learning and Approximate Dynamic Programming. IEEE Press, Wiley-Interscience, Hoboken (2004)
45. Shannon, T., Lendaris, G.: A new hybrid critic–training method for approximate dynamic programming. In: Proceedings of International Society for the System Sciences (2000)
46. Szuster, M., Hendzel, Z., Burghardt, A.: Fuzzy sensor-based navigation with neural tracking control of the wheeled mobile robot. Lect. Notes Artif. Int. **8468**, 302–313 (2014)
47. Szuster, M., Hendzel, Z.: Discrete globalised dual heuristic dynamic programming in control of the two-wheeled mobile robot. Math. Probl. Eng. **2014**, 1–16 (2014)
48. Szuster, M., Gierlak, P.: Approximate dynamic programming in tracking control of a robotic manipulator. Int. J. Adv. Robot. Syst. **13**, 1–18 (2016)
49. Szuster, M., Gierlak, P.: Globalised dual heuristic dynamic programming in control of robotic manipulator. AMM **817**, 150–161 (2016)
50. Szuster, M.: Globalised dual heuristic dynamic programming in tracking control of the wheeled mobile robot. Lect. Notes Artif. Int. **8468**, 290–301 (2014)
51. Syam, R., Watanabe, K., Izumi, K.: Adaptive actor-critic learning for the control of mobile robots by applying predictive models. Soft. Comput. **9**, 835–845 (2005)
52. Syam, R., Watanabe, K., Izumi, K., Kiguchi, K.: Control of nonholonomic mobile robot by an adaptive-critic method with simulated experience based value functions. In: Proceedings of the IEEE International Conference of Robotics and Automation, vol. 4, pp. 3960–3965 (2002)
53. Vamvoudakis, K.G., Lewis, F.L.: Online actor-critic algorithm to solve the continuous-time infinite horizon optimal control problem. Automatica **46**, 878–888 (2010)
54. Vamvoudakis, K.G., Lewis, F.L.: Multi-player non-zerosum games: online adaptive learning solution of coupled Hamilton-Jacobi equations. Automatica **47**, 1556–1569 (2011)
55. Venayagamoorthy, G.K., Harley, R.G., Wunsch, D.C.: Comparison of heuristic dynamic programming and dual heuristic programming adaptive critics of a turbogenerator. IEEE Trans. Neural Netw. **13**, 764–773 (2002)
56. Venayagamoorthy, G.K., Wunsch, D.C., Harley, R.G.: Adaptive critic based neurocontroller for turbogenerators with global dual heuristic programming. In: Proceedings of the IEEE Power Engineering Society Winter Meeting, vol. 1, pp. 291–294 (2000)
57. Visnevski, N., Prokhorov, D.: Control of a nonlinear multivariable system with adaptive critic designs. In: Proceedings of Artificial Neural Networks in Engineering, vol. 6, pp. 559–565 (1996)
58. Vrabie, D., Lewis, F.: Neural network approach to continuous-time direct adaptive optimal control for partially unknown nonlinear systems. Neural Netw. **22**, 237–246 (2009)
59. Wang, D., Liu, D., Wei, Q.: Finite-horizon neuro-optimal tracking control for a class of discrete-time nonlinear systems using adaptive dynamic programming approach. Neurocomputing **78**, 14–22 (2012)
60. Wang, D., Liu D., Wei, Q., Zhao D., Jin, N.: Optimal control of unknown nonaffine nonlinear discrete-time systems based on adaptive dynamic programming. Automatica **48**, 1825–1832 (2012)

61. Wang, D., Liu, D., Zhao, D., Huang, Y., Zhang, D.: A neural network-based iterative GDHP approach for solving a class of nonlinear optimal control problems with control constraints. Meural Comput. Appl. **22**, 219–227 (2013)
62. Wang, F.-Y., Zhang H., Liu D.: Adaptive dynamic programming: an introduction. IEEE Comput. Intell. Mag. **4**, 39–47 (2009)
63. Xu, X., Hou, Z., Lian, C., He, H.: Online learning control using adaptive critic designs with sparse kernel machines. IEEE Trans. Neural Netw. Learn. Syst. **24**, 762–775 (2013)
64. Xu, X., Wang, X., Hu, D.: Mobile robot path-tracking using an adaptive critic learning PD controller. Lect. Notes Comput. Sci. **3174**, 25–34 (2004)
65. Xu, X., Zuo, L., Huang, Z.: Reinforcement learning algorithms with function approximation: recent advances and applications. Inform. Sci. **261**, 1–31 (2014)
66. Zhang, H., Cui, L., Zhang, X., Luo, Y.: Data-driven robust approximate optimal tracking control for unknown general nonlinear systems using adaptive dynamic programming method. IEEE Trans. Neural Netw. **22**, 2226–2236 (2011)
67. Zelinsky, A., Gaskett, C., Wettergreen, D.: Q-learning in continous state and action spaces. In: Proceedings of Australian Joint Conference on Artificial Intelligence, pp. 417–428. Springer (1999)
68. Zhang, X., Zhang, H., Luo, Y.: Adaptive dynamic programming-based optimal control of unknown nonaffine nonlinear discrete-time systems with proof of convergence. Neurocomputing **91**, 48–55 (2012)
69. Zhong, X., Ni, Z., He, H.: A theoretical foundation of goal representation heuristic dynamic programming. IEEE Trans. Neural Netw. Learn. Syst. **PP**, 1–13 (2105)

# Chapter 7
# Control of Mechatronic Systems

The following part of this work discusses selected control methods for nonlinear systems with the application of intelligent algorithms, with an indication of their significant advantages and drawbacks. The criteria for consideration in the accuracy of the tracking motion executed by nonlinear systems constitute a comparison of yielded results to the values obtained from tests, wherein the control system was based only on a PD controller. The structures of the currently applied control algorithms for nonlinear objects are mostly composed of a controller and a compensator, and optionally of a robust control term, which was discussed in Sect. 3.1. The compensator compensates for nonlinearities of a controlled object, while the stabilizer (e.g. PD controller) minimizes the tracking error that results from inaccurate compensation. The robust term, however, ensures stability of the control system in a presence of major disturbances. The execution of compensation control may be seen as an inverse dynamics problem where a control signal is generated on the basis of a mathematical description of the controlled object, thereby ensuring the execution of the desired parameters of motion. This is not easy to implement, given that the algorithm for nonlinearity compensation shall meet several requirements. It is crucial to ensure stability of the control system and take account of variable operating conditions of the controlled object that result from e.g. load variations, mass distributions and, in consequence, mass moments of inertia, or variable operating conditions stemming from the movement on various types of surface, which is the case for a mobile robot.

The relevant literature provides examples of control algorithms that fail to meet the last requirement. For example, the algorithm of the computed moment [15, 64, 82, 91, 95] is based on a mathematical model. It's drawback is the requirement for availability of a model structure and its parameters to synthesize the control system. The definition of a mathematical model structure is not that difficult as there are a number of mathematical formalisms for the creation of motion equations that can describe the controlled object, whereas the definition of model's parameters is not an easy task. Furthermore, in the case of the computed moment algorithm it is assumed

© Springer International Publishing AG 2018
M. Szuster and Z. Hendzel, *Intelligent Optimal Adaptive Control
for Mechatronic Systems*, Studies in Systems, Decision and Control 120,
https://doi.org/10.1007/978-3-319-68826-8_7

that the parametric values are constant, thus making it impossible for the control algorithm to adapt to variable operating conditions of the controlled object.

These disadvantages are to a certain extent eliminated in robust control methods that draw on the theory of variable-structure systems [3, 15, 30, 80–82, 95]. These algorithms are used for tracking control of inaccurately described systems with explicit modeling uncertainties. It is assumed that each of the object's parameters takes on a value from a defined interval whose endpoints are defined based on the available information on the controlled object. The control based on a robust algorithm is properly executed when parametric disturbances meet the assumed constraints. Introduction of a switching component into the robust algorithm minimizes errors resulting from the lack of knowledge of the object's model. The disadvantage of the algorithm is the frequent use of overestimated values of parametric constraints which is related to high uncertainty.

These drawbacks do not apply to algorithms that have structures able to adapt their parameters in the event of disturbances, thereby ensuring proper implementation of the control process. Examples of such solutions include the adaptive control algorithms [15, 30, 50, 81, 82, 91], whose design allows for the adaptation of algorithm's parameters to variable operating conditions of an object according to the assumed adaptive algorithm. Their synthesis only requires the structure of the controlled object's model, where the parametric estimates are determined in real-time during the movement of an object. Their values do not need to converge to real parameters to ensure correct control execution.

The application of modern AI methods to control algorithms has proved positive given the nonlinearities occurring in the dynamic description of controlled objects. In particular, it is NNs that have become an attractive tool used in control algorithms for nonlinear objects, due to their ability to approximate any nonlinear function, learning capability and parametric adaptation [12, 14, 22, 27, 30, 40, 50, 56, 58, 68, 69, 83, 94]. Fuzzy logic systems (FLS) [16, 17, 19, 20, 33, 37, 83, 89, 92] are also widely used for dynamic objects' control systems. The reason for this is that these algorithms can approximate any nonlinear function and may be applied to system control, where models provide only an approximate description of the controlled object's properties. Another significant factor is clear representation of information included in a fuzzy model by means of a knowledge base, and the possibility to acquire expert knowledge on object control when creating the rules for a fuzzy model. The drawback of fuzzy logic systems is that they fail to adapt sets of fuzzy premises and conclusions to object's variable operating conditions. Neuro-fuzzy systems (NFS) are free from this disadvantage [32, 34, 67, 74, 76, 78] as their selected fuzzy sets' parameters can be adapted in real-time.

The development of artificial intelligence methods allowed the emergence of NDP algorithms, being a combination of the classical optimization theory based on Bellman's DP and intelligent algorithms such as NNs. The algorithms allow for the generation of suboptimal control law for a dynamic object in a forward process and are implemented by means of actor-critic structures, and used in control systems for nonlinear objects [4, 5, 13, 18, 21, 24, 25, 35, 46, 49, 66, 70, 72, 77, 84, 88]. The structure of NDP-based control systems is more complex than that of the neural control algorithm, however NDP algorithms often allow to reach the control objective in tasks where the application of other systems has not delivered the expected results.

This chapter discusses selected policies for motion control of mechatronic systems. It provides the synthesis of control algorithms for dynamic objects and simulation test concerned with the motion of selected systems. The studies include the application of a PD controller (Sect. 7.1) to tracking control, wherein the controller is used as a stabilizer and constitutes the key component of the control system. The subsections that follow present control algorithms in which the control system structure was complemented with a compensation algorithm for the controlled object's nonlinearities. Section 7.2 discusses the application of an adaptive algorithm for motion control. A neural control algorithm (Sect. 7.3) was used as an alternative solution to the problem. Further subchapters cover the applicability of NDP algorithms to control systems of dynamic objects. Section 7.4 presents a control system with an HDP algorithm, Sect. 7.5 focuses on the application of a DHP type algorithm, Sect. 7.6 includes a description of a GDHP algorithm implementation into a control system of a dynamic object. The application of an ADHDP algorithm to a control system is discussed in Sect. 7.7. The final Sect. 7.8 involves behavioral control methods for a WMR with the use of NDP algorithms.

## 7.1 Tracking Control of a WMR and a RM with a PD Controller

Proper execution of the tracking requires that the controlled object moves according to the parameters resulting from the solution of an inverse kinematics problem. For dynamic objects described by complex nonlinear equations of motion such as RMs or WMRs, proper execution of the desired trajectory depends upon the application of an appropriate control system. A typical structure of a control system for nonlinear objects is shown in Fig. 3.1 in Sect. 3.1. The key component of a control system is the PD/PID controller that stabilizes the tracking error. The relevant literature provides a number of publications on the PD/PID controller application to control of dynamic objects such as RMs [15, 82, 91, 95], WMRs [30, 50, 65] or underactuated systems [59, 75, 93], where the PD/PID controller generates the overall control signal or constitutes one of the components of a complex control algorithm. The PD controller delivers good control performance if the controlled object is linear, which is a theoretical assumption, and provided there are no rapid movements of the object. This chapter focuses on the tracking control of a WMR and RM with the application of a PD controller.

### 7.1.1   Synthesis of PD-Type Control

The synthesis of a WMR's tracking control system with the application of a PD controller is detailed in [30], while an adequate description of a RM's PD control system is included in [95].

The tracking problem is defined as a search for the control law $\mathbf{u}$ whereby the executed object's trajectory follows the desired trajectory, given the variables $\mathbf{q}_d$, $\dot{\mathbf{q}}_d$, $\ddot{\mathbf{q}}_d$ are constrained. The tracking error $\mathbf{e}$ is defined as

$$\mathbf{e} = \mathbf{q} - \mathbf{q}_d \ . \tag{7.1}$$

The filtered tracking error $\mathbf{s}$ is introduced being a linear combination of the tracking error and its derivative, defined on the basis of (7.1) as

$$\mathbf{s} = \dot{\mathbf{e}} + \mathbf{\Lambda}\mathbf{e} \ , \tag{7.2}$$

where $\mathbf{\Lambda}$ – constant, positive - definite, diagonal design matrix.

The control signal of a proportional-derivative controller (PD) is expressed by equation

$$\mathbf{u} = \mathbf{u}_{PD} = \mathbf{K}_D\mathbf{s} \ , \tag{7.3}$$

where $\mathbf{K}_D$ – constant, positive - definite design matrix of controller gains. Expression $\mathbf{K}_D\mathbf{s}$ is equivalent to the description of a conventional PD controller with derivative (D) and proportional (P) terms whose respective gains are $\mathbf{K}_D$ and $\mathbf{K}_P = \mathbf{K}_D\mathbf{\Lambda}$.

Figure 7.1 shows a schematic diagram of a nonlinear object's tracking control system with a PD controller in regard to a WMR.

The denotations shown in the diagram were introduced in Sect. 2.1, where $\boldsymbol{\alpha}$ is the vector of WMR's drive wheels rotation angles corresponding to the vector $\mathbf{q}$.

### 7.1.2   Simulation Tests

The simulation tests involved the tracking control of a WMR and a RM with the application of a PD controller.

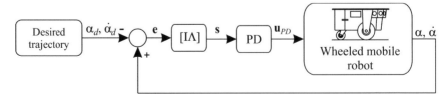

**Fig. 7.1** Schematic diagram of a WMR's tracking control system with the application of a PD controller

### 7.1.2.1 Numerical Test of the WMR Motion Control with the Use of PD Controller

The simulation tests on the application of a PD controller to tracking control of the WMR's point $A$ were carried out in a virtual computing environment. Numerical tests were performed of which two results were selected. These were obtained on the basis of the trajectory with an 8-shaped path presented in Sect. 2.1.1.2. Two scenarios were investigated, one with a parametric disturbance, where the change of model's parameters was related to the WMR being weighed down with additional mass, and one that considered no such disturbance.

The numerical tests with a PD controller were conducted for subsequent comparison of tracking control accuracy with the application of other control algorithms.

The following parameters were used in all numerical tests:

- PD controller parameters:

    - $\mathbf{K}_D = \mathbf{I}$ – constant diagonal matrix of PD controller gains,
    - $\mathbf{\Lambda} = 0.5\,\mathbf{I}$ – constant diagonal matrix,

- $h = 0.01$ [s] – numerical test calculation step.

The applied PD controller parameters correspond to the coefficients of proportional gain $\mathbf{K}_P = \mathbf{K}_D \mathbf{\Lambda}$. The value of matrix $\mathbf{\Lambda}$ adopted in the simulation tests provides low sensitivity of the system to parametric changes of the WMR's model and to other disturbances, while allowing for control of a real object i.e. the Pioneer 2-DX WMR. Excessive values of matrix $\mathbf{\Lambda}$ elements generate control signals that cannot be applied to a real object, due to the characteristics of the actuators used. The values of matrix $\mathbf{K}_D$ were selected in such a way as to provide good performance of the WMR's point $A$ tracking control. It should be noted that these values cannot be too high, as that could result in system instability.

The following performance indexes were used for the quality assessment of the generated control signals and the execution of the tracking: – maximum absolute value of angular displacement error for the WMR's drive wheels 1 and 2, $e_{maxj}$ [rad],

$$e_{maxj} = \max_k \left( |e_{j\{k\}}| \right) , \tag{7.4}$$

where $j = 1, 2$,
– root mean square error for angular displacement of the $j$th wheel, $e_j$ [rad],

$$\varepsilon_j = \sqrt{\frac{1}{n} \sum_{k=1}^{n} e_{j\{k\}}^2} , \tag{7.5}$$

where $k$ – number of iteration step, $n$ – total number of steps,
– maximum absolute value of angular velocity error for the $j$th wheel $\dot{e}_{maxj}$ [rad/s],

$$\dot{e}_{maxj} = \max_k \left( |\dot{e}_{j\{k\}}| \right) \, , \tag{7.6}$$

– root mean square error for angular velocity of the $j$th wheel, $\dot{e}_j$ [rad/s],

$$\dot{\varepsilon}_j = \sqrt{\frac{1}{n} \sum_{k=1}^{n} \dot{e}_{j\{k\}}^2} \, , \tag{7.7}$$

– maximum absolute value of filtered tracking error for the $j$th wheel $s_{maxj}$ [rad/s],

$$s_{maxj} = \max_k \left( |s_{j\{k\}}| \right) \, , \tag{7.8}$$

– root mean square filtered tracking error for the $j$th wheel $s_j$ [rad/s],

$$\sigma_j = \sqrt{\frac{1}{n} \sum_{k=1}^{n} s_{j\{k\}}^2} \, , \tag{7.9}$$

and motion path errors for the robot's selected point:
– mean square error for distance $d_{\{k\}}$ [m] between the desired $\left( x_{dA\{k\}}, y_{dA\{k\}} \right)$ and actual $\left( x_{A\{k\}}, y_{A\{k\}} \right)$ position of the WMR's point $A$ on the $xy$ plane,

$$\rho_d = \sqrt{\frac{1}{n} \sum_{k=1}^{n} d_{\{k\}}^2} \, , \tag{7.10}$$

where $d_{\{k\}} = \sqrt{\left( x_{A\{k\}} - x_{dA\{k\}} \right)^2 + \left( y_{A\{k\}} - y_{dA\{k\}} \right)^2}$ [m],
– maximum distance $d_{max}$ [m],

$$d_{max} = \max_k \left( d_{\{k\}} \right) \, , \tag{7.11}$$

– distance between the desired $\left( x_{dA\{k\}}, y_{dA\{k\}} \right)$ and actual $\left( x_{A\{k\}}, y_{A\{k\}} \right)$ position of mobile robot's point $A$ after the completion of motion, when $k = n$, $d_n$ [m].

The volume of discrete measurement data for the WMR's trajectory with an 8-shaped path is $n = 4500$, while for the RM and the trajectory with a semicircle path $n = 4000$.

### 7.1.2.2   Examination of PD-Control, 8-Shaped Path, No-Disturbance Trial

The desired trajectory with an 8-shaped path discussed in Sect. 2.1.1.2 was used in the numerical test. The values of motion parameters obtained from the solution of the

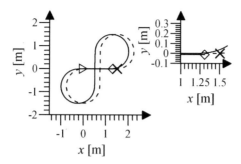

**Fig. 7.2**  Desired and executed motion path of the WMR's point $A$

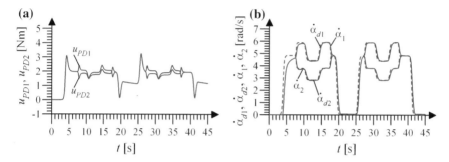

**Fig. 7.3**  **a** Control signals $u_{PD1}$ and $u_{PD2}$, **b** desired ($\dot{\alpha}_{d1}$, $\dot{\alpha}_{d2}$) and executed ($\dot{\alpha}_1$, $\dot{\alpha}_2$) angular velocities of drive wheels

inverse kinematics problem are denoted by index $d$ and used as the desired parameters of motion. Tracking control of the WMR was simulated with the application of a PD controller.

Figure 7.2a shows the desired and executed path of the WMR's point $A$. The left graph illustrates the desired (dashed line) and executed (solid line) path of the WMR's point $A$, the triangle shows the initial position of point $A$, "**X**" marks the desired final position, the rhombus indicates the actual final position of the WMR's point $A$. The graph on the right shows an enlarged view of the section of the graph on the left.

The executed path of the WMR's point $A$ is the result of sending signals generated by the tracking control algorithm to the actuators of the robot. The values of the control signals $u_{PD1}$ and $u_{PD2}$ for wheels 1 and 2 are shown in Fig. 7.3a, values of the desired (dashed line, $\dot{\alpha}_{d1}$, $\dot{\alpha}_{d2}$) and executed angular velocities (solid line, $\dot{\alpha}_1$, $\dot{\alpha}_2$) of the drive wheels are presented in Fig. 7.3b.

The difference between the desired and executed path of the WMR's point $A$ is due to the erroneous execution of the desired trajectory. The values of the errors for rotation angles ($e_1$, $e_2$), and errors for angular velocities ($\dot{e}_1$, $\dot{e}_2$), of the WMR drive wheels 1 and 2 are shown in Fig. 7.4a, b, respectively.

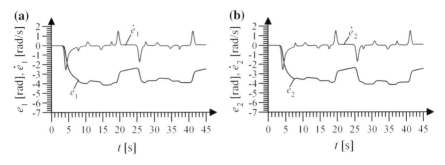

**Fig. 7.4**  **a** Tracking errors $e_1$ and $\dot{e}_1$, **b** tracking errors $e_2$ and $\dot{e}_2$

**Table 7.1**  Values of selected performance indexes

| Index | $e_{maxj}$ [rad] | $\varepsilon_j$ [rad] | $\dot{e}_{maxj}$ [rad/s] | $\dot{\varepsilon}_j$ [rad/s] | $s_{maxj}$ [rad/s] | $\sigma_j$ [rad/s] |
|---|---|---|---|---|---|---|
| Wheel 1, $j = 1$ | 4.16 | 3.31 | 2.56 | 0.47 | 3.25 | 1.74 |
| Wheel 2, $j = 2$ | 4.16 | 3.31 | 2.56 | 0.47 | 3.25 | 1.74 |

**Table 7.2**  Values of selected performance indexes of path execution

| Index | $d_{max}$ [m] | $d_n$ [m] | $\rho_d$ [m] |
|---|---|---|---|
| Value | 0.316 | 0.197 | 0.229 |

It should be noted that the goal of the tracking control algorithm concerned here is to minimize tracking errors. The values of individual performance indexes for the tracking control generated by the PD controller are shown in Tables 7.1 and 7.2.

The selected performance indexes were calculated by splitting the executed trajectory with an 8-shaped path in two phases, where phase I ($t \in \langle 0.01, 22.5 \rangle [s]$, $k = 1, \ldots, 2250$ discrete measurements) includes the motion along a curved path in the leftward direction, from the initial position at point $S$, to a complete stop at point $P$ (see Fig. 2.6a). Phase II includes the motion along a curved path in the rightward direction, from point $P$ to point $G$ ($t \in \langle 22.51, 45 \rangle [s]$, $k = 2251, \ldots, 4500$). The division of trajectory into two phases when calculating the performance indexes aims to examine the impact of the knowledge stored within the parameters of the adaptive structures on the quality of the generated control. The execution of motion with the application of the presented trajectory can be construed as two experiments, wherein adaptation of algorithm parameters occurs. In the first experiment the parameters adapted take on zero initial values, while in the second experiment the initial parametric values are those obtained from the first experiment. The values of selected performance indexes for individual motion phases are shown in Table 7.3.

The selected indexes show similar values in both parts of the numerical test when no adaptive algorithms are applied to the control system structure e.g. NNs. This occurs when a PD controller is used in the tracking control of the WMR's point $A$.

**Table 7.3** Values of selected performance indexes in I and II stage of motion

| Index | $e_{maxj}$ [rad] | $\varepsilon_j$ [rad] | $\dot{e}_{maxj}$ [rad/s] | $\dot{\varepsilon}_j$ [rad/s] | $s_{maxj}$ [rad/s] | $\sigma_j$ [rad/s] |
|---|---|---|---|---|---|---|
| Stage I, $k = 1, \ldots, 2250$ | | | | | | |
| Wheel 1, $j = 1$ | 4.16 | 3.22 | 2.56 | 0.55 | 3.11 | 1.74 |
| Wheel 2, $j = 2$ | 3.91 | 3.03 | 2.56 | 0.54 | 3.11 | 1.65 |
| Stage II, $k = 2251, \ldots, 4500$ | | | | | | |
| Wheel 1, $j = 1$ | 3.91 | 3.03 | 2.56 | 0.54 | 3.11 | 1.65 |
| Wheel 2, $j = 2$ | 4.16 | 3.22 | 2.56 | 0.55 | 3.11 | 1.74 |

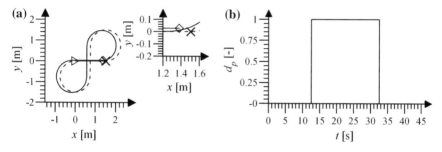

**Fig. 7.5** **a** Desired and executed path of the WMR's point $A$, **b** parametric disturbance signal $d_p$

### 7.1.2.3 Examination of PD-Control, 8-Shaped Path, Disturbance Trial

The desired trajectory with an 8-shaped path discussed in Sect. 2.1.1.2 was used in the numerical test. A parametric disturbance was introduced twice. The first one involved the change of the WMR model's parameters corresponding to the robot's frame being weighed down with a load of $m_{RL} = 4.0$ [kg] in time $t_{d1} = 12.5$ [s]. In the second disturbance scenario the additional load was removed in time $t_{d2} = 32.5$ [s].

Figure 7.5a shows the desired and executed path of the WMR's point $A$. The left graph illustrates the desired (dashed line) and executed (solid line) path of the WMR's point $A$, the triangle shows the initial position of point $A$, "**X**" marks the desired final position, the rhombus indicates the actual final position of the WMR's point $A$. The graph on the right shows an enlarged view of the section of the graph on the left. The graph in Fig. 7.5b presents the values of signal $d_p$ that indicates the occurrence of parametric disturbance. If $d_p = 0$, the nominal set of parameters **a** from Table 2.1 is applied to the WMR's dynamic model, if $d_p = 1$, the set of parameters $\mathbf{a}_d$ is used that corresponds to a parametric disturbance resulting from an additional mass carried by the WMR.

The values of the control signals $u_{PD1}$ and $u_{PD2}$ for wheels 1 and 2 are shown in Fig. 7.6a, values of the desired ($\dot{\alpha}_{d1}, \dot{\alpha}_{d2}$) and executed ($\dot{\alpha}_1, \dot{\alpha}_2$) angular velocities of the drive wheels are presented in Fig. 7.6b. The ellipses mark the graphs' sections which illustrate the impact of parametric disturbances on the obtained values. It can be noticed that the occurrence of the first disturbance in time $t_{d1}$ is followed by

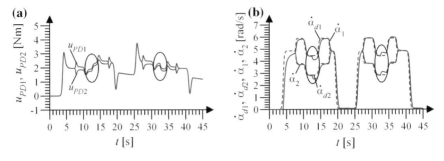

**Fig. 7.6 a** Control signals $u_{PD1}$ and $u_{PD2}$, **b** desired $(\dot{\alpha}_{d1}, \dot{\alpha}_{d2})$ and executed $(\dot{\alpha}_1, \dot{\alpha}_2)$ angular velocities of drive wheels

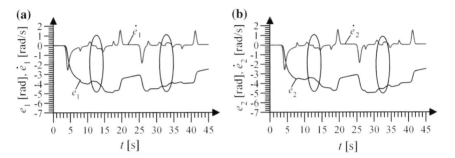

**Fig. 7.7 a** Tracking errors $e_1$ and $\dot{e}_1$, **b** tracking errors $e_2$ and $\dot{e}_2$

**Table 7.4** Values of selected performance indexes

| Index | $e_{maxj}$ [rad] | $\varepsilon_j$ [rad] | $\dot{e}_{maxj}$ [rad/s] | $\dot{\varepsilon}_j$ [rad/s] | $s_{maxj}$ [rad/s] | $\sigma_j$ [rad/s] |
|---|---|---|---|---|---|---|
| Wheel 1, $j = 1$ | 4.89 | 3.69 | 2.56 | 0.49 | 3.76 | 1.93 |
| Wheel 2, $j = 2$ | 4.98 | 3.7 | 2.56 | 0.49 | 3.76 | 1.93 |

an increase in the values of control signals due to a change in the dynamics of the controlled object. A momentary decrease in the values of actual angular velocities is observed. The reduction in mass of the load carried by the WMR in time $t_{d2}$ results in a decrease in the values of control signals and a momentary increase in the values of actual angular velocities of the drive wheels.

The values of errors for rotation angles $(e_1, e_2)$ and errors for angular velocities $(\dot{e}_1, \dot{e}_2)$ of the WMR drive wheels 1 and 2 are shown in Fig. 7.7a, b, respectively. An analysis of trajectory errors shows an increase in the values of errors after time $t_{d1}$ and their decrease after time $t_{d2}$, as compared to the no-disturbance simulation.

The values of individual performance indexes for the tracking control generated by the PD controller in the parametric disturbance simulation are shown in Tables 7.4 and 7.5.

**Table 7.5** Values of selected performance indexes of path execution

| Index | $d_{max}$ [m] | $d_n$ [m] | $\rho_d$ [m] |
|-------|---------------|-----------|--------------|
| Value | 0.301 | 0.13 | 0.188 |

**Table 7.6** Values of selected performance indexes in I and II stage of motion

| Index | $e_{maxj}$ [rad] | $\varepsilon_j$ [rad] | $\dot{e}_{maxj}$ [rad/s] | $\dot{\varepsilon}_j$ [rad/s] | $s_{maxj}$ [rad/s] | $\sigma_j$ [rad/s] |
|-------|------------------|------------------------|---------------------------|-------------------------------|--------------------|--------------------|
| Stage I, $k = 1, \ldots, 2250$ | | | | | | |
| Wheel 1, $j = 1$ | 4.89 | 3.54 | 2.56 | 0.57 | 3.11 | 1.92 |
| Wheel 2, $j = 2$ | 4.72 | 3.36 | 2.56 | 0.56 | 3.11 | 1.84 |
| Stage II, $k = 2251, \ldots, 4500$ | | | | | | |
| Wheel 1, $j = 1$ | 4.25 | 3.33 | 2.89 | 0.61 | 3.51 | 1.81 |
| Wheel 2, $j = 2$ | 4.67 | 3.53 | 2.89 | 0.63 | 3.51 | 1.91 |

The values of performance indexes obtained in the disturbance trial are higher than those yielded in the no-disturbance experiment. For example, the values of the performance index $\sigma_j$ are around 11% higher when a parametric disturbance occurs.

As in simulation Sect. 7.1.2.2 the values of selected performance indexes were calculated by dividing the executed trajectory with an 8-shaped path into two phases. The values of selected performance indexes for individual motion phases are presented in Table 7.6. The values of performance indexes in the I and II motion phase differ slightly.

The values obtained for the performance indexes are used further in this work for comparison of tracking control accuracy with the application of other control algorithms.

### 7.1.2.4 Numerical Test of the RM Motion Control with the Use of PD Controller

Subsequent simulation tests involved the application of a PD controller to tracking control of the RM's TCP and were carried out in a virtual computing environment. Numerical tests were performed of which one result was selected. It was obtained on the basis of the trajectory with a curved path presented in Sect. 2.2.1.2 and included a parametric disturbance, where the change of model's parameters corresponded to the mass of the load $m_{ML}$ applied to the manipulator's gripper.

The following parameters were used in the numerical test:

- PD controller parameters:

    - $\mathbf{K}_D = \mathbf{I}$ – constant diagonal matrix of PD controller gains,
    - $\mathbf{\Lambda} = \mathbf{I}$ – constant diagonal matrix,

- $h = 0.01$ [s] – numerical test calculation step.

The applied PD controller parameters correspond to the coefficients of proportional gain $\mathbf{K}_P = \mathbf{K}_D \mathbf{\Lambda}$. The value of matrix $\mathbf{\Lambda}$ assumed in the simulation tests provides low sensitivity of the system to parametric changes of the manipulator's model and to other disturbances, while allowing for control of a real object i.e. the Scorbot ER4pc RM. Excessive values of matrix $\mathbf{\Lambda}$ elements generate control signals that cannot be applied to a real object, due to the characteristics of the actuators used. The values of matrix $\mathbf{K}_D$ were selected in such a way as to provide good performance of the manipulator's TCP tracking control. It should be noted that these values cannot be too high, as not to cause system instability.

The performance indexes used for the quality assessment of the generated control and the execution of the tracking are similar to those of the WMR. However, the investigated RM has three degrees of freedom, hence $j = 1, 2, 3$, and the tracking errors are concerned with joint variables i.e. joints' rotation angles and angular velocities.

### 7.1.2.5  Examination of PD-Control, Semicircle Path, Disturbance Trial

The desired trajectory with a semicircle path discussed in Sect. 2.2.1.2 was used in the numerical test for the tracking control of a RM with the application of a PD controller. A parametric disturbance was introduced to change the parameters of the manipulator's model. The disturbance corresponds to the situation where the gripper carries a load of $m_{RL} = 1.0$ [kg] in time $t_{d1} \in < 21; 27 >$ [s], and $t_{d2} \in < 33; 40 >$ [s].

Figure 7.8a shows the desired (green line) and the executed (red line) path of the end-effector's point $C$. The triangle indicates the initial position (point $S$), "$\mathbf{X}$" marks the desired path's reversal position (point $G$), the rhombus shows the executed final position of the manipulator's point $C$. Figure 7.8b shows the path errors for the end-effector's point $C$, $e_x = x_{Cd} - x_C$, $e_y = y_{Cd} - y_C$, $e_z = z_{Cd} - z_C$, where $x_{Cd}, y_{Cd}, z_{Cd}$ are the pre-defined point $C$ coordinates. The rounded rectangles mark the graphs' sections which illustrate the impact of parametric disturbances on the obtained values. The graph in Fig. 7.8c presents the values of signal $d_p$ that indicates the occurrence of a parametric disturbance. If $d_p = 0$, the nominal set of parameters $\mathbf{p}$ from Table 2.3 is used in the manipulator's dynamics model, if $d_p = 1$, the set of parameters $\mathbf{p}_d$ is used that corresponds to a parametric disturbance resulting from an additional mass carried by the gripper.

The values of signals controlling the drive units of manipulator's individual links $u_{PD1}, u_{PD2}$ and $u_{PD3}$ are shown in Fig. 7.9a, the absolute value of the desired ($|v_{Cd}|$) and the executed ($|v_C|$) velocity of the end-effector's point $C$ is presented in Fig. 7.9b. During the parametric disturbance $t_{d1}$ and $t_{d2}$ there are noticeable changes in the values of individual signals as compared to the values yielded for a load-free manipulator. The most apparent are changes in the values of the control signal $u_{PD3}$.

The values of errors for rotation angles ($e_1, e_2, e_3$) and errors for angular velocities ($\dot{e}_1, \dot{e}_2, \dot{e}_3$) of the manipulator's links 1, 2 and 3 are shown in Fig. 7.10a, b, respectively.

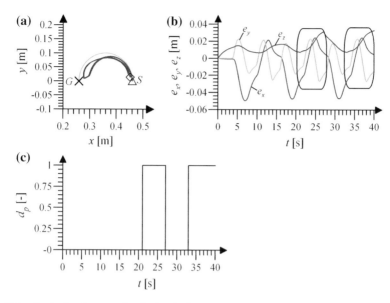

**Fig. 7.8** **a** Desired and executed path of end-effector's point $C$, **b** path errors $e_x, e_y, e_z$, **c** parametric disturbance signal $d_p$

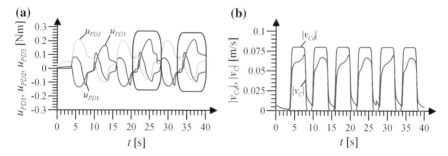

**Fig. 7.9** **a** Control signals $u_{PD1}$, $u_{PD2}$ and $u_{PD3}$, **b** absolute value of desired ($|v_{Cd}|$) and executed ($|v_C|$) velocity of end-effector's point $C$

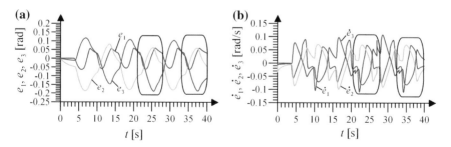

**Fig. 7.10** **a** Links' rotation angles errors $e_1$, $e_2$ and $e_3$, **b** links' angular velocities errors $\dot{e}_1$, $\dot{e}_2$ and $\dot{e}_3$

**Table 7.7**  Values of selected performance indexes

| Index | $e_{maxj}$ [rad] | $\varepsilon_j$ [rad] | $\dot{e}_{maxj}$ [rad/s] | $\dot{\varepsilon}_j$ [rad/s] | $s_{maxj}$ [rad/s] | $\sigma_j$ [rad/s] |
|---|---|---|---|---|---|---|
| Link 1, $j = 1$ | 0.064 | 0.04 | 0.11 | 0.045 | 0.122 | 0.06 |
| Link 2, $j = 2$ | 0.186 | 0.096 | 0.1 | 0.046 | 0.21 | 0.107 |
| Link 3, $j = 3$ | 0.188 | 0.094 | 0.11 | 0.048 | 0.213 | 0.106 |

**Table 7.8**  Values of selected performance indexes of path execution

| Index | $d_{max}$ [m] | $d_n$ [m] | $\rho_d$ [m] |
|---|---|---|---|
| Value | 0.0512 | 0.0362 | 0.0323 |

An analysis of trajectory errors shows an increase in the values of errors in time $t_{d1}$ and $t_{d2}$, as compared to the no-disturbance simulation.

The values of individual performance indexes for the tracking control generated by the PD controller in the simulation with a parametric disturbance are shown in Tables 7.7 and 7.8.

The values obtained for the performance indexes are used further in this work for comparison of tracking control accuracy with the application of other control algorithms.

### 7.1.3  Conclusions

The PD-based tracking control performance is low, which is reflected in high values of the assumed performance indexes. The errors of rotation angles and angular velocities take large values in the initial motion phase i.e. the acceleration. The occurrence of trajectory errors is related to the operation of the PD controller, and their values show little variation throughout the entire motion period. A significant increase in the values of trajectory errors can be observed at the time of the parametric disturbance. In the case where additional load was simulated, the increased error values occur throughout the entire motion of the WMR's and manipulator's model.

## 7.2  Adaptive Tracking Control of a WMR

In tracking control systems the purpose of the control system is to generate control signals that ensure the execution of the tracking motion construed as the motion of a robot's selected point along a trajectory with a given path. In adaptive control systems [3, 68, 81] the control signal that ensures the execution of the desired trajectory is generated by introducing the approximation of controlled object's nonlinearity. This enables the algorithm's parameters to adapt to the object's variable operating conditions. The synthesis of adaptive control algorithms requires the availability of the controlled object's mathematical model structure. The estimates of parameters are determined in real-time, while the object is in motion. The estimated parameters do not need to converge to the real ones in order for the control goal to be met.

## 7.2.1  Synthesis of an Adaptive Control Algorithm

In the synthesis of an adaptive control algorithm it is assumed that the function $f(\mathbf{x}_a)$ includes all nonlinearities of the controlled object and constitutes a combination of object's $a_i$ available parameters and signals $Y_i(\mathbf{x}_a)$, which can be written as

$$f(\mathbf{x}_a) = \sum_{i=1}^{m} Y_i(\mathbf{x}_a)\, a_i = \mathbf{Y}(\mathbf{x}_a)^T\, \mathbf{a}\,, \tag{7.12}$$

where $\mathbf{x}_a$ – vector of variables characterizing the object's motion. The synthesis of an adaptive control algorithm is aimed at the estimation of the controlled object's parameters $\hat{\mathbf{a}}$ that ensure the limitation of traking errors and values of the parameters being estimated. The following synthesis of an adaptive control algorithm is carried out for a WMR. It is assumed that the desired object's trajectory is known for which $\boldsymbol{\alpha}_d, \dot{\boldsymbol{\alpha}}_d, \ddot{\boldsymbol{\alpha}}_d$ are continuous and bounded.

Given the following WMR's dynamic equations of motion

$$\mathbf{M}\ddot{\boldsymbol{\alpha}} + \mathbf{C}(\dot{\boldsymbol{\alpha}})\,\dot{\boldsymbol{\alpha}} + \mathbf{F}(\dot{\boldsymbol{\alpha}}) + \boldsymbol{\tau}_d(t) = \mathbf{u}\,, \tag{7.13}$$

it is assumed that the values of the vector of disturbances $\boldsymbol{\tau}_d(t)$ are bounded, $\|\tau_{dj}(t)\| \le b_{dj}$, where $b_{dj} = 0, j = 1, 2$, hence there are no disturbances to the object.

When designing an adaptive tracking control algorithm for the WMR's point $A$ it is necessary to determine the control law $\mathbf{u}$ and adaptive algorithm for the vector of parameters $\hat{\mathbf{a}}$, which shall ensure that the executed motion parameters $\boldsymbol{\alpha}, \dot{\boldsymbol{\alpha}}$ converge to the desired values $\boldsymbol{\alpha}_d, \dot{\boldsymbol{\alpha}}_d$.

The desired trajectory errors are defined as

$$\begin{aligned} \mathbf{e} &= \boldsymbol{\alpha} - \boldsymbol{\alpha}_d\,, \\ \dot{\mathbf{e}} &= \dot{\boldsymbol{\alpha}} - \dot{\boldsymbol{\alpha}}_d\,, \end{aligned} \tag{7.14}$$

while the filtered tracking error $\mathbf{s}$ is given as

$$\mathbf{s} = \dot{\mathbf{e}} + \boldsymbol{\Lambda}\mathbf{e}\,, \tag{7.15}$$

where $\boldsymbol{\Lambda}$ – constant, positive-definite, diagonal design matrix.

Differentiating relation (7.15) and given (7.13), the description of a closed-loop system in the function of the filtered tracking error $\mathbf{s}$ was transformed to the following form

$$\mathbf{M}\dot{\mathbf{s}} = \mathbf{u} - \mathbf{C}\,(\dot{\boldsymbol{\alpha}})\,\mathbf{s} - \mathbf{M}\,(\ddot{\boldsymbol{\alpha}}_d + \boldsymbol{\Lambda}\dot{\mathbf{e}}) - \mathbf{C}\,(\dot{\boldsymbol{\alpha}})\,(\dot{\boldsymbol{\alpha}}_d + \boldsymbol{\Lambda}\mathbf{e}) - \mathbf{F}\,(\dot{\boldsymbol{\alpha}})\ . \tag{7.16}$$

Defining auxiliary signals

$$\begin{aligned}
\mathbf{v} &= \dot{\boldsymbol{\alpha}}_d - \boldsymbol{\Lambda}\mathbf{e}\ , \\
\dot{\mathbf{v}} &= \ddot{\boldsymbol{\alpha}}_d - \boldsymbol{\Lambda}\dot{\mathbf{e}}\ ,
\end{aligned} \tag{7.17}$$

and introducing them to (7.16), the following description of a closed-loop control system was derived

$$\mathbf{M}\dot{\mathbf{s}} = \mathbf{u} - \mathbf{C}\,(\dot{\boldsymbol{\alpha}})\,\mathbf{s} - \mathbf{f}\,(\mathbf{x}_v)\ , \tag{7.18}$$

where

$$\mathbf{f}\,(\mathbf{x}_v) = \mathbf{M}\dot{\mathbf{v}} + \mathbf{C}\,(\dot{\boldsymbol{\alpha}})\,\mathbf{v} + \mathbf{F}\,(\dot{\boldsymbol{\alpha}})\ , \tag{7.19}$$

and $\mathbf{x}_v = \left[\mathbf{v}^T,\ \dot{\mathbf{v}}^T,\ \dot{\boldsymbol{\alpha}}^T\right]^T$.

The adopted control signal includes compensation for the object's nonlinearity

$$\mathbf{u} = \mathbf{f}\,(\mathbf{x}_v) - \mathbf{u}_{PD}\ , \tag{7.20}$$

where

$$\mathbf{u}_{PD} = \mathbf{K}_D\mathbf{s}\ , \tag{7.21}$$

is the PD controller equation, where $\mathbf{K}_D$ – constant, positive-definite design matrix of controller gains. The expression $\mathbf{K}_D\mathbf{s}$ is equivalent to the description of a conventional PD controller with the derivative and proportional terms, whose gains are respectively $\mathbf{K}_D$ and $\mathbf{K}_P = \mathbf{K}_D\boldsymbol{\Lambda}$.

If the model describing the WMR's dynamics were known precisely, the control (7.20) would ensure the execution of the tracking. However, the parameters of the controlled object's model are usually not known or only their approximate values are available. If that is the case the function $\mathbf{f}\,(\mathbf{x}_v)$ shall be estimated.

Given the properties of the WMR's mathematical model, the nonlinear function $\mathbf{f}\,(\mathbf{x}_v)$ can be written in a linear form with respect to parameters

$$\mathbf{f}\,(\mathbf{x}_v) = \mathbf{M}\dot{\mathbf{v}} + \mathbf{C}\,(\dot{\boldsymbol{\alpha}})\,\mathbf{v} + \mathbf{F}\,(\dot{\boldsymbol{\alpha}}) = \mathbf{Y}_v\,(\mathbf{v}, \dot{\mathbf{v}}, \dot{\boldsymbol{\alpha}})\,\mathbf{a}\ , \tag{7.22}$$

where $\mathbf{Y}_v\,(\mathbf{v}, \dot{\mathbf{v}}, \dot{\boldsymbol{\alpha}})$ – matrix of signals defined previously, constructed in line with relation (7.19). By analogy with (7.22), the estimate of the nonlinear function $\mathbf{f}\,(\mathbf{x}_v)$ is given by relation

$$\hat{\mathbf{f}}\,(\mathbf{x}_v) = \hat{\mathbf{M}}\dot{\mathbf{v}} + \hat{\mathbf{C}}\,(\dot{\boldsymbol{\alpha}})\,\mathbf{v} + \hat{\mathbf{F}}\,(\dot{\boldsymbol{\alpha}}) = \mathbf{Y}_v\,(\mathbf{v}, \dot{\mathbf{v}}, \dot{\boldsymbol{\alpha}})\,\hat{\mathbf{a}}\ , \tag{7.23}$$

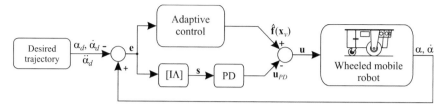

**Fig. 7.11** Structure of the WMR's adaptive tracking control system

where $\hat{\mathbf{f}}(\mathbf{x}_v)$ – function estimating the function $\mathbf{f}(\mathbf{x}_v)$. Matrices $\hat{\mathbf{M}}$, $\hat{\mathbf{C}}(\dot{\boldsymbol{\alpha}})$ and the vector $\hat{\mathbf{F}}(\dot{\boldsymbol{\alpha}})$ have the same structure as $\mathbf{M}$, $\mathbf{C}(\dot{\boldsymbol{\alpha}})$ and $\mathbf{F}(\dot{\boldsymbol{\alpha}})$, but include the estimated parameters $\hat{\mathbf{a}}$.

The modified control law (7.20) has the form

$$\mathbf{u} = \hat{\mathbf{f}}(\mathbf{x}_v) - \mathbf{u}_{PD} = \mathbf{Y}_v(\mathbf{v}, \dot{\mathbf{v}}, \dot{\boldsymbol{\alpha}})\,\hat{\mathbf{a}} - \mathbf{K}_D\mathbf{s}\,. \tag{7.24}$$

Substituting the control signal (7.24) to the description of a closed-loop control system (7.18) we obtain

$$\mathbf{M}\dot{\mathbf{s}} + \mathbf{C}(\dot{\boldsymbol{\alpha}})\,\mathbf{s} = \mathbf{Y}_v(\mathbf{v}, \dot{\mathbf{v}}, \dot{\boldsymbol{\alpha}})\,\tilde{\mathbf{a}} - \mathbf{K}_D\mathbf{s}\,, \tag{7.25}$$

where $\tilde{\mathbf{a}}$ – WMR's model parameter estimation error, $\tilde{\mathbf{a}} = \hat{\mathbf{a}} - \mathbf{a}$.

A schematic diagram of an adaptive tracking control system for the WMR's point $A$ is shown in Fig. 7.11.

The parameter adaptation law $\hat{\mathbf{a}}$ for the control algorithm was defined by applying the Lyapunov stability theory. The adopted positive-definite candidate Lyapunov function is expressed by relation

$$L = \frac{1}{2}\left[\mathbf{s}^T\mathbf{M}\mathbf{s} + \tilde{\mathbf{a}}^T\boldsymbol{\Gamma}^{-1}\tilde{\mathbf{a}}\right]\,, \tag{7.26}$$

where $\boldsymbol{\Gamma}$ – diagonal design matrix with positive elements.

Derivative of function $L$ is given by

$$\dot{L} = \mathbf{s}^T\mathbf{M}\dot{\mathbf{s}} + \frac{1}{2}\mathbf{s}^T\dot{\mathbf{M}}\mathbf{s} + \tilde{\mathbf{a}}^T\boldsymbol{\Gamma}^{-1}\dot{\tilde{\mathbf{a}}}\,. \tag{7.27}$$

Expression $\mathbf{M}\dot{\mathbf{s}}$ was derived from (7.25) and introduced to (7.27), thus

$$\dot{L} = -\mathbf{s}^T\mathbf{K}_D\mathbf{s} + \frac{1}{2}\mathbf{s}^T\left[\dot{\mathbf{M}} - 2\mathbf{C}(\dot{\boldsymbol{\alpha}})\right]\mathbf{s} + \mathbf{s}^T\mathbf{Y}_v(\mathbf{v}, \dot{\mathbf{v}}, \dot{\boldsymbol{\alpha}})\,\tilde{\mathbf{a}} + \tilde{\mathbf{a}}^T\boldsymbol{\Gamma}^{-1}\dot{\tilde{\mathbf{a}}}\,. \tag{7.28}$$

Matrix $\mathbf{D}(\dot{\boldsymbol{\alpha}}) = \dot{\mathbf{M}} - 2\mathbf{C}(\dot{\boldsymbol{\alpha}})$ is skew-symmetric based on the assumptions for the structural properties of the WMR's mathematical model described in Sect. 2.1.2. Given the above property, the expression (7.28) takes the form

$$\dot{L} = -\mathbf{s}^T \mathbf{K}_D \mathbf{s} + \tilde{\mathbf{a}}^T \left[ \mathbf{Y}_v^T (\mathbf{v}, \dot{\mathbf{v}}, \dot{\boldsymbol{\alpha}}) \mathbf{s} + \boldsymbol{\Gamma}^{-1} \dot{\tilde{\mathbf{a}}} \right] . \tag{7.29}$$

Selecting the parameter adaptation law **a** in the form of relation

$$\dot{\hat{\mathbf{a}}} = -\boldsymbol{\Gamma} \mathbf{Y}_v^T (\mathbf{v}, \dot{\mathbf{v}}, \dot{\boldsymbol{\alpha}}) \mathbf{s} , \tag{7.30}$$

we finally get

$$\dot{L} = -\mathbf{s}^T \mathbf{K}_D \mathbf{s} \le 0 . \tag{7.31}$$

Function $L$ is positive-definite in the space of variables $\left[ \mathbf{s}^T, \tilde{\mathbf{a}}^T \right]^T$, whereas $\dot{L}$ is negative-semi-definite, hence the function $L$ is a Lyapunov function. In line with the Lyapunov stability theory the signals **s** and $\tilde{\mathbf{a}}$ are bounded. Based on the Barbalat's lemma [81] it can be demonstrated that the filtered tracking error **s** converges to zero, which consequently makes the vector of tracking errors **e** and $\dot{\mathbf{e}}$ converge to zero. However, it cannot whether the model's parameter estimation error $\tilde{\mathbf{a}}$ converges to zero, which is not deemed necessary for the proper execution of the tracking motion.

### 7.2.2  Simulation Tests

An adaptive algorithm was used for the simulation tests of the WMR tracking control. The studies were carried out in a virtual computing environment. Numerical tests were performed of which one result was selected. It was obtained on the basis of the trajectory with an 8-shaped path presented in Sect. 2.1.1.2. A parametric disturbance was simulated, where the change of model's parameters was related to the WMR being weighed down with additional mass.

The following adaptive control system parameters were used in the numerical test:

- PD controller parameters:

    - $\mathbf{K}_D = \mathbf{I}$ – constant diagonal matrix of PD controller gains,
    - $\boldsymbol{\Lambda} = 0.5\,\mathbf{I}$ – constant diagonal matrix,

- $\boldsymbol{\Gamma} = \text{diag}\,\{0.00375, 0.0375, 0.0075, 0.0015, 0.75, 0.75\}$ – diagonal matrix of gain coefficients for the adaptation of parameters $\hat{\mathbf{a}}$,
- $\hat{\mathbf{a}}\,(t_0) = \mathbf{0}$ – zero initial values of parameter estimates $\hat{a}_1, \ldots, \hat{a}_6$, for $t_0 = 0$ [s],
- $h = 0.01$ [s] – numerical test calculation step.

The performance indexes from point Sect. 7.1.2 were adopted to assess the quality of the generated control and the execution of the tracking motion. The selected results, presented below, were yielded from numerical tests for an adaptive algorithm with the application of a trajectory with an 8-shaped path and a modeled parametric disturbance.

### 7.2.2.1 Examination of Adaptive Control, 8-Shaped Path, Disturbance Trial

Simulation tests of an adaptive algorithm were conducted on the basis of the trajectory with an 8-shaped path discussed in Sect. 2.1.1.2. A parametric disturbance was introduced twice. The first one involved the change of the WMR model's parameters corresponding to the robot's frame being weighed down with a load of $m_{RL} = 4.0$ in time $t_{d1} = 12.5$ [s]. In the second disturbance scenario the additional load was removed in time $t_{d2} = 33$ [s]. The denotations are identical to those used in the previous numerical tests.

Figure 7.12a shows the values of the overall control signals of the adaptive control system $u_1$ and $u_2$, obtained in the numerical test on the execution of the WMR's point $A$ tracking motion with the desired trajectory with an 8-shaped path. The overall control signals consist of the WMR's nonlinearity compensation signals ($\hat{f}_1$ and $\hat{f}_2$), and the control signals generated by the PD controller ($u_{PD1}$ and $u_{PD2}$), that are presented in Fig. 7.12b, c, respectively. The inverse values of the PD control signals ($-u_{PD1}$ and $-u_{PD2}$) are presented for ease of comparison to the PD control signals of other algorithms, and more convenient assessment of their impact on the overall control signals of the adaptive control system. The values of the WMR's parameter estimates are shown in Fig. 7.12d. Zero initial values of the parameter estimates were assumed. The highest values are observed for the parameters of the adaptive system $\hat{a}_5$ and $\hat{a}_6$. They are the estimates of parameters that model the WMR's resistance to

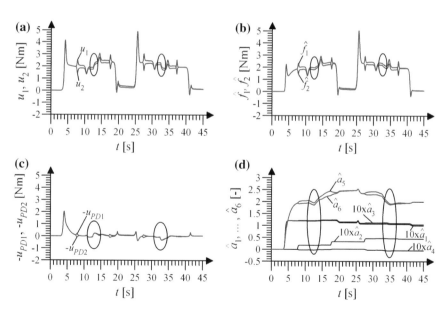

**Fig. 7.12** **a** Drive wheels 1 and 2 control signals $u_1$ and $u_2$, **b** WMR's nonlinearity compensation signals $\hat{f}_1$ and $\hat{f}_2$, **c** control signals $-u_{PD1}$ and $-u_{PD2}$, **d** estimates of parameters $\hat{a}_1, \ldots, \hat{a}_6$

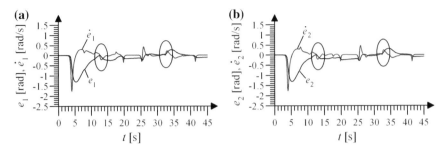

**Fig. 7.13  a** Trajectory errors $e_1$ and $\dot{e}_1$, **b** trajectory errors $e_2$ and $\dot{e}_2$

motion. At the time of parametric disturbances the values of these parameters change as to compensate for the changes in the controlled object's dynamics.

At the beginning of motion phase I i.e. the acceleration, it is the PD control signals that play a major role in the overall control signals. This is due to the zero initial values of the adaptive algorithm parameters. As the parameter adaptation process advances the values of the control signals generated by the adaptive algorithm increase, while the values of the PD control signals are minimized and close to zero. At the time of disturbance $t_{d1}$ the values of control signals increase due to the compensation for changes in the WMR's dynamics. When the second disturbance occurs, $t_{d2}$, a decrease in the values of control signals is observed, which is due to removal of additional load carried by the WMR.

The error values of the rotation angles ($e_1$, $e_2$) and angular velocities ($\dot{e}_1$, $\dot{e}_2$) of the WMR drive wheels 1 and 2 are shown in Fig. 7.13a, b. The largest values of tracking errors are observed at the outset of phase I, during the acceleration period, where the adaptive system's parameters have zero initial values. In that motion period the overall control signals are generated by the PD controller and the adaptation of the WMR's nonlinearities does not occur, and hence the highest values of tracking errors. The values of trajectory errors are being reduced during the parameter adaptation process when the system generates control signals that compensate for the WMR's nonlinearities. An analysis of tracking errors shows an increase in error values after time $t_{d1}$ and $t_{d2}$ due to the parametric disturbance. The impact the parametric disturbances have on the execution of the tracking motion is compensated by the adaptive system which generates control signals that include the change in the WMR's dynamics.

The left graph in Fig. 7.14 shows the desired (dashed line) an the executed motion path of the WMR's point $A$. The graph on the right shows an enlarged view of the section of the graph on the left. The denotations are identical to those used in the previous numerical tests.

The values of individual performance indexes for the tracking motion with the application of an adaptive control system are shown in Tables 7.9 and 7.10.

**Fig. 7.14** Desired and executed motion path of the WMR's point $A$

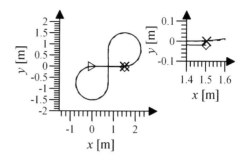

**Table 7.9** Values of selected performance indexes

| Index | $e_{maxj}$ [rad] | $\varepsilon_j$ [rad] | $\dot{e}_{maxj}$ [rad/s] | $\dot{\varepsilon}_j$ [rad/s] | $s_{maxj}$ [rad/s] | $\sigma_j$ [rad/s] |
|---|---|---|---|---|---|---|
| Wheel 1, $j = 1$ | 1.3 | 0.35 | 1.76 | 0.21 | 2.04 | 0.28 |
| Wheel 2, $j = 2$ | 1.3 | 0.35 | 1.76 | 0.21 | 2.04 | 0.28 |

**Table 7.10** Values of selected performance indexes of path execution

| Index | $d_{max}$ [m] | $d_n$ [m] | $\rho_d$ [m] |
|---|---|---|---|
| Value | 0.108 | 0.02 | 0.032 |

**Table 7.11** Values of selected performance indexes in I and II stage of motion

| Index | $e_{maxj}$ [rad] | $\varepsilon_j$ [rad] | $\dot{e}_{maxj}$ [rad/s] | $\dot{\varepsilon}_j$ [rad/s] | $s_{maxj}$ [rad/s] | $\sigma_j$ [rad/s] |
|---|---|---|---|---|---|---|
| Stage I, $k = 1, \ldots, 2250$ | | | | | | |
| Wheel 1, $j = 1$ | 1.3 | 0.49 | 1.76 | 0.29 | 2.04 | 0.37 |
| Wheel 2, $j = 2$ | 1.3 | 0.48 | 1.76 | 0.29 | 2.04 | 0.37 |
| Stage II, $k = 2251, \ldots, 4500$ | | | | | | |
| Wheel 1, $j = 1$ | 0.29 | 0.11 | 0.42 | 0.1 | 0.47 | 0.11 |
| Wheel 2, $j = 2$ | 0.33 | 0.12 | 0.42 | 0.1 | 0.47 | 0.12 |

All performance indexes demonstrate lower values than those obtained in the execution of trajectory with the application of the control system based only on the PD controller.

Table 7.11 shows the values of selected performance indexes in the I and II phase of motion along an 8-shaped path. They were calculated in accordance with the methodology outlined in the numerical test Sect. 7.1.2.3.

The values of the performance indexes in phase II are lower than those obtained in phase I. This is due to the fact that the information on the controlled object was accumulated in the parameters of the adaptive system and further applied at the outset of phase II to generate control signals. The controlled object's nonlinearity compensation algorithm generates control signals that ensure proper performance of the tracking motion in phase II, while the values of the PD control signals are close to zero.

### 7.2.3 Conclusions

The application of compensation for nonlinearities of the controlled object by means of an adaptive system results in a decrease in the values of the trajectory performance indexes compared with the results obtained in the simulation with a PD-based control system. The initial values of the adaptive system parameters have a significant impact on the performance of motion. In the presented example the worst-case scenario was selected with zero initial values of the parameters $\hat{a}_1, \ldots, \hat{a}_6$. That is why the largest tracking error values occur at the beginning of phase I of the WMR's motion. In the case where parametric disturbances occur the adaptive algorithm compensates for changes in the controlled object's dynamics caused by an additional load applied to the robot's frame.

## 7.3 Neural Tracking Control of a WMR

In neural tracking control systems for mechatronic systems the control signal that ensures the execution of the desired trajectory is generated by introducing the approximation of controlled object's nonlinearity into the adaptive algorithm, with the application of NNs. Due to their properties such as the ability to map any nonlinear function, parameter learning or adaptation capabilities, NNs are commonly used for the modeling and control of complex nonlinear systems [29, 40, 50, 68, 73, 83].

### 7.3.1 Synthesis of a Neural Control Algorithm

The synthesis of a neural control algorithm was carried out on a WMR. In the synthesis of a neural tracking control algorithm the following dynamic equations of the WMR's motion were assumed

$$\mathbf{M}\ddot{\boldsymbol{\alpha}} + \mathbf{C}(\dot{\boldsymbol{\alpha}})\dot{\boldsymbol{\alpha}} + \mathbf{F}(\dot{\boldsymbol{\alpha}}) + \boldsymbol{\tau}_d(t) = \mathbf{u} , \qquad (7.32)$$

where $\boldsymbol{\tau}_d(t)$ – vector of bounded disturbances, $\|\tau_{dj}(t)\| \le b_{dj}$, $b_{dj}$ – constant, positive value, $j = 1, 2$.

The definition of the tracking motion from Sect. 7.2.1 was assumed. The trajectory errors $\mathbf{e}$, $\dot{\mathbf{e}}$ and the filtered tracking error $\mathbf{s}$ were defined as

$$\begin{aligned}
\mathbf{e} &= \boldsymbol{\alpha}_d - \boldsymbol{\alpha} , \\
\dot{\mathbf{e}} &= \dot{\boldsymbol{\alpha}}_d - \dot{\boldsymbol{\alpha}} , \\
\mathbf{s} &= \dot{\mathbf{e}} + \boldsymbol{\Lambda}\mathbf{e} ,
\end{aligned} \qquad (7.33)$$

where $\boldsymbol{\Lambda}$ – constant, positive-definite, diagonal design matrix.

The description of a closed-loop system in the function of the filtered tracking error $\mathbf{s}$ was transformed to the following form

$$\mathbf{M\dot{s}} = -\mathbf{C}(\dot{\alpha})\,\mathbf{s} + \mathbf{M}(\ddot{\alpha}_d + \Lambda\dot{\mathbf{e}}) + \mathbf{C}(\dot{\alpha})(\dot{\alpha}_d + \Lambda\mathbf{e}) + \mathbf{F}(\dot{\alpha}) + \boldsymbol{\tau}_d(t) - \mathbf{u}\,.$$
$$(7.34)$$

Defining auxiliary signals

$$\mathbf{v} = \dot{\alpha}_d + \Lambda\mathbf{e}\,,$$
$$\dot{\mathbf{v}} = \ddot{\alpha}_d + \Lambda\dot{\mathbf{e}}\,,$$
$$(7.35)$$

and introducing them to (7.34), the following description of a closed control system was derived

$$\mathbf{M\dot{s}} = -\mathbf{C}(\dot{\alpha})\,\mathbf{s} + \mathbf{f}(\mathbf{x}_v) + \boldsymbol{\tau}_d(t) - \mathbf{u}\,,$$
$$(7.36)$$

where

$$\mathbf{f}(\mathbf{x}_v) = \mathbf{M}\dot{\mathbf{v}} + \mathbf{C}(\dot{\alpha})\,\mathbf{v} + \mathbf{F}(\dot{\alpha})\,,$$
$$(7.37)$$

and $\mathbf{x}_v = \left[\mathbf{v}^T,\ \dot{\mathbf{v}}^T,\ \dot{\alpha}^T\right]^T$.

Given the properties of the WMR's mathematical model, the nonlinear function $\mathbf{f}(\mathbf{x}_v)$ can be written in a linear form with respect to parameters

$$\mathbf{f}(\mathbf{x}_v) = \mathbf{M}\dot{\mathbf{v}} + \mathbf{C}(\dot{\alpha})\,\mathbf{v} + \mathbf{F}(\dot{\alpha}) = \mathbf{Y}_v(\mathbf{v}, \dot{\mathbf{v}}, \dot{\alpha})\,\mathbf{a}\,,$$
$$(7.38)$$

where $\mathbf{Y}_v(\mathbf{v}, \dot{\mathbf{v}}, \dot{\alpha})$ – matrix of signals defined previously, constructed in line with relation (7.37).

When designing a tracking control system providing for the WMR's nonlinearity compensation it is vital to determine an algorithm for parameter estimation $\hat{\mathbf{a}}$ that shall ensure the limitation of the estimate values for the estimated parameters and tracking errors. In the neural control algorithm, a single-layer RVFL NN was used for the approximation of the nonlinear function $\mathbf{f}(\mathbf{x}_v)$ (7.37), the object's parameters were denoted by $\mathbf{W}$, and their estimates, represented by NN weights, were labeled as $\hat{\mathbf{W}}$. The applied RVFL NN was described in detail in Sect. 3.2.1. It is a linear NN, due to the output layer weights, with constant input layer weights randomly selected in the NN initialization process and bipolar sigmoid neuron activation functions.

If the NN's neuron activation function vector $\mathbf{S}(\mathbf{x}_N)$ is selected as a set of basic functions, then the NN with functional extensions and ideal weights $\mathbf{W}$ shall approximate any continuous function defined on a compact set with the desired accuracy [61, 90].

On these assumptions the nonlinear function $\mathbf{f}(\mathbf{x}_v)$, featured in relation (7.38), is written as

$$\mathbf{f}(\mathbf{x}_v) = \mathbf{W}^T\mathbf{S}(\mathbf{x}_N) + \boldsymbol{\tau}_N\,,$$
$$(7.39)$$

where $\mathbf{x}_N$ – vector of NN input signals, scaled into the interval $x_{Ni} \in \langle -1, 1 \rangle$ according to the relation $\mathbf{x}_N = \kappa_N\mathbf{x}_v$, $\kappa_N$ – positive-definite, diagonal design matrix of scaling coefficients for the NN inputs, $i = 1, \ldots, M$, $M$ – size of the NN inputs vector. The components of the vector of errors for nonlinear function approximation with a NN $\boldsymbol{\tau}_N$ take on values that are limited to the interval $|\tau_{Ni}| \leq d_{Nj}$, where $d_{Nj}$ is a constant, positive value, $j = 1, 2$.

Assuming the ideal NN weights $\mathbf{W}$ are bounded, the estimate of function $\mathbf{f}(\mathbf{x}_\nu)$ is

$$\hat{\mathbf{f}}(\mathbf{x}_\nu) = \hat{\mathbf{W}}^T \mathbf{S}(\mathbf{x}_N) \ , \tag{7.40}$$

where $\hat{\mathbf{W}}$ – estimate of NN ideal weights, determined in the adaptation process under the adopted adaptation law.

The assumed control signal includes compensation for object's nonlinearities

$$\mathbf{u} = \mathbf{u}_N + \mathbf{u}_{PD} \ , \tag{7.41}$$

where

$$\mathbf{u}_N = \hat{\mathbf{W}}^T \mathbf{S}(\mathbf{x}_N) \ , \tag{7.42}$$

is a control signal generated by the NN, compensating for nonlinearities of the controlled object, while

$$\mathbf{u}_{PD} = \mathbf{K}_D \mathbf{s} \ , \tag{7.43}$$

is the PD controller equation, where $\mathbf{K}_D$ – constant, positive-definite design matrix of controller gains. Expression $\mathbf{K}_D \mathbf{s}$ is equivalent to the description of a conventional PD controller with derivative (D) and proportional (P) terms whose respective gains are $\mathbf{K}_D$ and $\mathbf{K}_P = \mathbf{K}_D \boldsymbol{\Lambda}$.

The approximation error of the nonlinear function $\mathbf{f}$ is due to a difference between the NN ideal weights $\mathbf{W}$ and their estimates $\hat{\mathbf{W}}$,

$$\tilde{\mathbf{W}} = \mathbf{W} - \hat{\mathbf{W}} \ , \tag{7.44}$$

which can be written as

$$\tilde{\mathbf{f}}(\mathbf{x}_\nu) = \mathbf{W}^T \mathbf{S}(\mathbf{x}_N) - \hat{\mathbf{W}}^T \mathbf{S}(\mathbf{x}_N) + \boldsymbol{\tau}_N \ , \tag{7.45}$$

where $\boldsymbol{\tau}_N$ – NN bounded estimation error of a nonlinear function.

Introducing the assumed control law (7.41) into relation (7.36), and given (7.45), the closed-loop control system is described as follows

$$\mathbf{M}\dot{\mathbf{s}} + \mathbf{C}(\dot{\boldsymbol{\alpha}})\mathbf{s} = -\mathbf{K}_D \mathbf{s} + \boldsymbol{\tau}_d(t) + \tilde{\mathbf{W}}^T \mathbf{S}(\mathbf{x}_N) + \boldsymbol{\tau}_N \ . \tag{7.46}$$

Figure 7.15 shows a schematic diagram of the WMR's neural tracking control system with the application of the RVFL NN to compensate for the WMR's nonlinearities.

NN weights adaptation law $\hat{\mathbf{W}}$ for the neural control algorithm was defined by means of the Lyapunov stability theory [30, 36, 68]. The adopted positive-definite candidate Lyapunov function is expressed by relation

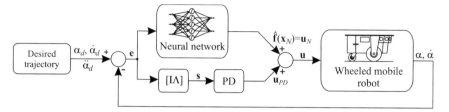

**Fig. 7.15** Structure of a neural tracking control system of the WMR's point $A$ with the application of the RVFL NN

$$L = \frac{1}{2} \left[ \mathbf{s}^T \mathbf{M} \mathbf{s} + \mathrm{tr} \left( \tilde{\mathbf{W}}^T \boldsymbol{\Gamma}_N^{-1} \tilde{\mathbf{W}} \right) \right] , \tag{7.47}$$

where $\boldsymbol{\Gamma}_N$ – diagonal design matrix with positive elements. Derivative of function $L$ is given by

$$\dot{L} = \mathbf{s}^T \mathbf{M} \dot{\mathbf{s}} + \frac{1}{2} \mathbf{s}^T \dot{\mathbf{M}} \mathbf{s} + \mathrm{tr} \left( \tilde{\mathbf{W}}^T \boldsymbol{\Gamma}_N^{-1} \dot{\tilde{\mathbf{W}}} \right) . \tag{7.48}$$

Expression $\mathbf{M}\dot{\mathbf{s}}$ was derived from the relation describing a closed-loop system (7.46) and introduced to (7.48), thus obtaining

$$\begin{aligned}
\dot{L} = &-\mathbf{s}^T \mathbf{K}_D \mathbf{s} + \tfrac{1}{2} \mathbf{s}^T \left[ \dot{\mathbf{M}} - 2\mathbf{C} \left( \dot{\boldsymbol{\alpha}} \right) \right] \mathbf{s} \\
&+ \mathrm{tr} \left[ \tilde{\mathbf{W}}^T \left( \boldsymbol{\Gamma}_N^{-1} \dot{\tilde{\mathbf{W}}} + \mathbf{S} \left( \mathbf{x}_N \right) \mathbf{s}^T \right) \right] + \mathbf{s}^T \left( \boldsymbol{\tau}_N + \boldsymbol{\tau}_d \left( t \right) \right) .
\end{aligned} \tag{7.49}$$

Matrix $\mathbf{D} \left( \dot{\boldsymbol{\alpha}} \right) = \dot{\mathbf{M}} - 2\mathbf{C} \left( \dot{\boldsymbol{\alpha}} \right)$ is skew-symmetric based on the assumptions for the structural properties of the WMR's mathematical model described in Sect. 2.1.2.

Selecting the NN weights adaptation law $\hat{\mathbf{W}}$ in the form of relation

$$\dot{\hat{\mathbf{W}}} = -\boldsymbol{\Gamma}_N \mathbf{S} \left( \mathbf{x}_N \right) \mathbf{s}^T , \tag{7.50}$$

we finally get

$$\dot{L} = -\mathbf{s}^T \mathbf{K}_D \mathbf{s} + \mathbf{s}^T \left( \boldsymbol{\tau}_N + \boldsymbol{\tau}_d \left( t \right) \right) \leq -K_{Dmin} \|\mathbf{s}\|^2 + \left( \|\mathbf{b}_d\| + \|\mathbf{b}_N\| \right) \|\mathbf{s}\| , \tag{7.51}$$

where $K_{Dmin}$ – the lowest eigenvalue of matrix KD.

Since the limit values for individual components of error vectors $\boldsymbol{\tau}_N$ and $\boldsymbol{\tau}_d \left( t \right)$ are constant, then (7.47) is a Lyapunov function whose derivative $\dot{L} \leq 0$, provided the following condition is satisfied

$$\Psi = \{ \mathbf{s} : \|\mathbf{s}\| > \left( \|\mathbf{b}_d\| + \|\mathbf{b}_N\| \right) / K_{Dmin} \equiv b_S \} . \tag{7.52}$$

From relation (7.52) it follows that the filtered tracking error $\mathbf{s}$ is uniformly finally bounded to the set $\Psi$, with the practical bound denoted by $b_S$. Furthermore, it results from that relation that the tracking errors are bounded and that increasing the PD controller gain coefficients by changing values of the design matrix $\mathbf{K}_D$ elements, leads to the reduction of the filtered tracking error $\mathbf{s}$ and consequently the reduction of the tracking errors $\mathbf{e}$ and $\dot{\mathbf{e}}$. Thus formulated problem related to the synthesis of the neural tracking control of the WMR's point $A$ generates the control law that ensures proper performance of the closed-loop control system until the neural compensator starts to generate a compensation control signal, by application a conventional PD controller. This allows to circumvent the NN weights pre-learning process to ensure the NN operation as a stabilizer in the initial NN weights adaptation process, hence the designed neural algorithm can be used for a real-time generation of a control signal.

In the neural control system a conventional PD controller was used, which has a simple design and provides control system stability in the absence of a compensation signal for the controlled object's nonlinearities. The drawback of the PD controller is the occurrence of an offset that can be corrected by adding an integral term to the PD controller. Thus designed PID controller indeed does reduce the offset but has a more complex structure, and the time the system needs to reach the state of full compensation is long. The PID-based control is characterized by system sensitivity to variable operating conditions and insufficient accuracy related to the control of nonlinear objects, such as a WMR.

## 7.3.2   Simulation Tests

The simulation tests on the application of a neural control system to tracking control of a WMR were carried out in a virtual computing environment. Numerical tests were performed of which one result was selected. It was obtained on the basis of the trajectory with an 8-shaped path presented in Sect. 2.1.1.2. The test scenario included a parametric disturbance, where the change of model's parameters was related to the WMR being weighed down with additional load.

The following neural control system parameters were used in the numerical test:

- PD controller parameters:

    - $\mathbf{K}_D = \mathbf{I}$ – constant diagonal matrix of PD controller gains,
    - $\mathbf{\Lambda} = 0.5\,\mathbf{I}$ – constant diagonal matrix,

- type of NNs applied: RVFL,

    - $m = 8$ – number of NN neurons,
    - $\mathbf{\Gamma}_N = \mathrm{diag}\,\{0.4\}$ – diagonal matrix of gain coefficients for the adaptation of neurons weights $\hat{\mathbf{W}}$,
    - $\mathbf{D}_N$ – NN input layer weights randomly selected from the interval $D_{Ni,i} \in \langle -0.1, 0.1 \rangle$,
    - $\hat{\mathbf{W}}_j\,(t_0) = \mathbf{0}$ – zero initial values of the output layer weights, $j = 1, 2, t_0 = 0$ [s],

– $\beta_N = 2$ – slope coefficient for bipolar sigmoid neuron activation functions,

• $h = 0.01$ [s] – numerical test calculation step.

The performance indexes from point Sect. 7.1.2 were adopted to assess the quality of the tracking motion execution.

The selected results, presented below, were obtained from neural control simulation test with the application of a trajectory with an 8-shaped path and a modeled parametric disturbance.

### 7.3.2.1  Examination of Neural Control, 8-Shaped Path, Disturbance Trial

A numerical test of a neural control algorithm was conducted on the basis of the trajectory with an 8-shaped path discussed in Sect. 2.1.1.2. A parametric disturbance was introduced twice. The first one involved the change of the WMR model's parameters corresponding to the robot's frame being weighed down with a load of $m_{RL} = 4.0$ [kg] in time $t_{d1} = 12.5$ [s]. In the second disturbance scenario the additional load was removed in time $t_{d2} = 33$ [s]. The denotations are identical to those used in the previous numerical tests.

Figure 7.16a shows the values of the overall control signals of the neural control system. They consist of the compensation control signals generated by the NNs

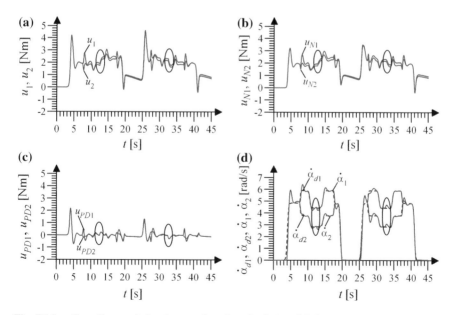

**Fig. 7.16  a** Overall control signals $u_1$ and $u_2$ for wheels 1 and 2, **b** neural control signals $u_{N1}$ and $u_{N2}$, **c** PD control signals $u_{PD1}$ and $u_{PD2}$, **d** desired ($\dot{\alpha}_{d1}$, $\dot{\alpha}_{d2}$) and executed ($\dot{\alpha}_1$, $\dot{\alpha}_2$) angular velocities of the WMR's drive wheels

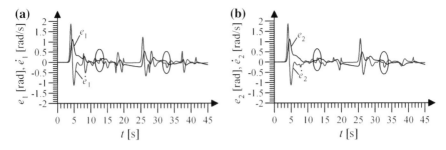

**Fig. 7.17** **a** Tracking errors $e_1$ and $\dot{e}_1$, **b** tracking errors $e_2$ and $\dot{e}_2$

(Fig. 7.16b) and the PD control signals (rys. Fig. 7.16c). The values of the desired $(\dot{\alpha}_{d1}, \dot{\alpha}_{d2})$ and executed $(\dot{\alpha}_1, \dot{\alpha}_2)$ angular velocities of the WMR drive wheels are presented in Fig. 7.16d.

At the beginning of the execution of the desired trajectory in phase I it is the PD control signals that play a major role in the overall control signals. This is due to the zero initial values of the NNs weights in the WMR nonlinearity compensating system. The NNs weights adaptation process leads to an increase in the values of neural control signals that become dominant in the overall control signals. At the time of disturbance $t_{d1}$ the values of control signals are observed to increase, due to the compensation for changes in the WMR's dynamics. When the second disturbance occurs, $t_{d2}$, a decrease in the values of control signals is observed, which is due to the removal of the additional load carried by the WMR. The removal of the load carried by the WMR causes a momentary increase in the values of tracking errors which is presented in Fig. 7.17. The largest values of tracking errors are observed at the outset of phase I, where there is no compensation for the WMR's nonlinearities and when the overall control signals have similar values to those produced by the PD controller. As a result of the NNs weights adaptation process the information on the WMR's dynamics is retained, which increases the importance of NNs generated control signals in the overall control signals and leads to a decrease in the values of tracking errors in subsequent phases of motion.

The values of the NNs output layers weights are shown in Fig. 7.18. The zero initial values of weights were selected in the NNs initialization process. The NNs weights remain bounded and stabilize around certain values. At the times of parametric disturbances the values of NNs weights change, as a consequence of the adaptive process. This is to compensate for the changes in the controlled object's dynamics.

Figure 7.19 shows the desired (dashed line) and the executed motion path of the WMR's point $A$. The denotations are identical to those used in the previous simulations. The graph on the right shows an enlarged view of the section of the graph on the left.

The values of individual performance indexes for the tracking motion with the application of a neural control system are shown in Tables 7.12 and 7.13.

The performance indexes demonstrate lower values than those obtained in the execution of trajectory with the application of the PD controller. The values of

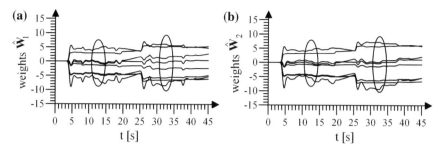

**Fig. 7.18** **a** NN 1 output layer weights $\hat{\mathbf{W}}_1$, **b** NN 2 output layer weights $\hat{\mathbf{W}}_2$

**Fig. 7.19** Desired and executed path of the WMR's point $A$

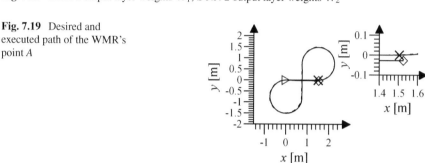

**Table 7.12** Values of selected performance indexes

| Index | $e_{maxj}$ [rad] | $\varepsilon_j$ [rad] | $\dot{e}_{maxj}$ [rad/s] | $\dot{\varepsilon}_j$ [rad/s] | $s_{maxj}$ [rad/s] | $\sigma_j$ [rad/s] |
|---|---|---|---|---|---|---|
| Wheel 1, $j = 1$ | 1.13 | 0.21 | 1.87 | 0.3 | 2.19 | 0.32 |
| Wheel 2, $j = 2$ | 1.13 | 0.2 | 1.87 | 0.27 | 2.19 | 0.29 |

**Table 7.13** Values of selected performance indexes of path execution

| Index | $d_{max}$ [m] | $d_n$ [m] | $\rho_d$ [m] |
|---|---|---|---|
| Value | 0.093 | 0.035 | 0.027 |

performance indexes are comparable to the ones produced in the numerical tests of the adaptive control algorithm.

Table 7.14 presents the values of selected performance indexes in the I and II phase of motion along an 8-shaped path, calculated in accordance with the methodology outlined in point Sect. 7.1.2.3.

The values of the performance indexes in phase II are lower than those obtained in phase I. This is due to the non-zero values of weights that were used at the beginning of phase II. The knowledge of the controlled object's dynamics, stored in the NNs weights, allows to generate a compensation signal for the WMR's nonlinearities, which leads to improved performance of the generated tracking control.

**Table 7.14**  Values of selected performance indexes in I and II stage of motion

| Index | $e_{maxj}$ [rad] | $\varepsilon_j$ [rad] | $\dot{e}_{maxj}$ [rad/s] | $\dot{\varepsilon}_j$ [rad/s] | $s_{maxj}$ [rad/s] | $\sigma_j$ [rad/s] |
|---|---|---|---|---|---|---|
| Stage I, $k = 1, \ldots, 2250$ | | | | | | |
| Wheel 1, $j = 1$ | 1.13 | 0.25 | 1.87 | 0.35 | 2.19 | 0.37 |
| Wheel 2, $j = 2$ | 1.13 | 0.25 | 1.87 | 0.33 | 2.19 | 0.35 |
| Stage II, $k = 2251, \ldots, 4500$ | | | | | | |
| Wheel 1, $j = 1$ | 0.4 | 0.11 | 0.65 | 0.19 | 0.76 | 0.2 |
| Wheel 2, $j = 2$ | 0.32 | 0.12 | 0.5 | 0.13 | 0.6 | 0.15 |

### 7.3.3  Conclusions

The application of compensation for the WMR's nonlinearities to a tracking control algorithm by means of linear NNs with respect to the output layer weights improves the performance of the generated control signals and the execution of the tracking motion, which is demonstrated by lower values of performance indexes. The selection of constant values for RVFL NNs input layer weights carried out in the initialization process has a significant effect on the quality of compensation for the WMR's nonlinearities. The adopted initial values of the output layers weights have a decisive impact on the performance of the tracking motion in its early phase. The presented numerical test explains the influence of weights stored information on the nonlinearity compensation process. The use of zero initial values for the NNs output layers weights applies to the least favorable case, yet it allows for the evaluation of a control algorithm, expressed by the values of performance indexes, and minimizes the influence the random weight selection has on the control generation process. The number of NNs neurons as well as the value of reinforcement learning coefficients have a significant impact on the NNs weights adaptation process as they are the algorithm's design parameters.

## 7.4  Heuristic Dynamic Programming in Tracking Control of a WMR

Adaptive critic design in HDP configuration [26, 42, 44, 77, 79] is the basic algorithm of the ADP family of algorithms. It is composed of actor and critic structures as well as a predictive model of the controlled object. There is a necessity for the availability of the controlled object's mathematical model to generate the weights adaptation law for the actor's NNs. In the HDP algorithm the value function is approximated by the critic. Since the function is a scalar quantity, the critic design is limited to one adaptive structure e.g. an NN. The actor generates the object's control law, hence its design results from the number of independent signals that control the actuators. For example, in a WMR's tracking control system with two degrees of freedom the actor consists of two adaptive structures e.g. NNs generating independent signals that control the robot's drive units. This chapter provides the synthesis of a discrete tracking control system for a dynamic object with the application of an NDP algorithm in HDP configuration. The object considered below is a WMR. The presented algorithm operates in a discrete-time domain, thus a discrete model of the WMR's dynamics was used for the synthesis of a control system. This chapter is based on the information relating to the HDP-type ACD whose description is included in Sect. 6.2.1, and on the description of the WMR's dynamics presented in Sect. 2.1.2.

## 7.4.1 Synthesis of HDP-Type Control

The tracking motion problem is defined as a search for the discrete control law $\mathbf{u}_{\{k\}}$ that minimizes the tracking errors $\mathbf{e}_{1\{k\}}$ and $\mathbf{e}_{2\{k\}}$ which are determined on the basis of (2.39), given by

$$
\begin{aligned}
\mathbf{e}_{1\{k\}} &= \mathbf{z}_{1\{k\}} - \mathbf{z}_{d1\{k\}} \, , \\
\mathbf{e}_{2\{k\}} &= \mathbf{z}_{2\{k\}} - \mathbf{z}_{d2\{k\}} \, ,
\end{aligned}
\tag{7.53}
$$

with the defined trajectory $\mathbf{z}_{d\{k\}} = \left[ \mathbf{z}_{d1\{k\}}^T, \mathbf{z}_{d2\{k\}}^T \right]^T$ $(\mathbf{z}_{\{k\}} \to \mathbf{z}_{d\{k\}}, k \to \infty)$, where the control algorithm remains stable.

The filtered tracking error $\mathbf{s}_{\{k\}}$ is defined in line with (7.53) as

$$
\mathbf{s}_{\{k\}} = \mathbf{e}_{2\{k\}} + \mathbf{\Lambda}\mathbf{e}_{1\{k\}} \, ,
\tag{7.54}
$$

where $\mathbf{\Lambda}$ – constant, positive-definite, diagonal design matrix.

Based on the discrete description of the WMR's dynamics (2.39) and relations (7.53) and (7.54), the filtered tracking error $\mathbf{s}_{\{k+1\}}$ is defined as follows

$$
\mathbf{s}_{\{k+1\}} = -\mathbf{Y}_f \left( \mathbf{z}_{2\{k\}} \right) + \mathbf{Y}_d \left( \mathbf{z}_{\{k\}}, \mathbf{z}_{d\{k+1\}} \right) - \mathbf{Y}_{\tau\{k\}} + h\mathbf{M}^{-1}\mathbf{u}_{\{k\}} \, ,
\tag{7.55}
$$

where

$$
\begin{aligned}
&\mathbf{Y}_f \left( \mathbf{z}_{2\{k\}} \right) = h\mathbf{M}^{-1} \left[ \mathbf{C} \left( \mathbf{z}_{2\{k\}} \right) \mathbf{z}_{2\{k\}} + \mathbf{F} \left( \mathbf{z}_{2\{k\}} \right) \right] \, , \quad \mathbf{Y}_{\tau\{k\}} = h\mathbf{M}^{-1} \boldsymbol{\tau}_{d\{k\}} \, , \\
&\mathbf{Y}_d \left( \mathbf{z}_{\{k\}}, \mathbf{z}_{d\{k+1\}} \right) = \mathbf{z}_{2\{k\}} - \mathbf{z}_{d2\{k+1\}} + \mathbf{\Lambda} \left[ \mathbf{z}_{1\{k\}} + h\mathbf{z}_{2\{k\}} - \mathbf{z}_{d1\{k+1\}} \right] = \\
&\mathbf{z}_{2\{k\}} - \mathbf{z}_{d2\{k\}} - h\mathbf{z}_{d3\{k\}} + \mathbf{\Lambda} \left[ \mathbf{z}_{1\{k\}} - \mathbf{z}_{d1\{k\}} + h\mathbf{z}_{2\{k\}} - h\mathbf{z}_{d2\{k\}} \right] = \\
&\mathbf{s}_{\{k\}} + \mathbf{Y}_e \left( \mathbf{z}_{\{k\}}, \mathbf{z}_{d\{k\}}, \mathbf{z}_{d3\{k\}} \right) \, , \\
&\mathbf{Y}_e \left( \mathbf{z}_{\{k\}}, \mathbf{z}_{d\{k\}}, \mathbf{z}_{d3\{k\}} \right) = h\mathbf{\Lambda}\mathbf{e}_{2\{k\}} - h\mathbf{z}_{d3\{k\}} \, ,
\end{aligned}
\tag{7.56}
$$

and $\mathbf{z}_{d3\{k\}}$ is the vector of desired angular accelerations obtained by writing out $\mathbf{z}_{d2\{k+1\}}$ applying the Euler's forward method. The vector $\mathbf{Y}_f \left( \mathbf{z}_{2\{k\}} \right)$ includes all the WMR's nonlinearities.

In the designed tracking control system for the WMR's point $A$ an NDP algorithm in HDP configuration was used, where the actor's and critic's structures were implemented in the form of NNs. The HDP algorithm requires the availability of the controlled object's model in order for the actor's NN weights adaptation to be performed.

The execution of the WMR's point $A$ tracking motion requires that the control signals are generated for two independent drive units, due to the robot's design. The control system design features two NDP algorithms, each of which consists of the actor and critic structure.

In the presented control algorithm the HDP structure shall generate the suboptimal control law $\mathbf{u}_{A\{k\}} = \left[u_{A1\{k\}}, u_{A2\{k\}}\right]^T$ to minimize the value functions $\mathbf{V}_{\{k\}}\left(\mathbf{s}_{\{k\}}\right) = \left[V_{1\{k\}}\left(\mathbf{s}_{\{k\}}\right), V_{2\{k\}}\left(\mathbf{s}_{\{k\}}\right)\right]^T$ given by relations

$$V_{j\{k\}}\left(\mathbf{s}_{\{k\}}\right) = \sum_{k=0}^{n} \gamma^k L_{Cj\{k\}}\left(\mathbf{s}_{\{k\}}\right) \ , \tag{7.57}$$

with the cost functions

$$L_{Cj\{k\}}\left(\mathbf{s}_{\{k\}}\right) = \frac{1}{2}R_j s_{j\{k\}}^2 \ , \tag{7.58}$$

where $j = 1, 2, R_1, R_2$ – positive design coefficients. The purpose of the NDP structure in HDP configuration, given the value function (7.57) and with the cost function (7.58), is to generate a control law that minimizes the values of filtered tracking errors $\mathbf{s}_{\{k\}}$.

A general schematic diagram of the ADP structure in HDP configuration is shown in Fig. 7.20, where $\mathbf{u}_{PD\{k\}}$, $\mathbf{u}_{S\{k\}}$ and $\mathbf{u}_{E\{k\}}$ are additional control signals, whose application stems from the analysis of system stability.

The HDP algorithm, described in Sect. 6.2.1, consists of a predictive model and two adaptive structures:

- **Predictive model** for the dynamics of the WMR's filtered tracking error generates the prediction of state $\mathbf{s}_{\{k+1\}}$ for step $k + 1$ of the discrete process, based on (7.55), according to relation

$$\mathbf{s}_{\{k+1\}} = -\mathbf{Y}_f\left(\mathbf{z}_{2\{k\}}\right) + \mathbf{Y}_d\left(\mathbf{z}_{\{k\}}, \mathbf{z}_{d\{k+1\}}\right) + h\mathbf{M}^{-1}\mathbf{u}_{\{k\}} \ . \tag{7.59}$$

- **Actor**, implemented in the form of two RVFL NNs, generates the suboptimal control law $\mathbf{u}_{A\{k\}}$ in the $k$th step according to relation

$$u_{Aj\{k\}}\left(\mathbf{x}_{Aj\{k\}}, \mathbf{W}_{Aj\{k\}}\right) = \mathbf{W}_{Aj\{k\}}^T \mathbf{S}\left(\mathbf{D}_A^T \mathbf{x}_{Aj\{k\}}\right) \ , \tag{7.60}$$

where $j = 1, 2$, $\mathbf{W}_{Aj\{k\}}$ – $j$th actor's RVFL NN output layer weights vector, $\mathbf{x}_{Aj\{k\}} = \kappa_A \left[1, s_{j\{k\}}, \mathbf{z}_{2\{k\}}^T, z_{d2j\{k\}}, z_{d3j\{k\}}\right]^T$ – $j$th actor's NN inputs vector, $\kappa_A$ – constant diagonal matrix of positive scaling coefficients for the actor's NN inputs values, $\mathbf{S}\left(\mathbf{D}_A^T \mathbf{x}_{Aj\{k\}}\right)$ – neuron activation function vector, $\mathbf{D}_A$ – matrix of constant weights in the actor's NN input layer, randomly selected in the initialization process. Bipolar sigmoid neuron activation functions were used.

The actor's NN weights adaptation procedure for the HDP algorithm of the designed control system is to minimize the performance index

$$
e_{Aj\{k\}} = \frac{\partial L_{Cj\{k\}}\left(\mathbf{s}_{\{k\}}\right)}{\partial u_{j\{k\}}} + \gamma \left[\frac{\partial s_{j\{k+1\}}}{\partial u_{j\{k\}}}\right]^T \frac{\partial \hat{V}_{j\{k+1\}}\left(\mathbf{x}_{Cj\{k+1\}}, \mathbf{W}_{Cj\{k\}}\right)}{\partial s_{j\{k+1\}}}, \qquad (7.61)
$$

where $\mathbf{x}_{Cj\{k+1\}} = \kappa_C\left[s_{j\{k+1\}}\right]$ – critic's NN inputs vector for step $k + 1$, calculated on the basis of the predictive model.

The NN output layer weights are updated according to the following relation

$$
\mathbf{W}_{Aj\{k+1\}} = \mathbf{W}_{Aj\{k\}} - e_{Aj\{k\}}\mathbf{\Gamma}_{Aj}\mathbf{S}\left(\mathbf{D}_A^T\mathbf{x}_{Aj\{k\}}\right), \qquad (7.62)
$$

where $\mathbf{\Gamma}_{Aj}$ – constant diagonal matrix of learning coefficients for the actor's $j$th NN neurons.

- **Critic** estimates the value function (7.57) and is implemented in the form of two NNs with Gaussian-type neuron activation functions. It generates the following signals

$$
\hat{V}_{j\{k\}}\left(\mathbf{x}_{Cj\{k\}}, \mathbf{W}_{Cj\{k\}}\right) = \mathbf{W}_{Cj\{k\}}^T\mathbf{S}\left(\mathbf{x}_{Cj\{k\}}\right), \qquad (7.63)
$$

where $j = 1, 2$, $\mathbf{W}_{Cj\{k\}}$ – output layer weights vector for the critic's $j$th NN, $\mathbf{x}_{Cj\{k\}} = \kappa_C\left[s_{j\{k\}}\right]$ – critic's NN inputs vector, $\kappa_C$ – constant positive scaling coefficient for the critic's NN inputs, $\mathbf{S}\left(\mathbf{x}_{Cj\{k\}}\right)$ – vector of Gaussian-type neuron activation functions. In the HDP algorithm the critic's NN output layer weights are adapted through backpropagation of the temporal difference error defined as

**Fig. 7.20** Schematic diagram of NDP structure in HDP configuration

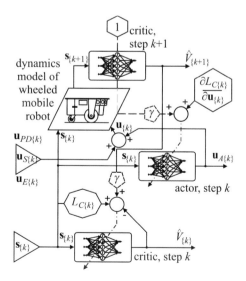

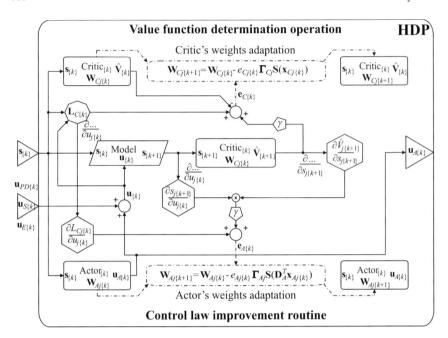

**Fig. 7.21** Schematic diagram of NNs weights adaptation process for NDP structures in HDP configuration

$$e_{Cj\{k\}} = L_{Cj\{k\}}\left(\mathbf{s}_k\right) + \gamma \, \hat{V}_{j\{k+1\}}\left(\mathbf{x}_{Cj\{k+1\}}, \mathbf{W}_{Cj\{k\}}\right) - \hat{V}_{j\{k\}}\left(\mathbf{x}_{Cj\{k\}}, \mathbf{W}_{Cj\{k\}}\right) , \quad (7.64)$$

where $\mathbf{s}_{\{k+1\}}$ – signal determined from the predictive model (7.59).

The critic's NN output layer weights are adapted through backpropagation of the temporal difference error (7.64) according to the following relation

$$\mathbf{W}_{Cj\{k+1\}} = \mathbf{W}_{Cj\{k\}} - e_{Cj\{k\}} \mathbf{\Gamma}_{Cj} \mathbf{S}\left(\mathbf{x}_{Cj\{k\}}\right) , \quad (7.65)$$

where $\mathbf{\Gamma}_{Cj}$ – constant diagonal matrix of positive learning coefficients for the critic's $j$th NN.

A schematic diagram of the NN weights adaptation procedure for the NDP structure in HDP configuration is shown in Fig. 7.21, where $\mathbf{u}_{PD\{k\}}$, $\mathbf{u}_{S\{k\}}$ and $\mathbf{u}_{E\{k\}}$ are additional control signals, whose application stems from the analysis of system stability.

The stability analysis of the proposed discrete tracking control system was carried out in line with the Lyapunov stability theory. In the designed control system the overall control signals $\mathbf{u}_{\{k\}}$ consist of the actor's NN generated control signals $\mathbf{u}_{A\{k\}}$, the PD control signals $\mathbf{u}_{PD\{k\}}$, additional control signals $\mathbf{u}_{E\{k\}}$ whose structure results from the discretization of the WMR's mathematical model, and supervisory term's control signals $\mathbf{u}_{S\{k\}}$, whose structure was derived from the Lyapunov stability

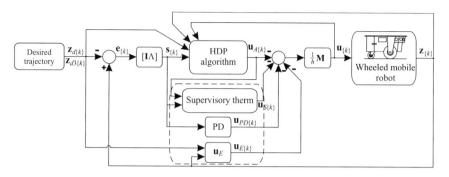

**Fig. 7.22** Schematic diagram of a WMR's tracking control system with the application of an HDP algorithm

theorem. Figure 7.22 shows a schematic diagram of a tracking control system for the WMR's point $A$ with the application of HDP algorithms.

The supervisory term's control signals $\mathbf{u}_{S\{k\}}$ provide stability of the closed-loop system which means that the filtered tracking error $\mathbf{s}_{\{k\}}$ is bounded. The overall control signals are given in the form

$$\mathbf{u}_{\{k\}} = -\frac{1}{h}\mathbf{M}\left[\mathbf{u}_{A\{k\}} - \mathbf{u}_{S\{k\}} + \mathbf{u}_{PD\{k\}} + \mathbf{u}_{E\{k\}}\right] , \qquad (7.66)$$

where

$$\begin{aligned} \mathbf{u}_{PD\{k\}} &= \mathbf{K}_D \mathbf{s}_{\{k\}} , \\ \mathbf{u}_{E\{k\}} &= h\left[\mathbf{\Lambda}\mathbf{e}_{2\{k\}} - \mathbf{z}_{d3\{k\}}\right] , \\ \mathbf{u}_{S\{k\}} &= \mathbf{I}_S \mathbf{u}_{S\{k\}}^* , \end{aligned} \qquad (7.67)$$

where $\mathbf{K}_D$ – constant, positive-definite design matrix, $\mathbf{I}_S$ – diagonal switching matrix, $I_{Sj,j} = 1$ when $\left|s_{j\{k\}}\right| \geq \rho_j$, $I_{Sj,j} = 0$ when $\left|s_{j\{k\}}\right| < \rho_j$, $\rho_j$ – constant design value, $\mathbf{u}_{S\{k\}}^*$ – supervisory term's control signal derived further in this subchapter, $j = 1, 2$. Given $I_{Sj,j} = 1$, by substituting (7.66) into (7.55), we get

$$\mathbf{s}_{\{k+1\}} = \mathbf{Y}_d\left(\mathbf{z}_{\{k\}}, \mathbf{z}_{d\{k+1\}}\right) - \mathbf{Y}_f\left(\mathbf{z}_{2\{k\}}\right) - \mathbf{Y}_{\tau\{k\}} - \mathbf{u}_{A\{k\}} - \mathbf{K}_D\mathbf{s}_{\{k\}} + \mathbf{u}_{S\{k\}}^* - \mathbf{u}_{E\{k\}} . \qquad (7.68)$$

Assuming a positive-definite Lyapunov candidate function

$$L_{\{k\}} = \mathbf{s}_{\{k\}}^T \mathbf{s}_{\{k\}} , \qquad (7.69)$$

its difference

$$\Delta L_{\{k\}} = \mathbf{s}_{\{k+1\}}^T \mathbf{s}_{\{k+1\}} - \mathbf{s}_{\{k\}}^T \mathbf{s}_{\{k\}} < 0 , \qquad (7.70)$$

can be written as

$$\left\|\mathbf{s}_{\{k+1\}}\right\|^2 < \left\|\mathbf{s}_{\{k\}}\right\|^2 . \qquad (7.71)$$

Based on (7.71) [63] is written as

$$\|\mathbf{s}_{\{k\}}\| \, \|\mathbf{s}_{\{k+1\}}\| \, < \, \|\mathbf{s}_{\{k\}}\|^2 \, , \tag{7.72}$$

from the Cauchy-Schwarz inequality it follows

$$|\mathbf{s}_{\{k\}}^T \mathbf{s}_{\{k+1\}}| \leq \|\mathbf{s}_{\{k\}}\| \, \|\mathbf{s}_{\{k+1\}}\| \, , \tag{7.73}$$

hence the inequality (7.72) is equivalent to

$$|\mathbf{s}_{\{k\}}^T \mathbf{s}_{\{k+1\}}| \, < \, \|\mathbf{s}_{\{k\}}\|^2 \, . \tag{7.74}$$

Relation (7.74) is equivalent to

$$\mathbf{s}_{\{k\}}^T \left[ \mathbf{s}_{\{k+1\}} - \mathbf{s}_{\{k\}} \right] < 0 \, , \tag{7.75}$$

thus

$$\Delta L_{\{k\}} = \mathbf{s}_{\{k\}}^T \left[ \mathbf{s}_{\{k+1\}} - \mathbf{s}_{\{k\}} \right] < 0 \, . \tag{7.76}$$

Substituting (7.68) into (7.76) we get

$$\Delta L_{\{k\}} = \mathbf{s}_{\{k\}}^T \left[ -\mathbf{K}_D \, \mathbf{s}_{\{k\}} - \mathbf{Y}_f \left( \mathbf{z}_{2\{k\}} \right) - \mathbf{Y}_{\tau\{k\}} - \mathbf{u}_{A\{k\}} + \mathbf{u}_{S\{k\}}^* \right] \, . \tag{7.77}$$

If assumed that all the elements of the vector of disturbances are bounded $\left| Y_{\tau j\{k\}} \right| < b_{dj}$, $b_{dj} > 0$, we may write the following relation

$$\Delta L_{\{k\}} \leq -\mathbf{s}_{\{k\}}^T \mathbf{K}_D \, \mathbf{s}_{\{k\}} + \sum_{j=1}^{2} |s_{j\{k\}}| \left[ \left| Y_{f\{k\}} \left( \mathbf{z}_{2\{k\}} \right) \right| + b_{dj} + \left| u_{Aj\{k\}} \right| \right] + \sum_{j=1}^{2} s_{j\{k\}} u_{Sj\{k\}}^* \, . \tag{7.78}$$

Given the supervisory term's control signals $\mathbf{u}_{S\{k\}}^*$ as

$$u_{Sj\{k\}}^* = -\mathrm{sgn} \, s_{j\{k\}} \left[ F_j + \left| u_{Aj\{k\}} \right| + b_{dj} + \sigma_j \right] \, , \tag{7.79}$$

where $\left| Y_{fj} \left( \mathbf{z}_{2\{k\}} \right) \right| \leq F_j$, $F_j > 0$ and $\sigma_j$ is a small positive constant; from the inequality (7.78) it follows

$$\Delta L_{\{k\}} \leq 0 \, . \tag{7.80}$$

The difference of the Lyapunov function is negative-definite. The designed control signals provide the reduction of the filtered tracking errors $\mathbf{s}_{\{k\}}$ where $\left| s_{j\{k\}} \right| \geq \rho_j$. For the initial condition $\left| s_{j\{k=0\}} \right| < \rho_j$, the filtered tracking errors remain bounded to $\left| s_{j\{k\}} \right| < \rho_j$, $\forall k \geq 0$, $j = 1, 2$.

The synthesis and results of studies on the WMR's tracking control with the application of HDP algorithms are presented in [42, 44].

## 7.4.2 Simulation Tests

The simulation tests were carried out in a virtual computing environment for the WMR's tracking control with the application of a discrete control algorithm with ADP structure in HDP configuration. The selected numerical test results are based on the trajectory with an 8-shaped path presented in Sect. 2.1.1.2. A parametric disturbance was simulated, where the change of model's parameters was related to the WMR being weighed down with additional mass.

The following parameters were used for an HDP-based control system:

- PD controller parameters:

  - $\mathbf{K}_D = 3.6h\mathbf{I}$ – constant diagonal matrix of PD controller gains,
  - $\mathbf{\Lambda} = 0.5\,\mathbf{I}$ – constant diagonal matrix,

- type of actor's NNs applied: RVFL,

  - $m_A = 8$ – number of the actor's NN neurons,
  - $\mathbf{\Gamma}_{Aj} = \mathrm{diag}\,\{0.1\}$ – diagonal matrix of gain coefficients for the adaptation of weights $\mathbf{W}_{Aj}, j = 1, 2$,
  - $\mathbf{D}_A$ – NN input layer weights randomly selected from the interval $D_{A[i,g]} \in \langle -0.1, 0.1 \rangle$,
  - $\mathbf{W}_{Aj\{k=0\}} = \mathbf{0}$ – zero initial values of the output layer weights,

- type of critic's NNs applied: NN with Gaussian-type neuron activation functions,

  - $m_C = 8$ – number of the critic's NN neurons,
  - $\mathbf{\Gamma}_{Cj} = \mathrm{diag}\,\{0.01\}$ – diagonal matrix of gain coefficients for the adaptation of weights $\mathbf{W}_{Cj}$,
  - $\mathbf{c}_C = [-1.0, -0.6, -0.3, -0.1, 0.1, 0.3, 0.6, 1.0]^T$ – positions of the centers of Gaussian curves in the input space,
  - $\mathbf{r}_C = [0.3, 0.175, 0.1, 0.085, 0.085, 0.1, 0.175, 0.3]^T$ – widths of the Gaussian curves,
  - $\mathbf{W}_{Cj\{k=0\}} = \mathbf{0}$ – zero initial values of the output layer weights in test Sect. 7.4.2.1,
  - $\mathbf{W}_{Cj\{k=0\}} = \mathbf{1}$ – non-zero initial values of the output layer weights in test Sect. 7.4.2.2,

- $\beta_N = 2$ – slope coefficient of bipolar sigmoid neuron activation functions,
- $R_j = 1$ – scaling coefficient in the cost function,
- $l_{AC} = 3$ – number of inner loop iteration steps for the actor's and critic's structure adaptation in the $k$th step of the iterative process,
- $\rho_j = 2.5$ – supervisory algorithm parameter,
- $\sigma_j = 0.01$ – supervisory algorithm parameter,
- $h = 0.01$ [s] – numerical test calculation step.

In the studies on the designed control systems with NN based algorithms the components of the inputs vectors were scaled to the interval $\langle -1, 1 \rangle$.

The selection of PD controller parameters for a discrete control system with an NDP structure was based on the results of numerical tests and verification studies. The parameters were picked in such a way as to correspond with those of the continuous controller from Sect. 7.1.2 and provide the exact same tracking control performance.

The following performance indexes were used for the quality assessment of the generated control and the execution of the tracking motion: – maximum absolute value of angular displacement error for the WMR's drive wheels 1 and 2, $e_{max1j}$ [rad],

$$e_{max1j} = \max_k \left( |e_{1j\{k\}}| \right) ,$$  (7.81)

where $j = 1, 2$,
– root mean square error for angular displacement of the $j$th wheel, $e_{1j}$ [rad],

$$\varepsilon_{1j} = \sqrt{\frac{1}{n} \sum_{k=1}^{n} e_{1j\{k\}}^2} ,$$  (7.82)

where $k$ – number of iteration step, $n$ – total number of steps,
– maximum absolute value of angular velocity error for the $j$th wheel $e_{max2j}$ [rad/s],

$$e_{max2j} = \max_k \left( |e_{2j\{k\}}| \right) ,$$  (7.83)

– root mean square error for angular velocity of the $j$th wheel, $e_{2j}$ [rad/s],

$$\varepsilon_{2j} = \sqrt{\frac{1}{n} \sum_{k=1}^{n} e_{2j\{k\}}^2} .$$  (7.84)

The other performance indexes were adopted as in point Sect. 7.1.2. The volume of discrete measurement data for the WMR's trajectory with an 8-shaped path is $n = 4500$.

In description of the numerical tests of discrete-time algorithms the notation of variables was simplified by removing the process $k$th step index. To facilitate analysis of results and comparison of continuous functions for the previously presented WMR's control systems, the axis of abscissae of the graphs representing discrete functions was scaled in time $t$.

The selected results, presented below, were yielded from numerical tests for the tracking motion execution with the application of a trajectory with an 8-shaped path and a modeled parametric disturbance.

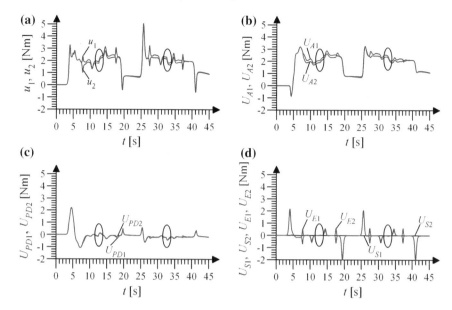

**Fig. 7.23** **a** Overall control signals $u_1$ and $u_2$ for wheels 1 and 2, **b** actor's NNs control signals $U_{A1}$ and $U_{A2}$, $\mathbf{U}_A = -\frac{1}{h}\mathbf{Mu}_A$, **c** PD control signals $U_{PD1}$ and $U_{PD2}$, $\mathbf{U}_{PD} = -\frac{1}{h}\mathbf{Mu}_{PD}$, **d** supervisory term's control signals $U_{S1}$ and $U_{S2}$, $\mathbf{U}_S = -\frac{1}{h}\mathbf{Mu}_S$, and additional control signals $u_{E1}$ and $u_{E2}$, $\mathbf{U}_E = -\frac{1}{h}\mathbf{Mu}_E$.

### 7.4.2.1 Examination of HDP-Type Control, 8-Shaped Path, Disturbance Trial with Zero Initial Values of NNs Weights

Zero initial values were assumed for the actor's and critic's NN output layer weights in a numerical test of a tracking control algorithm for the WMR's point $A$ with the application of ADP structures in HDP configuration. The goal of the system was to generate control signals that would ensure the execution of the desired WMR's point $A$ trajectory with an 8-shaped path under two parametric disturbances. The first disturbance involved the change of the WMR model's parameters related to the robot's frame being weighed down with a load of $m_{RL} = 4.0$ [kg] in time $t_{d1} = 12.5$ [s]. In the second disturbance scenario the additional load was removed in time $t_{d2} = 32.5$ [s].

The values of the tracking control signals for individual structures of the control system are shown in Fig. 7.23. The overall control signals $u_1$ and $u_2$ of wheels 1 and 2, respectively, are shown in Fig. 7.23a, according to relation (7.66) they consist of the control signals generated by the HDP structures $u_{A1}$ and $u_{A2}$ (rys. Fig. 7.23b), PD control signals $u_{PD1}$ and $u_{PD2}$ (rys. Fig. 7.23c), supervisory term's control signals $u_{S1}$ and $u_{S2}$, and additional control signals $u_{E1}$ and $u_{E2}$ (rys. Fig. 7.23d).

At the beginning of motion phase I, the acceleration, it is the PD control signals $\mathbf{u}_{PD}$ and signals $\mathbf{u}_E$ that play a major role in the overall control signals. The low

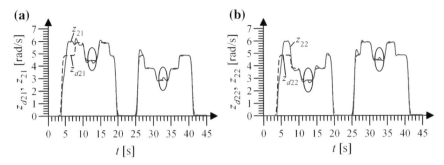

**Fig. 7.24** **a** Desired ($z_{d21}$) and executed ($z_{21}$) angular velocity of the WMR's wheel 1, **b** desired ($z_{d22}$) and executed ($z_{22}$) angular velocity of the WMR's wheel 2

values of the control signals generated by the HDP structures result from the zero initial values of the actor's NN output layer weights whose adaptation begins at the time when the different-from-zero filtered tracking errors occur. In phase II certain information on the dynamics of the controlled object is retained in the NN output layer weights of the ADP structures, while the actor's NN control signals play a dominant role in the overall tracking control signals for the WMR's point $A$. The occurrence of the first parametric disturbance in time $t_{d1}$ results in an increase in the values of control signals generated by the adaptive critic structures that compensate for the change in the dynamics of the controlled object. Similarly, the occurrence of the second disturbance in time $t_{d2}$ causes a decrease in the values of the HDP structures' control signals. The values of control signals $\mathbf{u}_E$ depend mainly on $\mathbf{z}_{d3}$, changes in the values of tracking errors have a negligible effect on those signals. The values of supervisory term's control signals $\mathbf{u}_S$ remain at zero, as the values of filtered tracking errors $\mathbf{s}$ do not exceed the established allowable threshold.

The values of the desired and executed angular velocities of the WMR's drive wheels 1 and 2 are presented in Fig. 7.24a, b respectively. The difference in the values of the desired and executed angular velocities of the WMR's drive wheels is the most evident at the beginning of phase I. This is due to the insufficient compensation for the WMR's nonlinearities performed by the neural critic system in the initial phase of adaptation, caused by the zero initial values of the NN output layer weights. At time $t_{d1}$ when the disturbance simulates the WMR being weighed down with additional load, the values of angular velocities decrease, since the generated control signals are not sufficient for the execution of the WMR motion with the desired velocity due to an increased resistance to motion. The increase in the value of the nonlinearity compensation signals generated by the HDP structures leads to an increase of the angular velocities and further execution of the trajectory according to the established parameters of motion. The occurrence of the second parametric disturbance in time $t_{d2}$ simulates the removal of additional load applied to the WMR which means return of the model parameters to the nominal values. At this point the generated control signals exceed the required values, which is due to the lower resistance to motion of the load-free object. Thus the value of the WMR's point $A$ actual velocity is higher

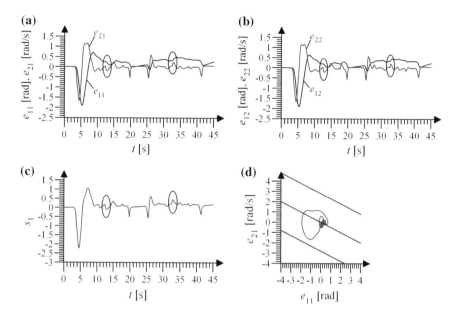

**Fig. 7.25** **a** Tracking errors $e_{11}$ and $e_{21}$, **b** tracking errors $e_{12}$ and $e_{22}$, **c** filtered tracking error $s_1$, **d** sliding manifold of wheel 1

than the desired one. As a result of weights adaptation in the HDP structures' NNs the values of control signals $u_{A1}$ and $u_{A2}$ are reduced, the velocity decreases and reaches the desired value.

The tracking error values of the WMR's wheel 1 and 2 are shown in Fig. 7.25a, b. Figure 7.25c presents the values of the filtered tracking error $s_1$, Fig. 7.25d sliding manifold of wheel 1. The largest values of tracking errors are observed in the initial phase of the motion. They are minimized during the motion as a result of weights adaptation in the HDP structures' NNs. At the time of parametric disturbances the absolute values of tracking errors increase, but due to the NNs weights adaptation process the NDP structures' control signals compensate for the change in the controlled object's dynamics. At the beginning of phase II, in the rightward motion along a curved path, the tracking error values are lower than those recorded at the beginning of phase I, as the information on the controlled object's dynamics is retained in the actor's NNs weights.

The values of the actor's and critic's NNs output layer weights for the first HDP structure are shown in Fig. 7.26a, b. The values of the NNs output layer weights for the second HDP structure, which generates the control signal for wheel 2, are similar. Zero initial values of the NN weights were assumed. The weights stabilize around certain values and remain bounded. At the times of parametric disturbances the values of the actor's NN weights change, as the generated compensation signal for the WMR's nonlinearities adjusts to the modified parameters of the controlled object.

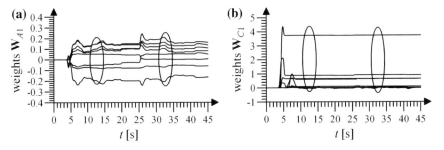

**Fig. 7.26  a** Actor's 1 NN output layer weights $\mathbf{W}_{A1}$, **b** critic's 1 NN output layer weights $\mathbf{W}_{C1}$

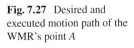

**Fig. 7.27**  Desired and executed motion path of the WMR's point $A$

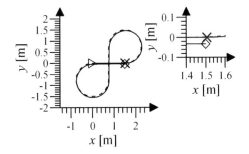

**Table 7.15**  Values of selected performance indexes

| Index | $e_{max1j}$ [rad] | $\varepsilon_{1j}$ [rad] | $e_{max2j}$ [rad/s] | $\varepsilon_{2j}$ [rad/s] | $s_{maxj}$ [rad/s] | $\sigma_j$ [rad/s] |
|---|---|---|---|---|---|---|
| Wheel 1, $j = 1$ | 1.9 | 0.46 | 1.66 | 0.35 | 2.22 | 0.42 |
| Wheel 2, $j = 2$ | 1.9 | 0.44 | 1.66 | 0.35 | 2.22 | 0.41 |

**Table 7.16**  Values of selected performance indexes of path execution

| Index | $d_{max}$ [m] | $d_n$ [m] | $\rho_d$ [m] |
|---|---|---|---|
| Value | 0.157 | 0.033 | 0.042 |

RVFL NNs were applied to the actor's structures, while the critic's structures were implemented with the use of NNs with Gaussian-type neuron activation functions.

The desired (dashed line) and executed (solid line) motion path of the WMR's point $A$ is shown in Fig. 7.27. The initial position of the WMR's point $A$ is marked with the triangle, "**X**" marks the target final position, while the rhombus indicates the executed final position of point $A$.

Tables 7.15 and 7.16 show the values of individual performance indexes for the execution of the tracking motion with the application of a control system with ADP algorithms in HDP configuration.

**Table 7.17** Values of selected performance indexes in I and II stage of motion

| Index | $e_{max1j}$ [rad] | $\varepsilon_{1j}$ [rad] | $e_{max2j}$ [rad/s] | $\varepsilon_{2j}$ [rad/s] | $s_{maxj}$ [rad/s] | $\sigma_j$ [rad/s] |
|---|---|---|---|---|---|---|
| Stage I, $k = 1, \ldots, 2250$ | | | | | | |
| Wheel 1, $j = 1$ | 1.9 | 0.55 | 1.66 | 0.48 | 2.22 | 0.55 |
| Wheel 2, $j = 2$ | 1.9 | 0.55 | 1.66 | 0.48 | 2.22 | 0.55 |
| Stage II, $k = 2251, \ldots, 4500$ | | | | | | |
| Wheel 1, $j = 1$ | 0.48 | 0.3 | 0.65 | 0.14 | 0.64 | 0.21 |
| Wheel 2, $j = 2$ | 0.45 | 0.29 | 0.65 | 0.14 | 0.6 | 0.2 |

The performance indexes demonstrate lower values than those obtained in the execution of trajectory with the application of the control system based only on the PD controller. However, the indexes values are higher than those yielded from numerical tests carried out for the adaptive and neural systems.

Table 7.17 presents the values of selected performance indexes in the I and II phase of motion along an 8-shaped path, calculated in accordance with the methodology described in numerical test Sect. 7.1.2.2.

The values of the performance indexes in phase II are lower than those obtained in phase I. This is due to the non-zero values of the NN output layer weights that were used for the numerical test at the beginning of phase II. The knowledge of the controlled object's dynamics, stored in the output layer weights of the HDP structures' NNs, allows to generate a compensation signal for the WMR's nonlinearities, which leads to improved performance of the tracking motion.

### 7.4.2.2 Examination of HDP-Type Control, 8-Shaped Path, Disturbance Trial with Non-zero Initial Values of NNs Weights

A numerical test of a tracking control algorithm was carried out for the WMR's point $A$ with the application of ADP structures in HDP configuration, non-zero initial values were assumed for the critic's NN output layer weights ($\mathbf{W}_{Cj\{k=0\}} = 1, j = 1, 2$). The initial values of the actor's NN output layer weights were assumed to equal zero. The goal of the system was to generate control signals that would ensure the execution of the desired WMR's point $A$ trajectory with an 8-shaped path. A parametric disturbance was simulated twice. The first disturbance involved the change of the WMR model's parameters corresponding to the robot's frame being weighed down with a load of $m_{RL} = 4.0$ [kg] in time $t_{d1} = 12.5$ [s]. In the second disturbance scenario the additional load was removed in time $t_{d2} = 32.5$ [s].

The values of the tracking control signals for individual structures of the control system are shown in Fig. 7.28. The overall control signals $u_1$ and $u_2$ of wheels 1 and 2, respectively, are shown in Fig. 7.28a, according to relation (7.66) they consist of the control signals generated by the HDP structures $u_{A1}$ and $u_{A2}$ (Fig. 7.28b), PD

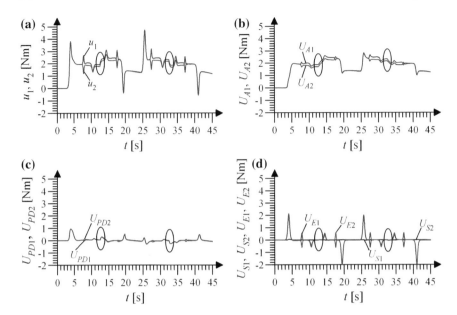

**Fig. 7.28**  **a** Overall control signals $u_1$ and $u_2$ for wheels 1 and 2, **b** actor's NNs control signals $U_{A1}$ and $U_{A2}$, $\mathbf{U}_A = -\frac{1}{h}\mathbf{Mu}_A$, **c** PD control signals $U_{PD1}$ and $U_{PD2}$, $\mathbf{U}_{PD} = -\frac{1}{h}\mathbf{Mu}_{PD}$, **d** supervisory term's control signals $U_{S1}$ and $U_{S2}$, $\mathbf{U}_S = -\frac{1}{h}\mathbf{Mu}_S$, and additional control signals $u_{E1}$ and $u_{E2}$, $\mathbf{U}_E = -\frac{1}{h}\mathbf{Mu}_E$

control signals $u_{PD1}$ and $u_{PD2}$ (Fig. 7.28c), supervisory term's control signals $u_{S1}$ and $u_{S2}$, and additional control signals $u_{E1}$ and $u_{E2}$ (Fig. 7.28d).

The use of non-zero initial values for the critic's NN output layer weights accelerated the HDP structures adaptive process. The improvement of the quality of compensation for the WMR's nonlinearities by means of control signals $u_{A1}$ and $u_{A2}$ causes a decrease in the values of the PD control signals, $u_{PD1}$ and $u_{PD2}$, at the beginning of phase I. In phase II, which starts at $t = 22.5$ [s], the PD control signals take close-to-zero values. Slight changes in the values of the PD control signals are observed when disturbances occur. The impact the change in the WMR's model parameters has on the object's dynamics is compensated by changes in the values of control signals generated by the actor's NN in the HDP structures. The values of control signals $\mathbf{u}_E$ depend mainly on $\mathbf{z}_{d3}$, changes in the values of tracking errors have a negligible effect on those signals. The values of supervisory term's control signals $\mathbf{u}_S$ remain at zero, as the filtered tracking errors $\mathbf{s}$ do not exceed the established allowable threshold.

The values of the executed angular velocities of the WMR's drive wheels, shown in Fig. 7.29, are close to the desired ones. At time $t_{d1}$ when the disturbance simulates the WMR being weighed down with additional load, the values of angular velocities decrease, since the generated control signals are not sufficient for the execution of the WMR motion with the desired velocity due to an increased resistance to motion. The increase in the value of the WMR's nonlinearity compensation signals generated by

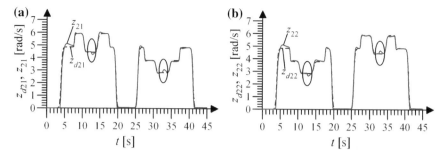

**Fig. 7.29** **a** Desired ($z_{d21}$) and executed ($z_{21}$) angular velocity of the WMR's wheel 1, **b** desired ($z_{d22}$) and executed ($z_{22}$) angular velocity of the WMR's wheel 2

the HDP structures leads to an increase of the angular velocities and further execution of the trajectory according to the desired parameters of motion. The occurrence of the second parametric disturbance in time $t_{d2}$ simulates the removal of additional load applied to the WMR which means return of the model parameters to the nominal values. At this point the generated control signals exceed the required values, which is due to the lower resistance to motion of the load-free object. Thus the value of the WMR's point $A$ velocity is higher than the desired one. As a result of output layer weights adaptation in the HDP structures' NNs the values of control signals $u_{A1}$ and $u_{A2}$ are reduced, the velocity decreases and reaches the desired value.

The tracking error values are shown in Fig. 7.30. The values of tracking errors were reduced as a result of the use of non-zero initial values for the critics's NN output layer weights in HDP algorithms. The parametric disturbances have only a limited impact on the performance of the tracking motion and the increase in the absolute values of tracking errors caused by changes in the controlled object's dynamics is compensated through the adaptation of the NN output layer weights in the NDP structures.

The values of the actor's and critic's NNs output layer weights for the first HDP structure are shown in Fig. 7.31. The values of the NNs output layer weights for the second HDP structure, that generates the control signal for wheel 2, are similar. Zero initial values of the actor's NN weights were assumed, while the initial values of the critic's NN output layer weights equal one. The weights stabilize around certain values and remain bounded. At the times of parametric disturbances the values of the actor's NN weights change, as the generated compensation signal for the WMR's nonlinearities adjusts to the modified parameters of the controlled object. RVFL NNs were applied to the actor's structures, while the critic's structures were implemented with the use of NNs with Gaussian-type neuron activation functions.

The desired (dashed line) and executed (solid line) motion path of the WMR's point $A$ is shown in Fig. 7.32. The initial position of the WMR's point $A$ is marked with the triangle, "**X**" marks the target final position, while the rhombus indicates the executed final position of point $A$.

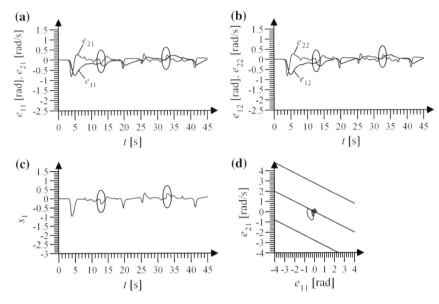

**Fig. 7.30** **a** Tracking errors $e_{11}$ and $e_{21}$, **b** tracking errors $e_{12}$ and $e_{22}$, **c** filtered tracking error $s_1$, **d** sliding manifold of wheel 1

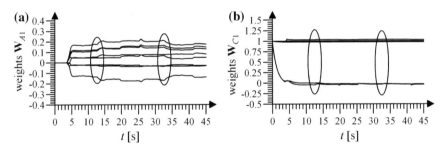

**Fig. 7.31** **a** Actor's 1 NN output layer weights $\mathbf{W}_{A1}$, **b** critic's 1 NN output layer weights $\mathbf{W}_{C1}$

**Fig. 7.32** Desired and executed motion path of the WMR's point $A$

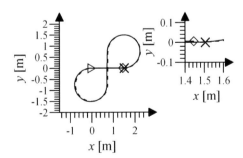

**Table 7.18**  Values of selected performance indexes

| Index | $e_{max1j}$ [rad] | $\varepsilon_{1j}$ [rad] | $e_{max2j}$ [rad/s] | $\varepsilon_{2j}$ [rad/s] | $s_{maxj}$ [rad/s] | $\sigma_j$ [rad/s] |
|---|---|---|---|---|---|---|
| Wheel 1, $j = 1$ | 0.75 | 0.2 | 0.81 | 0.14 | 0.96 | 0.17 |
| Wheel 2, $j = 2$ | 0.75 | 0.21 | 0.81 | 0.14 | 0.96 | 0.18 |

**Table 7.19**  Values of selected performance indexes of path execution

| Index | $d_{max}$ [m] | $d_n$ [m] | $\rho_d$ [m] |
|---|---|---|---|
| Value | 0.083 | 0.061 | 0.04 |

**Table 7.20**  Values of selected performance indexes in I and II stage of motion

| Index | $e_{max1j}$ [rad] | $\varepsilon_{1j}$ [rad] | $e_{max2j}$ [rad/s] | $\varepsilon_{2j}$ [rad/s] | $s_{maxj}$ [rad/s] | $\sigma_j$ [rad/s] |
|---|---|---|---|---|---|---|
| Stage I, $k = 1, \ldots, 2250$ | | | | | | |
| Wheel 1, $j = 1$ | 0.75 | 0.26 | 0.81 | 0.17 | 0.96 | 0.22 |
| Wheel 2, $j = 2$ | 0.75 | 0.27 | 0.81 | 0.18 | 0.96 | 0.22 |
| Stage II, $k = 2251, \ldots, 4500$ | | | | | | |
| Wheel 1, $j = 1$ | 0.24 | 0.11 | 0.41 | 0.1 | 0.47 | 0.12 |
| Wheel 2, $j = 2$ | 0.23 | 0.11 | 0.41 | 0.1 | 0.47 | 0.12 |

Tables 7.18 and 7.19 show the values of individual performance indexes for the execution of the tracking motion with the application of a control system with ACDs in HDP configuration.

As revealed by the numerical test the performance indexes demonstrate lower values than those obtained only with the application of the PD controller to the tracking control algorithm. The use of non-zero initial values for the critic's NN output layer weights in HDP structures improved the performance of the tracking motion, which was reflected in the lower values of performance indexes. Furthermore, the indexes values are lower than those yielded from the numerical tests carried out for the adaptive and neural algorithms.

Table 7.20 presents the values of selected performance indexes in the I and II phase of motion along an 8-shaped path, calculated in accordance with the methodology described in numerical test Sect. 7.1.2.2.

The values of the performance indexes in phase II are lower than those obtained in phase I. This is due to the non-zero values of the NNs output layer weights that were used for the numerical test at the beginning of phase II. The knowledge of the controlled object's dynamics, stored in the NNs weights of the HDP structures, allows to generate a compensation signal for the WMR's nonlinearities, which leads to improved performance of the generated tracking control.

## *7.4.3   Conclusions*

The application of NDP structures in HDP configuration to a tracking control system improved the quality of the generated control action that was reflected in the lower values of the assumed performance indexes as compared with those produced by the PD controller. The adoption of zero initial values for the actor's and critic's NNs output layer weights results in slower weights adaptation process and reduced performance of the tracking motion (numerical test Sect. 7.4.2.1), as opposed to the numerical test which adopted non-zero initial values for the critic's NNs output layer weights. It is also particularly relevant that the critic's NNs in HDP structures do not generate a control signal but only produce an estimate of the value function. The performance indexes values obtained from the numerical test Sect. 7.4.2.2 are lower than those yielded from the tracking control test carried out for the adaptive and neural algorithms.

The pace of output layer weights adaptation within the actor's RVFL NNs depends on the selection of constant values for the input layer weights in the random drawing during the NNs initialization. The rate of weights adaptation is also affected by the number of adaptation cycles for HDP structures within the inner loop $l_{AC}$ in the $k$th step of a discrete process and by the values of NNs learning coefficients. Too small values of learning coefficients slow down the adaptive process, whereas high values of coefficients may lead to process instability. A similar effect on the performance of the generated control signals has the number of inner loop calculation cycles $l_{AC}$ in the actor's and critic's NN output layer weights adaptation process in a given $k$th discrete step. However, increasing the parameter results in higher computational complexity of the executed algorithm. As the process advances the $l_{AC}$ value may change, owing to the adopted criterion e.g. a difference in the value of the cost function approximation generated by the critic's NN in two consecutive inner loop calculations. Another important factor in the performance of the generated control signals is the number of NN neurons and type of neuron activation function. The present studies involved a NNs with 8 neurons in the hidden layer, as further development of NNs structures produced no significant reduction in the values of the adopted performance indexes, but caused a proportional increase of computational cost. NNs linear with respect to the output layer weights were used. These NNs are characterized by a simple structure and uncomplicated adaptive process, which is especially important for the application of real-time algorithms to control systems of dynamic objects. The application of HDP algorithms requires the availability of the controlled object's mathematical model.

## 7.5 Dual-Heuristic Dynamic Programming in Tracking Control of a WMR and a RM

The adaptive critic structure in DHP configuration [26, 31, 41, 44, 46, 53, 77, 79, 86] is a more complex algorithm than the HDP algorithm. It is composed of actor and critic structures as well as a predictive model of the controlled object. The mathematical model of the controlled object must be known so as to generate the weights adaptation law for the actor's and critic's NNs. The difference to the HDP algorithm is in the critic function. In the DHP algorithm the critic approximates the derivative of the value function with respect to the controlled object's state, hence the critic's complex design results from the size of the state vector. In the case of a WMR with two degrees of freedom, the critic consists of two adaptive structures e.g. NNs. The actor generates the object's control law, hence its design results from the number of independent signals that control the actuators. For example, in a WMR's tracking control system with two degrees of freedom the actor consists of two adaptive structures e.g. NNs generating independent signals that control the robot's drive units. This chapter provides the synthesis of a discrete tracking control system for a dynamic object with the application of an NDP algorithm in HDP configuration. The objects considered below are a WMR and a RM with three degrees of freedom. The presented algorithms operate in a discrete-time domain, thus discrete models of the WMR's and 3-DoF RM's dynamics were used for the synthesis of a controlled system. This chapter is based on the information relating to the DHP-type adaptive critic design whose description is included in Sect. 6.2.2 as well as the description of the WMR's dynamics presented in Sect. 2.1.2 and the description of the RM's dynamics from Sect. 2.2.2.

### 7.5.1 Synthesis of DHP-Type Control

The synthesis of a control algorithm for a dynamic object with the application of an NDP structure in HDP configuration was carried out in a similar manner as for the derivation of the control system with an HDP algorithm described in Sect. 7.4.1. The synthesis involved the WMR's and RM's control systems. Since the dynamic description differs for each of the controlled objects, their control algorithms shall be also different. The following synthesis of the WMR's control algorithm with a DHP structure is presented in a detailed manner, while the analysis of the tracking control algorithm for the RM points to the differences in the synthesis and results obtained, that follow from the specific character of the controlled object.

### 7.5.1.1  Synthesis of WMR's Tracking Control with the Application of a DHP Algorithm

In the synthesis of a discrete tracking control algorithm for a WMR the tracking errors $\mathbf{e}_{1\{k\}}$ and $\mathbf{e}_{2\{k\}}$ were defined in line with relation (7.53), the filtered tracking errors $\mathbf{s}_{\{k\}}$ were specified in relation (7.54), and $\mathbf{s}_{\{k+1\}}$ according to relation (7.55) derived in Sect. 7.4.1.

In the designed tracking control system for the WMR's point $A$ an NDP algorithm in DHP configuration was used, where the actor's and critic's structures were implemented in the form of two RVFL NNs, which results from the dimension of the state vector for the WMR's model dynamics. The DHP algorithm requires the availability of the controlled object's model in order for the actor's and critic's NN weights adaptation to be performed.

The goal of the NDP algorithm in DHP configuration is to generate the suboptimal control law $\mathbf{u}_{A\{k\}} = \left[ u_{A1\{k\}}, u_{A2\{k\}} \right]^T$ that minimizes the value function given by relation

$$V_{\{k\}} \left( \mathbf{s}_{\{k\}} \right) = \sum_{k=0}^{n} \gamma^k L_{C\{k\}} \left( \mathbf{s}_{\{k\}} \right) \ , \tag{7.85}$$

with the cost function

$$L_{C\{k\}} \left( \mathbf{s}_{\{k\}} \right) = \frac{1}{2} \mathbf{s}_{\{k\}}^T \mathbf{R} \mathbf{s}_{\{k\}} \ , \tag{7.86}$$

where $\mathbf{R}$ – positive-definite, constant diagonal matrix.

A general schematic diagram of the NDP structure in DHP configuration is shown in Fig. 7.33.

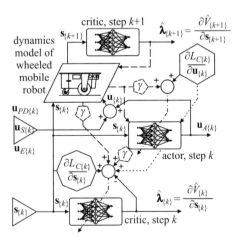

**Fig. 7.33** Schematic diagram of NDP structure in DHP configuration

The DHP algorithm, described in Sect. 6.2.2, consists of a predictive model and two adaptive structures:

- **Predictive model** for the dynamics of the WMR's filtered tracking error generates the prediction of state $\mathbf{s}_{\{k+1\}}$ for step $k+1$ of the discrete process, based on (7.55), according to relation

$$\mathbf{s}_{\{k+1\}} = \mathbf{s}_{\{k\}} - \mathbf{Y}_f\left(\mathbf{z}_{2\{k\}}\right) + \mathbf{Y}_e\left(\mathbf{z}_{\{k\}}, \mathbf{z}_{d\{k\}}, \mathbf{z}_{d3\{k\}}\right) + h\mathbf{M}^{-1}\mathbf{u}_{\{k\}} . \tag{7.87}$$

- **Actor**, implemented in the form of two RVFL NNs, generates the suboptimal control law $\mathbf{u}_{A\{k\}}$ in the $k$th step according to relation

$$u_{Aj\{k\}}\left(\mathbf{x}_{Aj\{k\}}, \mathbf{W}_{Aj\{k\}}\right) = \mathbf{W}_{Aj\{k\}}^T \mathbf{S}\left(\mathbf{D}_A^T \mathbf{x}_{Aj\{k\}}\right) , \tag{7.88}$$

where $j = 1, 2$, the actor's NNs of the DHP algorithm are designed in a similar manner as in the HDP structure presented in Sect. 7.4.1.

The actor's NNs weights adaptation procedure for the DHP algorithm is to minimize the performance index

$$\mathbf{e}_{A\{k\}} = \frac{\partial L_{C\{k\}}\left(\mathbf{s}_{\{k\}}\right)}{\partial \mathbf{u}_{\{k\}}} + \gamma \left[\frac{\partial \mathbf{s}_{\{k+1\}}}{\partial \mathbf{u}_{\{k\}}}\right]^T \hat{\boldsymbol{\lambda}}_{\{k+1\}}\left(\mathbf{x}_{C\{k+1\}}, \mathbf{W}_{C\{k\}}\right) , \tag{7.89}$$

according to relation

$$\mathbf{W}_{Aj\{k+1\}} = \mathbf{W}_{Aj\{k\}} - e_{Aj\{k\}} \boldsymbol{\Gamma}_{Aj} \mathbf{S}\left(\mathbf{D}_A^T \mathbf{x}_{Aj\{k\}}\right) , \tag{7.90}$$

where $\hat{\boldsymbol{\lambda}}_{\{k\}}\left(\mathbf{x}_{C\{k+1\}}, \mathbf{W}_{C\{k\}}\right)$ – critic's NN output for step $k+1$, $\boldsymbol{\Gamma}_{Aj}$ – constant diagonal matrix of positive learning coefficients for the actor's $j$th NN output layer weights.

- **Critic** estimates the derivatives of value functions (7.85) with respect to the state vector of the controlled object

$$\boldsymbol{\lambda}_{\{k\}} = \begin{bmatrix} \frac{\partial V_{\{k\}}\left(\mathbf{s}_{\{k\}}\right)}{\partial s_{1\{k\}}} \\ \frac{\partial V_{\{k\}}\left(\mathbf{s}_{\{k\}}\right)}{\partial s_{2\{k\}}} \end{bmatrix} . \tag{7.91}$$

Therefore it was implemented in the form of two RVFL NNs, each of which estimates one component of the vector $\boldsymbol{\lambda}_{\{k\}}$ (7.91). The critic's NN output is expressed by the following relation

$$\hat{\lambda}_{j\{k\}}\left(\mathbf{x}_{Cj\{k\}}, \mathbf{W}_{Cj\{k\}}\right) = \mathbf{W}_{Cj\{k\}}^T \mathbf{S}\left(\mathbf{D}_C^T \mathbf{x}_{Cj\{k\}}\right) , \tag{7.92}$$

where $j = 1, 2$, $\mathbf{W}_{Cj\{k\}}$ – output layer weights vector for the $j$th critic's RVFL NN, $\mathbf{x}_{Cj\{k\}} = \kappa_C\left[1, s_{j\{k\}}\right]^T$ – critic's NN inputs vector, $\kappa_C$ – constant diagonal matrix of positive scaling coefficients for the critic's NN inputs, $\mathbf{S}\left(\mathbf{D}_C^T \mathbf{x}_{Cj\{k\}}\right)$ – vector of bipolar sigmoid neuron activation functions, $\mathbf{D}_C$ – matrix of constant weights in the critic's NN input layer, randomly selected in the initialization process.

In the DHP algorithm the critic's NNs output layer weights are adapted through backpropagation of the performance index defined as

$$
\begin{aligned}
\mathbf{e}_{C\{k\}} = {} & \frac{\partial L_{C\{k\}}\left(\mathbf{s}_{\{k\}}\right)}{\partial \mathbf{s}_{\{k\}}} + \left[\frac{\partial \mathbf{u}_{\{k\}}}{\partial \mathbf{s}_{\{k\}}}\right]^T \frac{\partial L_{C\{k\}}\left(\mathbf{s}_{\{k\}}\right)}{\partial \mathbf{u}_{\{k\}}} + \\
& + \gamma \left[\frac{\partial \mathbf{s}_{\{k+1\}}}{\partial \mathbf{s}_{\{k\}}} + \left[\frac{\partial \mathbf{u}_{\{k\}}}{\partial \mathbf{s}_{\{k\}}}\right]^T \frac{\partial \mathbf{s}_{\{k+1\}}}{\partial \mathbf{u}_{\{k\}}}\right]^T \hat{\boldsymbol{\lambda}}_{\{k+1\}}\left(\mathbf{x}_{C\{k+1\}}, \mathbf{W}_{C\{k\}}\right) + \\
& - \hat{\boldsymbol{\lambda}}_{\{k\}}\left(\mathbf{x}_{C\{k\}}, \mathbf{W}_{C\{k\}}\right) ,
\end{aligned}
\tag{7.93}
$$

where $\mathbf{s}_{\{k+1\}}$ – signal determined from the predictive model (7.87).

The critic's NN weights are adapted through backpropagation of the performance index (7.93) according to the following relation

$$
\mathbf{W}_{Cj\{k+1\}} = \mathbf{W}_{Cj\{k\}} - e_{Cj\{k\}} \boldsymbol{\Gamma}_{Cj} \mathbf{S}\left(\mathbf{D}_C^T \mathbf{x}_{Cj\{k\}}\right) ,
\tag{7.94}
$$

where $\boldsymbol{\Gamma}_{Cj}$ – constant diagonal matrix of positive learning coefficients for the output layer weights of the critic's $j$th NN.

A schematic diagram of the NNs weights adaptation procedure for the NDP structure in DHP configuration is shown in Fig. 7.34, where $\mathbf{u}_{PD\{k\}}$, $\mathbf{u}_{S\{k\}}$ and $\mathbf{u}_{E\{k\}}$ are additional control signals, whose application stems from the analysis of system stability.

In the designed tracking control system for the WMR with the application of the NDP structure in DHP configuration the overall control signals $\mathbf{u}_{\{k\}}$ consist of the control signals generated by the actor's NNs $\mathbf{u}_{A\{k\}}$, the PD control signals $\mathbf{u}_{PD\{k\}}$, signals $\mathbf{u}_{E\{k\}}$ and additional supervisory term's control signals $\mathbf{u}_{S\{k\}}$, whose structure was derived by applying the Lyapunov stability theorem to the stability analysis of the designed control system. Figure 7.35 shows a schematic structure of a neural tracking control system for the WMR's point $A$ with the application of a DHP structure. The stability analysis of a closed-loop system was carried out in a similar manner as in Sect. 7.4.1 by assuming the overall control signal in the form of relation (7.66) and the supervisory term's control signal $\mathbf{u}_{S\{k\}}$ as defined in relation (7.79).

The synthesis and results of experimental studies on the tracking control system for a selected WMR's point with the application of DHP algorithm can be found in [41, 44, 46].

### 7.5.1.2   Synthesis of Robotic Manipulator's Tracking Control with the Application of a DHP Algorithm

In the synthesis of a discrete tracking control algorithm for a RM the tracking errors $\mathbf{e}_{1\{k\}}$ and $\mathbf{e}_{2\{k\}}$ were defined in line with relation (7.53), the filtered tracking errors $\mathbf{s}_{\{k\}}$ were specified in relation (7.54) derived in Sect. 7.4.1.

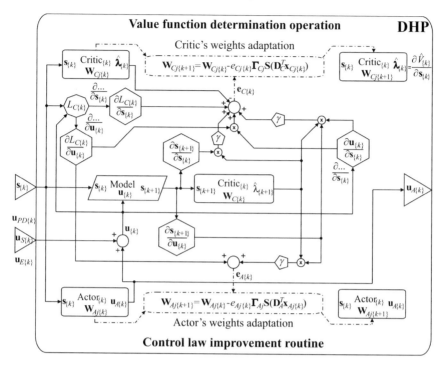

**Fig. 7.34** Schematic diagram of NNs weights adaptation process for NDP structures in DHP configuration

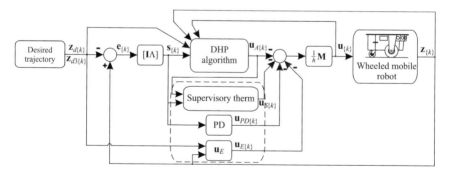

**Fig. 7.35** Schematic diagram of a WMR's tracking control system with the application of a DHP algorithm

Based on the discrete description of the RM's dynamics (2.91) and relation (7.53) and (7.54), the filtered tracking error $\mathbf{s}_{\{k+1\}}$ is defined as

$$\mathbf{s}_{\{k+1\}} = -\mathbf{Y}_f\left(\mathbf{z}_{1\{k\}}, \mathbf{z}_{2\{k\}}\right) + \mathbf{Y}_d\left(\mathbf{z}_{\{k\}}, \mathbf{z}_{d\{k+1\}}\right) - \mathbf{Y}_{\tau\{k\}}\left(\mathbf{z}_{1\{k\}}\right) + h\left[\mathbf{M}\left(\mathbf{z}_{1\{k\}}\right)\right]^{-1}\mathbf{u}_{\{k\}},$$

$$(7.95)$$

where

$$
\begin{aligned}
\mathbf{Y}_f\left(\mathbf{z}_{1\{k\}}, \mathbf{z}_{2\{k\}}\right) &= h\left[\mathbf{M}\left(\mathbf{z}_{1\{k\}}\right)\right]^{-1}\left[\mathbf{C}\left(\mathbf{z}_{1\{k\}}, \mathbf{z}_{2\{k\}}\right)\mathbf{z}_{2\{k\}} + \mathbf{F}\left(\mathbf{z}_{2\{k\}}\right) + \mathbf{G}\left(\mathbf{z}_{1\{k\}}\right)\right] , \\
\mathbf{Y}_{\tau\{k\}}\left(\mathbf{z}_{1\{k\}}\right) &= h\left[\mathbf{M}\left(\mathbf{z}_{1\{k\}}\right)\right]^{-1}\boldsymbol{\tau}_{d\{k\}} , \\
\mathbf{Y}_d\left(\mathbf{z}_{\{k\}}, \mathbf{z}_{d\{k+1\}}\right) &= \mathbf{z}_{2\{k\}} - \mathbf{z}_{d2\{k+1\}} + \boldsymbol{\Lambda}\left[\mathbf{z}_{1\{k\}} + h\mathbf{z}_{2\{k\}} - \mathbf{z}_{d1\{k+1\}}\right] = \\
\mathbf{z}_{2\{k\}} &- \mathbf{z}_{d2\{k\}} - h\mathbf{z}_{d3\{k\}} + \boldsymbol{\Lambda}\left[\mathbf{z}_{1\{k\}} - \mathbf{z}_{d1\{k\}} + h\mathbf{z}_{2\{k\}} - h\mathbf{z}_{d2\{k\}}\right] = \\
\mathbf{s}_{\{k\}} &+ \mathbf{Y}_e\left(\mathbf{z}_{\{k\}}, \mathbf{z}_{d\{k\}}, \mathbf{z}_{d3\{k\}}\right) , \\
\mathbf{Y}_e\left(\mathbf{z}_{\{k\}}, \mathbf{z}_{d\{k\}}, \mathbf{z}_{d3\{k\}}\right) &= h\left[\boldsymbol{\Lambda}\mathbf{e}_{2\{k\}} - \mathbf{z}_{d3\{k\}}\right] ,
\end{aligned}
$$

$$(7.96)$$

and $\mathbf{z}_{d3\{k\}}$ – the vector of desired angular accelerations obtained by writing out $\mathbf{z}_{d2\{k+1\}}$ applying the Euler's method. The vector $\mathbf{Y}_f\left(\mathbf{z}_{2\{k\}}\right)$ includes all the manipulator's nonlinearities.

In the designed tracking control system for the manipulator end-effector's point $C$ an ADP algorithm in DHP configuration was used. It consists of the actor's and critic's structures, each of which is implemented in the form of three RVFL NNs, which results from the dimension of the state vector for the dynamics model of a 3-DoF manipulator. The DHP algorithm requires the availability of the controlled object's model in order for the actor's and critic's NNs weights adaptation to be performed.

The value function is assumed as in (7.85) and the local cost is defined by relation (7.86), given the dimensionality of vector $\mathbf{s}_{\{k\}}$, which in the considered case equals three.

The DHP algorithm consists of a predictive model and two adaptive structures:
- **Predictive model** for the dynamics of the manipulator's filtered tracking error generates the prediction of state $\mathbf{s}_{\{k+1\}}$ for step $k + 1$ of the discrete process, based on (7.95), according to relation

$$
\mathbf{s}_{\{k+1\}} = -\mathbf{Y}_f\left(\mathbf{z}_{1\{k\}}, \mathbf{z}_{2\{k\}}\right) + \mathbf{Y}_d\left(\mathbf{z}_{\{k\}}, \mathbf{z}_{d\{k+1\}}\right) + h\left[\mathbf{M}\left(\mathbf{z}_{1\{k\}}\right)\right]^{-1}\mathbf{u}_{\{k\}} . \quad (7.97)
$$

- **Actor**, implemented in the form of three RVFL NNs, generates the suboptimal control law $\mathbf{u}_{A\{k\}}$ in the $k$th step according to relation (7.88).

The actor's NNs weights adaptation procedure for the DHP algorithm is to minimize the performance index (7.89) according to the following relation

$$
\mathbf{W}_{Aj\{k+1\}} = \mathbf{W}_{Aj\{k\}} - e_{Aj\{k\}}\boldsymbol{\Gamma}_{Aj}\mathbf{S}\left(\mathbf{D}_A^T\mathbf{x}_{Aj\{k\}}\right) - k_A|e_{Aj\{k\}}|\boldsymbol{\Gamma}_{Aj}\mathbf{W}_{Aj\{k\}} , \quad (7.98)
$$

where $k_A$ – positive constant. The last term in equation (7.98) is a NNs weights regularization mechanism [94] that prevents the over-fitting of the NNs weights.
- **Critic** estimates the derivatives of the value functions (7.85) with respect to the state vector of the controlled object

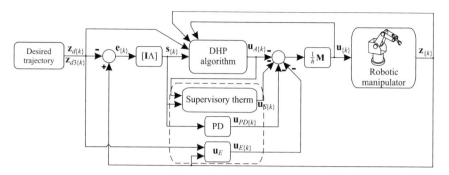

**Fig. 7.36** Schematic diagram of a robotic manipulator's tracking control system with the application of a DHP algorithm

$$\boldsymbol{\lambda}_{\{k\}} = \begin{bmatrix} \dfrac{\partial V_{\{k\}}\left(\mathbf{s}_{\{k\}}\right)}{\partial s_{1\{k\}}} \\[2ex] \dfrac{\partial V_{\{k\}}\left(\mathbf{s}_{\{k\}}\right)}{\partial s_{2\{k\}}} \\[2ex] \dfrac{\partial V_{\{k\}}\left(\mathbf{s}_{\{k\}}\right)}{\partial s_{3\{k\}}} \end{bmatrix} . \tag{7.99}$$

Hence, the critic was implemented in the form of three RVFL NNs, each of which estimates one component of the vector $\boldsymbol{\lambda}_{\{k\}}$ (7.99). The critic's NN output is expressed by relation (7.92), and its weights adaptation is carried out by minimizing the performance index (7.93) with the use of the backpropagation method, according to the following relation

$$\mathbf{W}_{Cj\{k+1\}} = \mathbf{W}_{Cj\{k\}} - e_{Cj\{k\}}\boldsymbol{\Gamma}_{Cj}\mathbf{S}\left(\mathbf{D}_C^T \mathbf{x}_{Cj\{k\}}\right) - k_C|e_{Cj\{k\}}|\boldsymbol{\Gamma}_{Cj}\mathbf{W}_{Cj\{k\}} , \tag{7.100}$$

where $k_C$ – positive constant.

The stability analysis of a closed-loop system was carried out in a similar manner as in Sect. 7.4.1. In the designed tracking control system for the manipulator end-effector's point $C$ with the application of the NDP structure in DHP configuration the overall control signals $\mathbf{u}_{\{k\}}$ consist of the control signals generated by the actor's NNs $\mathbf{u}_{A\{k\}}$, the PD control signals $\mathbf{u}_{PD\{k\}}$, signals $\mathbf{u}_{E\{k\}}$ and additional supervisory term's control signals $\mathbf{u}_{S\{k\}}$. Figure 7.36 shows the structure of a neural tracking control system for the manipulator end-effector's point $C$ with the application of a DHP structure.

The overall control signal is assumed as

$$\mathbf{u}_{\{k\}} = -\frac{1}{h}\mathbf{M}\left(\mathbf{z}_{1\{k\}}\right)\left[\mathbf{u}_{A\{k\}} - \mathbf{u}_{S\{k\}} + \mathbf{u}_{PD\{k\}} + \mathbf{u}_{E\{k\}}\right] , \tag{7.101}$$

where

$$\begin{aligned} \mathbf{u}_{PD\{k\}} &= \mathbf{K}_D \mathbf{s}_{\{k\}} , \\ \mathbf{u}_{E\{k\}} &= h\left[\boldsymbol{\Lambda}\mathbf{e}_{2\{k\}} - \mathbf{z}_{d3\{k\}}\right] , \\ \mathbf{u}_{S\{k\}} &= \mathbf{I}_S \mathbf{u}_{S\{k\}}^* , \end{aligned} \tag{7.102}$$

where $\mathbf{K}_D$ – constant, positive-definite design matrix, $\mathbf{I}_S$ – diagonal switching matrix, $I_{Sj,j} = 1$ when $|s_{j\{k\}}| \geq \rho_j$, $I_{Sj,j} = 0$ when $|s_{j\{k\}}| < \rho_j$, $\rho_j$ – constant design value, $\mathbf{u}^*_{S\{k\}}$ – supervisory term's control signal derived further in this subchapter, $j = 1, 2, 3$. Given $I_{Sj,j} = 1$, by substituting (7.101) into (7.95), we get

$$\mathbf{s}_{\{k+1\}} = \mathbf{Y}_d \left(\mathbf{z}_{\{k\}}, \mathbf{z}_{d\{k+1\}}\right) - \mathbf{Y}_f \left(\mathbf{z}_{1\{k\}}, \mathbf{z}_{2\{k\}}\right) - \mathbf{Y}_{\tau\{k\}} \left(\mathbf{z}_{1\{k\}}\right) - \mathbf{u}_{A\{k\}} + \mathbf{u}^*_{S\{k\}} - \mathbf{u}_{PD\{k\}} - \mathbf{u}_{E\{k\}} \, .$$
(7.103)

The stability analysis of a closed-loop system was carried out in a similar manner as in Sect. 7.4.1, by assuming a positive-definite Lyapunov candidate function (7.69), whose difference can be written in the form of relation (7.76). Substituting (7.103) into (7.76) we get

$$\Delta L_{\{k\}} = \mathbf{s}_{\{k\}}^T \left[ -\mathbf{K}_D \, \mathbf{s}_{\{k\}} - \mathbf{Y}_f \left(\mathbf{z}_{1\{k\}}, \mathbf{z}_{2\{k\}}\right) - \mathbf{Y}_{\tau\{k\}} \left(\mathbf{z}_{1\{k\}}\right) - \mathbf{u}_{A\{k\}} + \mathbf{u}^*_{S\{k\}} \right] \, .$$
(7.104)

If assumed that all components of the disturbance vector are bounded $\left| \mathbf{Y}_{\tau\{k\}} \left(\mathbf{z}_{1\{k\}}\right) \right| < b_{dj}$, $b_{dj} > 0$, we may write the following relation

$$\Delta L_{\{k\}} \leq -\mathbf{s}_{\{k\}}^T \mathbf{K}_D \, \mathbf{s}_{\{k\}} + \sum_{j=1}^{3} |s_{j\{k\}}| \left[ \left| Y_{f\{k\}} \left(\mathbf{z}_{1\{k\}}, \mathbf{z}_{2\{k\}}\right) \right| + b_{dj} + |u_{Aj\{k\}}| \right] + \sum_{i=1}^{3} s_{j\{k\}} u^*_{Sj\{k\}} \, .$$
(7.105)

Given the supervisory term's control signals $\mathbf{u}^*_{S\{k\}}$ as

$$u^*_{Sj\{k\}} = -\text{sgn} \, s_{j\{k\}} \left[ F_j + \left|u_{Aj\{k\}}\right| + b_{dj} + \sigma_j \right] \, ,$$
(7.106)

where $\left| Y_{fj} \left(\mathbf{z}_{1\{k\}}, \mathbf{z}_{2\{k\}}\right) \right| \leq F_j$, $F_j > 0$, and $\sigma_j$ is a small positive constant; from the inequality (7.78) it follows that the difference of the Lyapunov function is negative-definite. The designed control signal provides the reduction of the filtered tracking error $\mathbf{s}_{\{k\}}$ when $|s_{j\{k\}}| \geq \rho_j$. For the initial condition $|s_{j\{k=0\}}| < \rho_j$ the filtered tracking error $|s_{j\{k\}}| < \rho_j$ remains bounded to $\forall k \geq 0, j = 1, 2$.

The synthesis and results of studies on the manipulator's tracking control system with the application of a DHP algorithm are presented in [31, 86].

### 7.5.2 Simulation Tests

Simulation tests of tracking control systems of dynamic objects with the use of ADP algorithm in DHP configuration were performed. The first subchapter presents parameters of the WMR's control system with the DHP structure, the second results obtained by the WMR motion simulation. The third subchapter presents parameters of the RM's control system with the DHP structure, the fourth results obtained during the simulation of the RM motion.

### 7.5.2.1 Numerical Test of the WMR Motion Control with the Use of DHP Algorithm

The simulation tests of WMR tracking control with the use of a control algorithm with the NPD structure in DHP configuration were carried out in a virtual computing environment. From the carried out numerical tests the sequence resulting from executed trajectory with path in the shape of digit 8, as described in Sect. 2.1.1.2, was selected. A parametric disturbance was simulated, where the change of model's parameters was related to the WMR being weighed down with additional mass.

In the numerical test, the following parameters of the control system with the DHP structure were used:

- PD controller parameters:

  - $\mathbf{K}_D = 3.6h\mathbf{I}$ – fixed diagonal matrix of PD controller gain,
  - $\mathbf{\Lambda} = 0.5\,\mathbf{I}$ – fixed diagonal matrix,

- type of actor's NNs applied: RVFL,

  - $m_A = 8$ – number of actor's NN neurons,
  - $\mathbf{\Gamma}_{Aj} = \text{diag}\{0.25\}$ – diagonal matrix of gain coefficients for the adaptation of weights $\mathbf{W}_{Aj}, j = 1, 2$,
  - $\mathbf{D}_A$ – NN input layer weights randomly selected from the interval $D_{A[i,g]} \in \langle -0.1, 0.1\rangle$,
  - $\mathbf{W}_{Aj\{k=0\}} = \mathbf{0}$ – zero initial values of the output layer weights,

- type of critic's NNs applied: RVFL,

  - $m_C = 8$ – number of critic's NN neurons,
  - $\mathbf{\Gamma}_{Cj} = \text{diag}\{0.5\}$ – diagonal matrix of gain coefficients for the adaptation of weights $\mathbf{W}_{Cj}$,
  - $\mathbf{D}_C$ – NN input layer weights randomly selected from the interval $D_{C[i,g]} \in \langle -0.2, 0.2\rangle$,
  - $\mathbf{W}_{Cj\{k=0\}} = \mathbf{0}$ – zero initial values of output layer weights,

- $\beta_N = 2$ – slope coefficient of bipolar sigmoid neuron activation functions,
- $\mathbf{R} = \text{diag}\{1\}$ – matrix of scaling coefficients in the cost function,
- $l_{AC} = 3$ – number of inner loop iteration steps for the actor's and critic's structure adaptation in the $k$th step of the iterative process,
- $\rho_j = 2.5$ – supervisory algorithm parameter,
- $\mathbf{F} = [0.08, 0.08]^T$ – vector of constant values of assumed limitations of $Y_{fj}\left(\mathbf{z}_{2\{k\}}\right)$ function,
- $\sigma_j = 0.01$ – supervisory algorithm parameter,
- $h = 0.01$ [s] – numerical test calculation step.

In the studies on the designed control systems with NN based algorithms the components of the inputs vectors were scaled to the interval $\langle -1, 1\rangle$.

The same parameters of PD controller and performance indexes, as described in Sect. 7.4.2, were used. The number of discrete measurement data of WMR motion with path in the shaped of digit 8 amounts to $n = 4500$.

In the description of the numerical tests of algorithms operating in discrete time domain the notation of variables was simplified by resigning from the $k$th index step of the process. In order to facilitate the analysis of presented earlier results of tracking control systems of point $A$ of the WMR, and their comparison to continuous sequences, the abscissa axis of charts of discreet sequences were calibrated in time $t$.

The following data present selected results of conducted numerical tests of tracking motion implementation with the use of a trajectory in the shape of digit 8, with the parametric disturbance.

### 7.5.2.2   Examination of DHP-Type Control, 8-Shaped Path, Disturbance Trial

The numerical test of tracking control of point $A$ of the WMR was carried out with the use of NPD structure in DHP configuration, assuming zero initial values of actor's and critic's output layer weights. The task of the control system was to generate control signals to ensure the implementation of the desired trajectory of point $A$ of the WMR along the path in the shape of digit 8, with double parametric disturbance in the form of parameters changes in the WMR model corresponding to the loading of robot's frame with $m_{RL} = 4.0$ [kg] mass in time $t_{d1} = 12.5$ [s] and the removal of the additional mass in time $t_{d2} = 32.5$ [s].

The values of the tracking control signals for individual structures of the control system are shown in Fig. 7.37.

The overall control signals $u_1$ and $u_2$ controlling, respectively, wheel 1 and 2 are shown in Fig. 7.37a, accordingly to the equation (7.66) they consist of control signals generated by the DHP structure $u_{A1}$ and $u_{A2}$ (Fig. 7.37b), control signals of PD controller $u_{PD1}$ and $u_{PD2}$ (Fig. 7.37c), supervisory control signals $u_{S1}$ and $u_{S2}$ and additional control signals $u_{E1}$ and $u_{E2}$ (Fig. 7.37d).

The adaptation process of DHP structure parameters progresses faster in the initial phase of the I stage of motion than when using HDP structures with zero initial values of NN output layers weights, and generates control signals that have a large share in overall control signals from the beginning of the WMR point $A$ motion. The control signals of PD controller have small values in the initial phase of motion and then are reduced to values close to zero. During disturbance in time $t_{d1}$ and $t_{d2}$ the control signals generated by the DHP structure compensate for the changes in parameters of the controlled object. In the II stage of motion, after time $t = 22.5$ [s], the compensation for the WMR nonlinearity performed by NDP structure provides good quality of tracking, due to the information about the controlled object dynamics included in the values of actor's NNs output layer weights. The values of the $\mathbf{u}_E$ control signals depend mainly on $\mathbf{z}_{d3}$. Changes in values of tracking errors have a small influence on the sequences of these signals. The values of the $\mathbf{u}_S$ supervisory

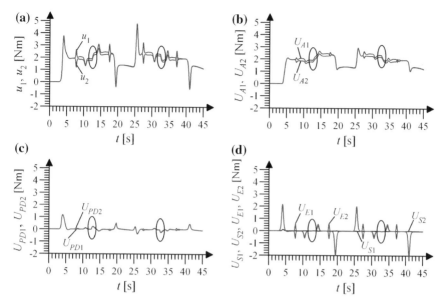

**Fig. 7.37** **a** Overall control signals $u_1$ and $u_2$ for wheels 1 and 2, **b** actor's NNs control signals $U_{A1}$ and $U_{A2}$, $\mathbf{U}_A = -\frac{1}{h}\mathbf{M}u_A$, **c** PD control signals $U_{PD1}$ and $U_{PD2}$, $\mathbf{U}_{PD} = -\frac{1}{h}\mathbf{M}u_{PD}$, **d** supervisory term's control signals $U_{S1}$ and $U_{S2}$, $\mathbf{U}_S = -\frac{1}{h}\mathbf{M}u_S$, and additional control signals $u_{E1}$ and $u_{E2}$, $\mathbf{U}_E = -\frac{1}{h}\mathbf{M}u_E$

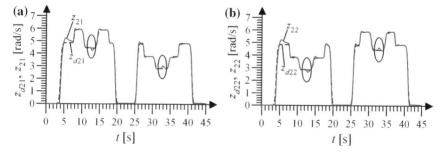

**Fig. 7.38** **a** Desired ($z_{d21}$) and executed ($z_{21}$) angular velocity of the WMR's wheel 1, **b** desired ($z_{d22}$) and executed ($z_{22}$) angular velocity of the WMR's wheel 2

control signals remain equal to zero, because the filtered tracking errors **s** do not exceed the acceptable level.

The values of executed angular velocities of WMR drive wheels rotation shown in Fig. 7.38 are similar to the desired ones. Only in the initial phase of the first stage of motion a clear difference between $z_{d21}$ and $z_{d22}$, $z_{21}$ and $z_{22}$ sequences resulting from assuming zero initial values of NNs output layer weights of DHP structure can be noticed. During the disturbance occurrence, in $t_{d1}$, simulating loading of the WMR with additional mass, the values of angular velocities of motion execution are

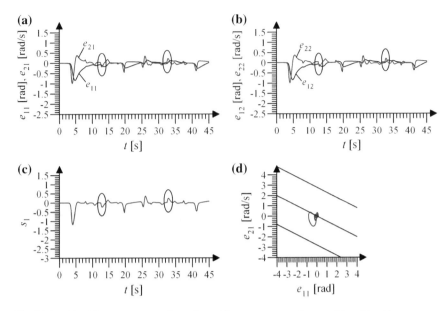

**Fig. 7.39** **a** Tracking errors $e_{11}$ and $e_{21}$, **b** tracking errors $e_{12}$ and $e_{22}$, **c** filtered tracking error $s_1$, **d** sliding manifold of wheel 1

decreasing, because the generated control signals are not sufficient for the execution of the WMR motion with the desired velocity due to increased resistance to motion. The increase in signal values compensating for the WMR nonlinearity, which are generated by the DHP structure, increases the values of angular velocities and causes the further execution of the trajectory with desired parameters of motion. Similarly, at the time $t_{d2}$ of the second parametric disturbance occurrence simulating the removal of additional load applied to the WMR which means return of parameters of the model to the nominal values, the generated values of control signals are too large in relation to the required ones due to smaller resistance to motion of the object with reduced mass, hence the value of executed velocity of point $A$ of the WMR is greater than the desired value. As a result of the NNs output layer weights adaptation process of the DHP structure the values of $u_{A1}$ and $u_{A2}$ control signals are reduced, the velocity of motion execution decreases and reaches the desired value.

The values of tracking errors of wheel 1 and 2 of the WMR are presented, respectively, in Fig. 7.39a, b. Figure 7.39c presents the values of filtered tracking error $s_1$, the values of filtered tracking error $s_2$ is similar. Figure 7.39d shows the sliding manifold of wheel 1. The largest values of tracking errors occur in the initial phase of motion; they are minimized due to the adaptation process of NNs output layer weights of DHP structure. The influence of the parametric disturbance on the quality of tracking motion execution is small, and the increase in the absolute error values of the trajectory execution, caused by changes in the dynamics of the controlled object, is eliminated by adopting NNs output layer weights of NDP structure.

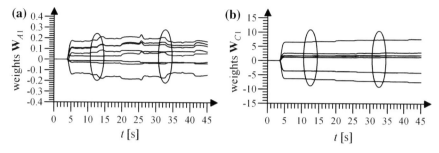

**Fig. 7.40** **a** Actor's 1 NN output layer weights $\mathbf{W}_{A1}$, **b** critic's 1 NN output layer weights $\mathbf{W}_{C1}$

**Fig. 7.41** Desired and executed motion path of the WMR's point $A$

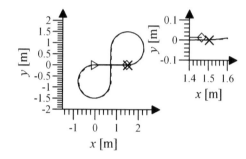

**Table 7.21** Values of selected performance indexes

| Index | $e_{max1j}$ [rad] | $\varepsilon_{1j}$ [rad] | $e_{max2j}$ [rad/s] | $\varepsilon_{2j}$ [rad/s] | $s_{maxj}$ [rad/s] | $\sigma_j$ [rad/s] |
|---|---|---|---|---|---|---|
| Wheel 1, $j = 1$ | 0.83 | 0.18 | 0.98 | 0.16 | 1.18 | 0.18 |
| Wheel 2, $j = 2$ | 0.83 | 0.19 | 0.98 | 0.16 | 1.18 | 0.18 |

Values of actor's and critic's NNs output layer weights, related to wheel 1 are shown in Fig. 7.40. Values of actor's and critic's NNs output layer weights related to wheel 2 are similar. Zero initial values of actor's and critic's NNs output layer weights were assumed. The values of weights stabilize around certain values or remain limited during the numerical test. During the parametric disturbance it is possible to notice the changes in values of actor's NN output layer weights resulting from the adjustments of generated control signal compensating for the WMR nonlinearity to the changed parameters of the controlled object.

The desired (dashed line) and executed (continuous line) motion path of WMR point $A$ is shown in Fig. 7.41. The initial position of point $A$ of the WMR is marked by a triangle, the desired final position by "**X**" and the final position of point $A$ by a rhombus.

Tables 7.21 and 7.22 show the values of individual performance indexes for the execution of the tracking motion with the application of a control system with ACD in DHP configuration.

**Table 7.22** Values of selected performance indexes of path execution

| Index | $d_{max}$ [m] | $d_n$ [m] | $\rho_d$ [m] |
|-------|---------------|-----------|--------------|
| Value | 0.069         | 0.043     | 0.034        |

**Table 7.23** Values of selected performance indexes in I and II stage of motion

| Index | $e_{max1j}$ [rad] | $\varepsilon_{1j}$ [rad] | $e_{max2j}$ [rad/s] | $\varepsilon_{2j}$ [rad/s] | $s_{maxj}$ [rad/s] | $\sigma_j$ [rad/s] |
|-------|-------------------|--------------------------|---------------------|----------------------------|--------------------|--------------------|
| Stage I, $k = 1, \ldots, 2250$ | | | | | | |
| Wheel 1, $j = 1$ | 0.83 | 0.25 | 0.98 | 0.2 | 1.18 | 0.23 |
| Wheel 2, $j = 2$ | 0.83 | 0.25 | 0.98 | 0.2 | 1.18 | 0.24 |
| Stage II, $k = 2251, \ldots, 4500$ | | | | | | |
| Wheel 1, $j = 1$ | 0.26 | 0.08 | 0.41 | 0.1 | 0.45 | 0.11 |
| Wheel 2, $j = 2$ | 0.26 | 0.08 | 0.41 | 0.1 | 0.45 | 0.11 |

As a result of the executed numerical test the obtained values of performance indexes were lower than in the case of PD regulator implementation in the algorithm of motion control. The obtained values of performance indexes are lower than in the case of the use of adaptive, neural control algorithm or control algorithm with HDP structure and zero initial values of actor's and critic's NNs output layer weights described in Sect. 7.4.2.1, and comparable to those obtained in the numerical test Sect. 7.4.2.2, where the non-zero initial values of critic's NNs output layer weights in HDP algorithm were implemented.

Table 7.23 presents the values of selected performance indexes in first and second stage of motion along the path in the shape of digit 8, calculated in accordance with the methodology described in the numerical test Sect. 7.1.2.2.

In the second stage of motion lower values of performance indexes were obtained than in the first stage. This results from the application of non-zero values of NNs output layer weights in the initial phase of the motion in the second stage of the numerical test. The knowledge of the controlled object dynamics, included in the NNs output layer weights of the DHP structure enables generation of signals compensating for the WMR nonlinearities, which improves the quality of the execution of the tracking motion.

### 7.5.2.3 Numerical Test of the RM Motion Control with the Use of DHP Algorithm

In a virtual computing environment simulation tests on the tracking control of RM with the use of control algorithm with NDP structure in DHP configuration were carried out. From the numerical tests selected were values obtained within implementation of trajectory with path in the shape of a semicircle, as described in Sect. 2.2.1.2. The numerical test in the version with the parametric disturbance in the form of changes in the parameters of the model corresponding to the loading of the RM's end-effector with transported mass was conducted.

In the numerical test, the following parameters of the control system with the DHP structure were applied:

- PD controller parameters:

  – $\mathbf{K}_D = \text{diag}\,\{0.4, 0.7, 0.7\}$ – constant diagonal matrix of PD controller gains,
  – $\mathbf{\Lambda} = \mathbf{I}$ – constant diagonal matrix,

- type of actor's NNs applied: RVFL,

  – $m_A = 8$ – number of the actor's NN neurons,
  – $\mathbf{\Gamma}_{A1} = \text{diag}\,\{0.26\}$ – diagonal matrix of gain coefficients for the adaptation of weights $\mathbf{W}_{A1}$,
  – $\mathbf{\Gamma}_{A2} = \mathbf{\Gamma}_{A3} = \text{diag}\,\{0.2\}$ – diagonal matrix of gain coefficients for the adaptation of weights $\mathbf{W}_{A2}$ and $\mathbf{W}_{A3}$,
  – $\mathbf{D}_A$ – NN input layer weights randomly selected from the interval $D_{A[i,g]} \in \langle -0.2, 0.2 \rangle$,
  – $\mathbf{W}_{Aj\{k=0\}} = \mathbf{0}$ – zero initial values of the output layer weights, $j = 1, 2, 3$,
  – $k_A = 0.1$ – positive constant,

- type of critic's NNs applied: RVFL,

  – $m_C = 8$ – number of the critic's NN neurons,
  – $\mathbf{\Gamma}_{C1} = \text{diag}\,\{2\}$ – diagonal matrix of gain coefficients for the adaptation of weights $\mathbf{W}_{C1}$,
  – $\mathbf{\Gamma}_{C2} = \mathbf{\Gamma}_{C3} = \text{diag}\,\{1\}$ – diagonal matrix of gain coefficients for the adaptation of weights $\mathbf{W}_{C2}$ and $\mathbf{W}_{C3}$,
  – $\mathbf{D}_C$ – NN input layer weights randomly selected from the interval $D_{C[i,g]} \in \langle -0.2, 0.2 \rangle$,
  – $\mathbf{W}_{Cj\{k=0\}} = \mathbf{0}$ – zero initial values of output layer weights,
  – $k_C = 0.1$ – positive constant,

- $\beta_N = 2$ – slope coefficient of bipolar sigmoid neuron activation functions,
- $\mathbf{R} = \text{diag}\,\{1\}$ – matrix of scaling coefficient in the cost function,
- $l_{AC} = 5$ – number of inner loop iteration steps for the actor's and critic's structure adaptation in the $k$th step of the iterative process,
- $\rho_j = 0.2$ – supervisory algorithm parameter,
- $\mathbf{F} = [0.25, 0.25, 0.35]^T$ – vector of constant values of assumed restrictions of function $Y_{fj}\left(\mathbf{z}_{1\{k\}}, \mathbf{z}_{2\{k\}}\right)$,
- $\sigma_j = 0.01$ – supervisory algorithm parameter,
- $h = 0.01$ [s] – numerical test calculation step.

The same performance indexes as described in Sect. 7.4.2, were applied. The number of discrete measurement data of the trajectory of RM with path in the shape of a semicircle amounts to $n = 4000$.

In the description of numerical tests of algorithms operating in discrete time domain the notation of variables was simplified by resigning from the $k$th step of the process. To facilitate analysis of results and comparison of continuous functions for the previously presented WMR's control systems, the axis of abscissae of the graphs representing discrete functions was scaled in time $t$.

The following are selected results of carried out numerical tests on tracking motion execution with the use of trajectory with path in the shape of a semicircle with the parametric disturbance.

### 7.5.2.4 Examination of DHP-Type Control, Semicircle Path, Disturbance Trial

In the numerical test of the RM control with NDP algorithm in DHP configuration the desired trajectory with path in the shape of a semicircle, discussed in Sect. 2.2.1.2, was applied. The parametric disturbance resulting in a change in parameters of the RM corresponding to the load of the gripper with a $m_{RL} = 1.0$ [kg] mass in time $t_{d1} \in < 21; 27 >$ [s] and $t_{d2} \in < 33; 40 >$ [s] was introduced.

In the Fig. 7.42a the green line presents the desired and the red line the executed path of point $C$ of RM's end-effector. The triangle, on the graph, indicates the initial

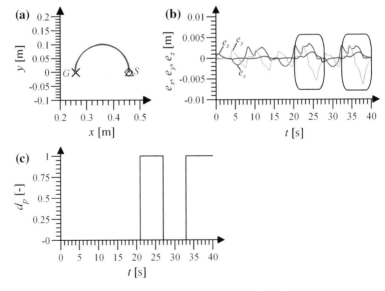

**Fig. 7.42** **a** Desired and executed path of end-effector's point $C$, **b** path errors $e_x, e_y, e_z$, **c** parametric disturbance signal $d_p$

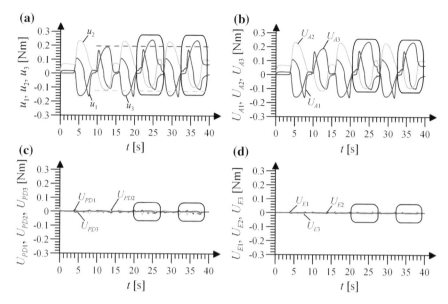

**Fig. 7.43** **a** The overall signals $u_1$, $u_2$ and $u_3$ controlling the motion of the RM's links, **b** the control signals generated by the actor NNs $U_{A1}$, $U_{A2}$ and $U_{A3}$, $\mathbf{U}_A = -\frac{1}{h}\mathbf{Mu}_A$, **c** control signals of PD controller $U_{PD1}$, $U_{PD2}$ and $U_{PD3}$, $\mathbf{U}_{PD} = -\frac{1}{h}\mathbf{Mu}_{PD}$, **d** control signals $U_{E1}$, $U_{E2}$ and $U_{E3}$, $\mathbf{U}_E = -\frac{1}{h}\mathbf{Mu}_E$

position (point $S$), the "**X**" represents desired reverse position of the path (point $G$), rhombus represents the reached final position of point $C$ of the RM. The graph Fig. 7.42b presents the error values of the desired path of point $C$ of the RM end-effector, $e_x = x_{Cd} - x_C, e_y = y_{Cd} - y_C, e_z = z_{Cd} - z_C$. The graph Fig. 7.42c presents the values of the $d_p$ signal informing about the occurrence of parametric disturbances. If $d_p = 0$, in the model of the RM dynamics the nominal set of **p** parameters provided in the Table 2.3 is applied and if $d_p = 1$, the set of $\mathbf{p}_d$ parameters corresponding to the occurrence of parametric disturbance resulting from loading the gripper with additional mass is applied.

The path of point $C$ of the RM end-effector results from introducing into the actuator systems tracking control signals, whose sequences are presented in Fig. 7.43. The overall control signals $u_1$, $u_2$ and $u_3$ of link 1, 2 and 3 are presented in Fig. 7.43a. According to the relation (7.101) they consist of control signals generated by the DHP structure $u_{A1}$, $u_{A2}$ and $u_{A3}$ (Fig. 7.43b), control signals of the PD controller $u_{PD1}$, $u_{PD2}$ and $u_{PD3}$ (Fig. 7.43c), supervisory control signals $u_{S1}$, $u_{S2}$ and $u_{S3}$ and additional control signals $u_{E1}$, $u_{E2}$ and $u_{E3}$ (Fig. 7.43d). The supervisory control signals are not shown on any of the diagrams, because their values during the numerical tests amounted to zero, which results from the assumed values of the $\rho_j$ parameter of supervisory control algorithm.

In the initial position of the arm of the RM, when $t = 0$ [s], the rotation angles of individual links amount to $z_{11} = 0, z_{12} = 0.16533, z_{13} = -0.16533$ [rad], and the point $C$ of the end-effector is at the maximum distance from the robot's base (point $S$). Despite the fact that in the initial stage of the numerical test, when $t \in \langle 0, 3.5 \rangle$, the values of desired angles of individual links of RM are constant; on the diagrams of control signals sequences occur nonzero values of signals controlling the motion of links 2 and 3, which results from the configuration of robotic arm and interaction of moments from gravity forces in individual joints. Dominant, among the overall control signals, are control signals generated by the actor's structure of the DHP algorithm and control signals of other elements of the control system have values close to zero. Loading the robotic manipulator end-effector with additional mass reduces the absolute value of the signal controlling link 2 motion, $u_2$, because the moment of the gravity forces resulting from the occurrence of additional mass helps to achieve the desired rotation of the link to the right. The opposite situation occurs in the case of link 3 and the control signal $u_3$. Loading the end-effector during the rotation to the left impedes the desired motion by generating the moment originating from the gravity forces acting on the transported mass. Thus, in the diagram of the control signal of link 3 motion during the parametric disturbance occurrences, the increased absolute values of the control signal $u_3$, which compensates for the increased moment of load in the link joint, can be noticed. Change of the controlled object dynamics resulting from the load transfer by the RM gripper is compensated by appropriate adjustment of the control signals of actor's NNs of the DHP algorithm.

The value of desired (red line) and executed (green line) angular velocity of link 2 rotation is shown in Fig. 7.44a. The desired ($|v_{Cd}|$) and the executed ($|v_C|$) absolute value of velocity of point $C$ of the RM gripper is shown in Fig. 7.44b. The small differences between desired and executed motion parameters can be noticed during simulation of loading of the end-effector with an additional mass ($t = 21$ [s], $t = 33$ [s]) and when the additional mass is removed ($t = 27$ [s]).

The tracking errors of the RM's link 1, 2 and 3 motion are presented in Fig. 7.45a, b. Figure 7.45c presents the sequences of filtered tracking error $s_1$, $s_2$ and $s_3$. Figure 7.45d demonstrates the sliding manifold of link 1. The largest values of tracking errors occur during the motion of the RM's arm loaded with additional mass.

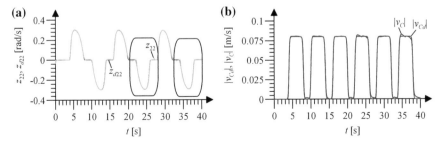

**Fig. 7.44** **a** Desired ($z_{d22}$) and executed ($z_{22}$) angular velocity of the RM's link 2 rotation, **b** desired ($|v_{Cd}|$) and executed ($|v_C|$) absolute value of velocity of point $C$ of the manipulator's end-effector

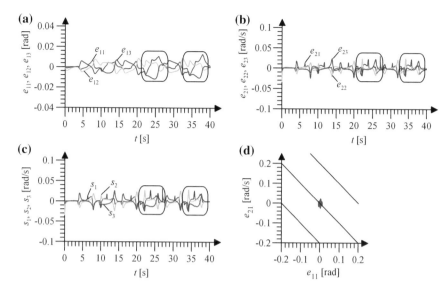

**Fig. 7.45** **a** Tracking errors of rotation angles of links, $e_{11}$, $e_{12}$ and $e_{13}$ **b** tracking errors of angular velocities, $e_{21}$, $e_{22}$ and $e_{23}$, **c** filtered tracking errors $s_1$, $s_2$ and $s_3$, **d** sliding manifold of link 1

However, the influence of parametric disturbance on the quality of tracking is small and the increase in the absolute value of tracking errors caused by changes in dynamics of the controlled object is minimized by adopting the NNs output layer weights of the DHP structure.

The values of the actor's and the critic's NNs output layer weights of the DHP structure are shown in Fig. 7.46. Zero initial values of actor's and critic's NNs output layer weights were applied. Values of weights stabilize around certain values and are limited during numerical test. When the parametric disturbance occurs it is possible to notice changes in the actor's NNs output layer weights. This results from the adjustments of generated signals compensating for nonlinearities of the RM to the changed parameters of the controlled object. Adaptation of NNs output layer weights of actor's and critic's structure, generating a signal controlling link 1, begins when the non-zero value of filtered tracking error $s_1$ occurs after time $t = 3$ [s]. In contrast, the NN weights of other DHP algorithm structures, generating signals controlling link 2 and 3, are adopted from the beginning of the numerical test. This results from the necessity of generating signals controlling the motion of link 2 and 3 which are also eliminating the influence of moments originating from the gravity forces and ensuring placement of point $C$ of the RM's end-effector in a desired position.

The values of individual performance indexes of executed tracking motion with the use of control system with ACD in DHP configuration are shown in Tables 7.24 and 7.25.

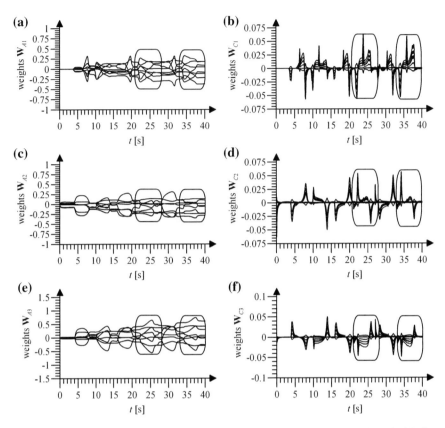

**Fig. 7.46** **a** Value of weights of actor first NN output layer $\mathbf{W}_{A1}$, **b** value of weights of critic first NN output layer $\mathbf{W}_{C1}$, **c** value of weights of actor second NN output layer $\mathbf{W}_{A2}$, **d** value of weights of critic second NN output layer $\mathbf{W}_{C2}$, **e** value of weights of actor third NN output layer $\mathbf{W}_{A3}$, **f** value of weights of critic third NN output layer $\mathbf{W}_{C3}$

**Table 7.24** Values of selected performance indexes

| Index | $e_{max1j}$ [rad] | $\varepsilon_{1j}$ [rad] | $e_{max2j}$ [rad/s] | $\varepsilon_{2j}$ [rad/s] | $s_{maxj}$ [rad/s] | $\sigma_j$ [rad/s] |
|---|---|---|---|---|---|---|
| Link 1, $j = 1$ | 0.0128 | 0.004 | 0.0246 | 0.0063 | 0.0276 | 0.0075 |
| Link 2, $j = 2$ | 0.0088 | 0.0039 | 0.0343 | 0.0062 | 0.0344 | 0.0074 |
| Link 3, $j = 3$ | 0.0117 | 0.0053 | 0.0302 | 0.006 | 0.0321 | 0.0081 |

**Table 7.25** Values of selected performance indexes of path execution

| Index | $d_{max}$ [m] | $d_n$ [m] | $\rho_d$ [m] |
|---|---|---|---|
| Value | 0.0064 | 0.0005 | 0.0025 |

The obtained values of performance indexes, after carried out numerical test, were significantly lower than in the case of applying PD controller in tracking control algorithm of RM.

### 7.5.3 Conclusions

The use of NDP structure in DHP configuration in the control system provides high quality of motion execution expressed by low values of performance indexes. In the case of applying the worst case scenario of zero values of the initial weights of NN output layer of actor's and critic's structures in the DHP algorithm, the obtained values of performance indexes are lower than those obtained during execution of motion with the use of a PD controller, the adaptive control system or neural control system. In the DHP structure critic's NN approximate derivative of a value function with respect to the state vector of the system. This complicates the actor's and critic's NN output layer weights adaptation law, but provides better quality of motion, which is confirmed by the carried out numerical tests.

A great influence on the process of actor's and critic's RVFL NN output layer weights adaptation have constant values of input layer weights selected randomly in the process of network initialization. The convergence of the output layer weights adaptation process depends on the values of the applied learning coefficients. With too low values of learning coefficients the weights adaptation process is slow. Assuming high values, on the other hand, can lead to instability of the process. An important constructional parameter of the DHP algorithm is the number of cycles of actor's and critic's structures adaptation in the inner loop, $l_{AC}$, carried out in each $k$th step of the process, which may change during the process due to the assumed criterion. The influence of other constructional parameters of NDP structures on the weights adaptation process of the NN output layer of the DHP algorithm is the same as in the HDP structure and was discussed earlier. Synthesis of DHP algorithm requires a knowledge of the mathematical description of the controlled object, whereas the application of the simplified model, which does not fully maps the behavior of the real object, does not result in a significant reduction of the generated control quality.

## 7.6 Globalised Dual-Heuristic Dynamic Programming in Tracking Control of a WMR and a RM

NDP algorithm in GDHP configuration [26, 77, 79, 85, 87, 88] has a similar structure as the structure of the HDP algorithm, simplified in comparison to the structure of DHP algorithm. This algorithm consists of actor's and critic's structures and predictive model of controlled object. Knowledge of the mathematical model of the controlled object is necessary to generate actor's and critic's NN weights adaptation law. Critic in GDHP algorithm approximates the value function as in the HDP algorithm, therefore, it is built with the use of only a single adaptive structure, e.g., NN. In the literature it possible to find information about other versions of GDHP algorithm, in which the critic approximates the value function and its derivative with respect to the state vector of the object [77]. Such an approach makes it necessary to use the NN with many outputs or several NNs with a single output, in order to implement the critic's structure. Since in the presented approach the critic's NN approximates the value function, the derivatives of the value function with respect to the state vector of the controlled object can be determined analytically. Therefore, the second version of the algorithm, with a more complex structure of the critic, will not be discussed. Thus, while the structure of critic of the GDHP algorithm is simplified in relation to the structure of critic of the DHP algorithm, the NN weights adaptation law of critic of the GDHP algorithm is more complex than in previously discussed algorithms. It is based on the minimization of performance indexes characteristic for NN weights adaptation algorithms of critic of the HDP and DHP structures. The actor generates the control law of the object, hence its construction results from the number of independent signals controlling actuator systems. For example, in a tracking control system of WMR with two degrees of freedom, the actor is comprised of two adaptive structures, e.g. NN, generating independent signals controlling the robot's driving modules. In this chapter, a synthesis of discrete tracking control system of dynamic object with the NDP algorithm in GDHP configuration was carried out based on the example of WMR and 3DoF RM. The presented algorithms operate in the discrete time domain and in the control system synthesis a discrete dynamics models of WMR and 3DoF RM were applied. This chapter is based on the information about the GDHP algorithm, whose description is provided in Sect. 6.2.3. Description of the WMR dynamics is provided in Sect. 2.1.2 and the description of the RM dynamics is presented in Sect. 2.2.2.

## 7.6.1  Synthesis of GDHP-Type Control

The synthesis of algorithm controlling the dynamic object with the use of NDP structure in GDHP configuration was carried out similarly to the implementation of the control system with HDP algorithm described in Sect. 7.4.1. Synthesis of WMR and RM tracking control systems was carried out. Because there are differences in the description of the dynamics of controlled objects listed, also control algorithms of the individual objects are different. The synthesis of algorithm controlling the WMR motion with GDHP structure is presented in detail, in the case of tracking control system of the RM the differences in the process of synthesis and obtained results due to the nature of controlled object were indicated.

### 7.6.1.1   Synthesis of WMR's Tracking Control with the Application of a GDHP Algorithm

In the synthesis of a discrete tracking control algorithm for a WMR the tracking errors $e_{1\{k\}}$ and $e_{2\{k\}}$ definitions were assumed according to the relation (7.53), the filtered tracking errors $s_{\{k\}}$ according to the relation (7.54), and $s_{\{k+1\}}$ according to relation (7.55) which was formulated in Sect. 7.4.1.

An NDP algorithm in GDHP configuration, in which the structure of the actor was implemented in the form of two RVFL NNs due to the number of independent signals controlling robot's actuator systems, was used in the designed tracking control system of point $A$ of the WMR. The structure of the critic was implemented with a single RVFL NN, because the critic approximates the value function.

The task of the NDP algorithm in GDHP configuration is to generate a suboptimal control law $\mathbf{u}_{A\{k\}} = \left[ u_{A1\{k\}}, u_{A2\{k\}} \right]^T$ that minimizes the value function defined by relation

$$V_{\{k\}}\left(\mathbf{s}_{\{k\}}\right) = \sum_{k=0}^{n} \gamma^k L_{C\{k\}}\left(\mathbf{s}_{\{k\}}\right) , \tag{7.107}$$

with the cost function

$$L_{C\{k\}}\left(\mathbf{s}_{\{k\}}\right) = \frac{1}{2}\mathbf{s}_{\{k\}}^T \mathbf{R}\mathbf{s}_{\{k\}} , \tag{7.108}$$

where $\mathbf{R}$ – positive-definite, fixed diagonal matrix. The task of the GDHP algorithm with the use of value function (7.107) with the local cost (7.108) is to generate a control law that will minimize the value of filtered tracking errors $s_{\{k\}}$.

The general scheme of NDP structure in GDHP configuration is presented in Fig. 7.47.

Described in Sect. 6.2.3 GDHP algorithm is composed of a predictive model and two adaptive structures:

- **Predictive model** of dynamics of the filtered tracking error of WMR generates a prediction of $s_{\{k+1\}}$ state of $k+1$ step of the discrete process based on the Eq. (7.55) according to the relation

$$\mathbf{s}_{\{k+1\}} = \mathbf{s}_{\{k\}} - \mathbf{Y}_f\left(\mathbf{z}_{2\{k\}}\right) + \mathbf{Y}_e\left(\mathbf{z}_{\{k\}}, \mathbf{z}_{d\{k\}}, \mathbf{z}_{d3\{k\}}\right) + h\mathbf{M}^{-1}\mathbf{u}_{\{k\}} . \tag{7.109}$$

- **Actor**, implemented in the form of two RVFL NNs, generates suboptimal control law $\mathbf{u}_{A\{k\}}$ in $k$th step according to the relation

$$u_{Aj\{k\}}\left(\mathbf{x}_{Aj\{k\}}, \mathbf{W}_{Aj\{k\}}\right) = \mathbf{W}_{Aj\{k\}}^T \mathbf{S}\left(\mathbf{D}_A^T \mathbf{x}_{Aj\{k\}}\right) , \tag{7.110}$$

where $j = 1, 2$, actor's NNs of GDHP algorithm are constructed similarly to the HDP structure shown in Sect. 7.4.1.

The procedure of actor's NN weights adaptation of the GDHP algorithm consists in minimization of performance index

**Fig. 7.47** Scheme of NDP
structure in GDHP
configuration

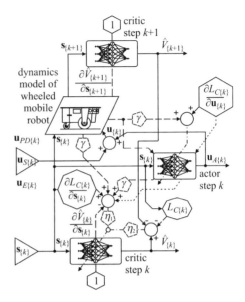

$$\mathbf{e}_{A\{k\}} = \frac{\partial L_{C\{k\}}\left(\mathbf{s}_{\{k\}}\right)}{\partial \mathbf{u}_{\{k\}}} + \gamma \left[\frac{\partial \mathbf{s}_{\{k+1\}}}{\partial \mathbf{u}_{\{k\}}}\right]^{T} \frac{\partial \hat{V}_{\{k+1\}}\left(\mathbf{x}_{C\{k+1\}}, \mathbf{W}_{C\{k\}}\right)}{\partial \mathbf{s}_{\{k+1\}}} , \qquad (7.111)$$

according to the relation

$$\mathbf{W}_{Aj\{k+1\}} = \mathbf{W}_{Aj\{k\}} - e_{Aj\{k\}}\boldsymbol{\Gamma}_{Aj}\mathbf{S}\left(\mathbf{D}_{A}^{T}\mathbf{x}_{Aj\{k\}}\right) , \qquad (7.112)$$

where $\hat{V}_{\{k+1\}}\left(\mathbf{x}_{C\{k+1\}}, \mathbf{W}_{C\{k\}}\right)$ – output from critic's NN for the $k+1$ step.
- **Critic**, implemented in the form of RVFL NN, approximates the value function
(7.107). Output of the critic NN is expressed by the following relation

$$\hat{V}_{\{k\}}\left(\mathbf{x}_{C\{k\}}, \mathbf{W}_{C\{k\}}\right) = \mathbf{W}_{C\{k\}}^{T}\mathbf{S}\left(\mathbf{D}_{C}^{T}\mathbf{x}_{C\{k\}}\right) . \qquad (7.113)$$

Critic's NN of GDHP algorithm in presented approach is constructed similarly to
the HDP structure shown in Sect. 7.4.1.

The procedure of weights adaptation of critic's NN of the GDHP algorithm consist
in minimization of two performance indexes. The first one is the temporal difference
error $e_{C1\{k\}}$ characteristic for the adaptation method of critic's algorithm of HDP
structure (6.7), expressed by the relation

$$e_{C1\{k\}} = L_{C\{k\}}\left(\mathbf{s}_{\{k\}}\right) + \gamma \hat{V}_{\{k+1\}}\left(\mathbf{x}_{C\{k+1\}}, \mathbf{W}_{C\{k\}}\right) - \hat{V}_{\{k\}}\left(\mathbf{x}_{C\{k\}}, \mathbf{W}_{C\{k\}}\right) , \quad (7.114)$$

while the second performance index is the derivative of the temporal difference error
$e_{C1\{k\}}$ with respect the state vector of the controlled object $\mathbf{s}_{\{k\}}$. This performance
index is used to adapt the structure of the critic in DHP algorithm (6.18) and in

GHDP algorithm is expressed by the relation

$$
e_{C2\{k\}} = \mathbf{I}_D^T \left\{ \frac{\partial L_{C\{k\}}\left(\mathbf{s}_{\{k\}}\right)}{\partial \mathbf{s}_{\{k\}}} + \left[ \frac{\partial \mathbf{u}_{\{k\}}}{\partial \mathbf{s}_{\{k\}}} \right]^T \frac{\partial L_{C\{k\}}\left(\mathbf{s}_{\{k\}}\right)}{\partial \mathbf{u}_{\{k\}}} + \right. \tag{7.115}
$$

$$
+ \gamma \left[ \frac{\partial \mathbf{s}_{\{k+1\}}}{\partial \mathbf{s}_{\{k\}}} + \left[ \frac{\partial \mathbf{u}_{\{k\}}}{\partial \mathbf{s}_{\{k\}}} \right]^T \frac{\partial \mathbf{s}_{\{k+1\}}}{\partial \mathbf{u}_{\{k\}}} \right]^T \frac{\partial \hat{V}_{\{k+1\}}\left(\mathbf{x}_{C\{k+1\}}, \mathbf{W}_{C\{k\}}\right)}{\partial \mathbf{s}_{\{k+1\}}} +
$$

$$
\left. - \frac{\partial \hat{V}_{\{k\}}\left(\mathbf{x}_{C\{k\}}, \mathbf{W}_{C\{k\}}\right)}{\partial \mathbf{s}_{\{k\}}} \right\} ,
$$

where $\mathbf{I}_D$ – constant vector, $\mathbf{I}_D = [1, 1]^T$, $\mathbf{s}_{\{k+1\}}$ – signal obtained from the predictive model (7.109).

Weights of critic's NNs are adopted with backpropagation method of performance indexes (7.114) and (7.115) according to the relation

$$
\mathbf{W}_{C\{k+1\}} = \mathbf{W}_{C\{k\}} - \eta_1 e_{C1\{k\}} \mathbf{\Gamma}_C \frac{\partial \hat{V}_{\{k\}}\left(\mathbf{x}_{C\{k\}}, \mathbf{W}_{C\{k\}}\right)}{\partial \mathbf{W}_{C\{k\}}} + \tag{7.116}
$$

$$
- \eta_2 e_{C2\{k\}} \mathbf{\Gamma}_C \frac{\partial^2 \hat{V}_{\{k\}}\left(\mathbf{x}_{C\{k\}}, \mathbf{W}_{C\{k\}}\right)}{\partial \mathbf{s}_{\{k\}} \partial \mathbf{W}_{C\{k\}}} ,
$$

where $\mathbf{\Gamma}_C$ – fixed diagonal matrix of positive learning coefficients of critic's NN output layer weights, $\eta_1$, $\eta_2$ – positive constants.

In the designed tracking control system of WMR with NDP structure in GDHP configuration the overall control signals $\mathbf{u}_{\{k\}}$ consists of control signals generated by the actor's NNs $\mathbf{u}_{A\{k\}}$, the signals generated by the PD controller $\mathbf{u}_{PD\{k\}}$, $\mathbf{u}_{E\{k\}}$ signals and additional supervisory control signals $\mathbf{u}_{S\{k\}}$ with a structure resulting from the application of Lyapunov stability theory to the analysis of the stability of designed control system. The scheme of the tracking control system of the WMR with the use of GDHP algorithm is presented in Fig. 7.48.

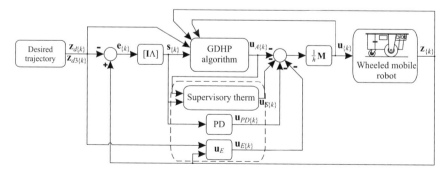

**Fig. 7.48** Scheme of the tracking control system of the WMR with the use of GDHP algorithm

The analysis of the closed-loop control system was conducted similarly as described in Sect. 7.4.1 by applying overall control signals in the form of relation (7.66) and the supervisory control signals $\mathbf{u}_{S\{k\}}$ in the form of relation (7.79).

Synthesis and the results of experimental tests carried out on tracking control system of WMR with the use of GDHP algorithm is presented in [85, 88].

### 7.6.1.2  Synthesis of RM's Tracking Control with the Application of a GDHP Algorithm

During the synthesis of discrete tracking control algorithm of RM the definition of tracking errors $\mathbf{e}_{1\{k\}}$ and $\mathbf{e}_{2\{k\}}$, according to the relation (7.53), and filtered tracking errors $\mathbf{s}_{\{k\}}$, according to the relation (7.54), which was introduced in Sect. 7.4.1, were accepted.

On the basis of the discrete description of RM dynamics (2.91), relation (7.53) and (7.54), the filtered tracking error $\mathbf{s}_{\{k+1\}}$ was defined as

$$\mathbf{s}_{\{k+1\}} = -\mathbf{Y}_f\left(\mathbf{z}_{1\{k\}}, \mathbf{z}_{2\{k\}}\right) + \mathbf{Y}_d\left(\mathbf{z}_{\{k\}}, \mathbf{z}_{d\{k+1\}}\right) - \mathbf{Y}_{\tau\{k\}}\left(\mathbf{z}_{1\{k\}}\right) + h\left[\mathbf{M}\left(\mathbf{z}_{1\{k\}}\right)\right]^{-1}\mathbf{u}_{\{k\}}, \tag{7.117}$$

where

$$\begin{aligned}
&\mathbf{Y}_f\left(\mathbf{z}_{1\{k\}}, \mathbf{z}_{2\{k\}}\right) = h\left[\mathbf{M}\left(\mathbf{z}_{1\{k\}}\right)\right]^{-1}\left[\mathbf{C}\left(\mathbf{z}_{1\{k\}}, \mathbf{z}_{2\{k\}}\right)\mathbf{z}_{2\{k\}} + \mathbf{F}\left(\mathbf{z}_{2\{k\}}\right) + \mathbf{G}\left(\mathbf{z}_{1\{k\}}\right)\right], \\
&\mathbf{Y}_{\tau\{k\}}\left(\mathbf{z}_{1\{k\}}\right) = h\left[\mathbf{M}\left(\mathbf{z}_{1\{k\}}\right)\right]^{-1}\boldsymbol{\tau}_{d\{k\}}, \\
&\mathbf{Y}_d\left(\mathbf{z}_{\{k\}}, \mathbf{z}_{d\{k+1\}}\right) = \mathbf{z}_{2\{k\}} - \mathbf{z}_{d2\{k+1\}} + \boldsymbol{\Lambda}\left[\mathbf{z}_{1\{k\}} + h\mathbf{z}_{2\{k\}} - \mathbf{z}_{d1\{k+1\}}\right] = \\
&\mathbf{z}_{2\{k\}} - \mathbf{z}_{d2\{k\}} - h\mathbf{z}_{d3\{k\}} + \boldsymbol{\Lambda}\left[\mathbf{z}_{1\{k\}} - \mathbf{z}_{d1\{k\}} + h\mathbf{z}_{2\{k\}} - h\mathbf{z}_{d2\{k\}}\right] = \\
&\mathbf{s}_{\{k\}} + \mathbf{Y}_e\left(\mathbf{z}_{\{k\}}, \mathbf{z}_{d\{k\}}, \mathbf{z}_{d3\{k\}}\right), \\
&\mathbf{Y}_e\left(\mathbf{z}_{\{k\}}, \mathbf{z}_{d\{k\}}, \mathbf{z}_{d3\{k\}}\right) = h\left[\boldsymbol{\Lambda}\mathbf{e}_{2\{k\}} - \mathbf{z}_{d3\{k\}}\right],
\end{aligned} \tag{7.118}$$

whereas $\mathbf{z}_{d3\{k\}}$ – a vector of desired angular accelerations. Vector $\mathbf{Y}_f\left(\mathbf{z}_{2\{k\}}\right)$ includes all nonlinearities of the RM.

In the designed tracking control system of point $C$ of the RM's arm the NDP algorithm in GDHP configuration was adopted, in which the structure of the actor was implemented in the form of three RVFL NNs due to the number of independent signals controlling the motion of robot's links. However, the critic was implemented with the use of a single RBF NN, because it approximates the value function in GDHP algorithm.

The value function (7.107) and the local cost determined by the relation (7.108) were assumed taking into account the dimensionality of vector $\mathbf{s}_{\{k\}}$, which in the analyzed case amounts to three.

The GDHP algorithm consists of a predictive model and two adaptive structures:
- **Predictive model** of dynamics of the filtered tracking error of RM generates a prediction of state $\mathbf{s}_{\{k+1\}}$ of $k + 1$ step of discrete process on the basis of (7.117) according to the relation

$$\mathbf{s}_{\{k+1\}} = -\mathbf{Y}_f\left(\mathbf{z}_{1\{k\}}, \mathbf{z}_{2\{k\}}\right) + \mathbf{Y}_d\left(\mathbf{z}_{\{k\}}, \mathbf{z}_{d\{k+1\}}\right) + h\left[\mathbf{M}\left(\mathbf{z}_{1\{k\}}\right)\right]^{-1}\mathbf{u}_{\{k\}} . \quad (7.119)$$

- **Actor**, implemented in the form of three RVFL NNs, generates suboptimal control law $\mathbf{u}_{A\{k\}}$ in $k$th step according to the relation (7.110), where $j = 1, 2, 3$.

The procedure of actor's NN weights adaptation of GDHP algorithm consist in minimization of performance index (7.111) according to the relation

$$\mathbf{W}_{Aj\{k+1\}} = \mathbf{W}_{Aj\{k\}} - e_{Aj\{k\}}\boldsymbol{\Gamma}_{Aj}\mathbf{S}\left(\mathbf{D}_A^T\mathbf{x}_{Aj\{k\}}\right) - k_A|e_{Aj\{k\}}|\boldsymbol{\Gamma}_{Aj}\mathbf{W}_{Aj\{k\}} , \quad (7.120)$$

where $k_A$ – a positive constant. The last element of the equation (7.120) is a regularization mechanism of NN weights [94] that prevents the over-fitting of the NNs weights.

- **Critic** approximates the value function (7.107) and was implemented in the form of one RBF NN, which output is expressed by the relation

$$\hat{V}_{\{k\}}\left(\mathbf{x}_{C\{k\}}, \mathbf{W}_{C\{k\}}\right) = \mathbf{W}_{C\{k\}}^T\mathbf{S}\left(\mathbf{x}_{C\{k\}}\right) , \quad (7.121)$$

where $\mathbf{W}_{C\{k\}}$ – vector of critic's RBF NN output layer weights, $\mathbf{x}_{C\{k\}} = \boldsymbol{\kappa}_C\mathbf{s}_{\{k\}}$ – vector of critic's NN inputs, $\boldsymbol{\kappa}_C$ – fixed diagonal matrix of positive scaling coefficients of critic's NN input values, $\mathbf{S}\left(\mathbf{x}_{C\{k\}}\right)$ – vector of radial neurons activation functions of Gaussian curve type described by the relation (3.5).

Adaptation process of critic's NN output layer weights of GDHP algorithm consists in minimization of performance indexes $e_{C1\{k\}}$ (7.114) and $e_{C2\{k\}}$ (7.115) discussed in Sect. 7.6.1.1, according to the relation

$$\mathbf{W}_{C\{k+1\}} = \mathbf{W}_{C\{k\}} - \eta_1 e_{C1\{k\}}\boldsymbol{\Gamma}_C\frac{\partial\hat{V}_{\{k\}}\left(\mathbf{x}_{C\{k\}}, \mathbf{W}_{C\{k\}}\right)}{\partial\mathbf{W}_{C\{k\}}} - k_{C1}|e_{C1\{k\}}|\boldsymbol{\Gamma}_C\mathbf{W}_{C\{k\}} +$$

$$(7.122)$$

$$-\eta_2 e_{C2\{k\}}\boldsymbol{\Gamma}_C\frac{\partial^2\hat{V}_{\{k\}}\left(\mathbf{x}_{C\{k\}}, \mathbf{W}_{C\{k\}}\right)}{\partial\mathbf{s}_{\{k\}}\partial\mathbf{W}_{C\{k\}}} - k_{C2}|e_{C2\{k\}}|\boldsymbol{\Gamma}_C\mathbf{W}_{C\{k\}} ,$$

where $k_{C1}, k_{C2}$ – positive constants.

The analysis of the closed-loop control system was carried out similarly as described in Sect. 7.5.1.2. The overall control signals $\mathbf{u}_{\{k\}}$, in the designed tracking control system of point $C$ of the RM's arm with NDP structure in GDHP configuration, consist of control signals generated by actor's NNs $\mathbf{u}_{A\{k\}}$, signals generated by the PD controller $\mathbf{u}_{PD\{k\}}$, signals $\mathbf{u}_{E\{k\}}$ and additional supervisory control signals $\mathbf{u}_{S\{k\}}$. The scheme of the structure of neural tracking control system of point $C$ of the RM's arm with GDHP structure is shown in Fig. 7.49.

The overall control signals were assumed according to the relation

$$\mathbf{u}_{\{k\}} = -\frac{1}{h}\mathbf{M}\left(\mathbf{z}_{1\{k\}}\right)\left[\mathbf{u}_{A\{k\}} - \mathbf{u}_{S\{k\}} + \mathbf{u}_{PD\{k\}} + \mathbf{u}_{E\{k\}}\right] , \quad (7.123)$$

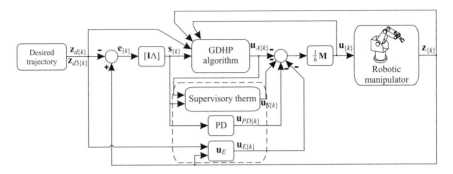

**Fig. 7.49** Scheme of tracking control system of the RM with GDHP algorithm

where

$$\mathbf{u}_{PD\{k\}} = \mathbf{K}_D\mathbf{s}_{\{k\}} ,$$
$$\mathbf{u}_{E\{k\}} = h\left[\mathbf{\Lambda e}_{2\{k\}} - \mathbf{z}_{d3\{k\}}\right] , \qquad (7.124)$$
$$\mathbf{u}_{S\{k\}} = \mathbf{I}_S\mathbf{u}_{S\{k\}}^* ,$$

where $\mathbf{K}_D$ – fixed positive-definite matrix, $\mathbf{I}_S$ – diagonal switching matrix, $I_{Sj,j} = 1$ when $\left|s_{j\{k\}}\right| \geq \rho_j$, $I_{Sj,j} = 0$ if $\left|s_{j\{k\}}\right| < \rho_j$, $\rho_j$ – constant design value, $j = 1, 2, 3$, $\mathbf{u}_{S\{k\}}^*$ – supervisory control signal adopted in the form of relation

$$u_{Sj\{k\}}^* = -\mathrm{sgn}\, s_{j\{k\}}\left[F_j + \left|u_{Aj\{k\}}\right| + b_{dj} + \sigma_j\right] , \qquad (7.125)$$

where $\left|\mathbf{Y}_{\tau\{k\}}\left(\mathbf{z}_{1\{k\}}\right)\right| < b_{dj}$, $b_{dj} > 0$, $\left|Y_{fj}\left(\mathbf{z}_{1\{k\}}, \mathbf{z}_{2\{k\}}\right)\right| \leq F_j$, $F_j > 0$, and $\sigma_j$ – a small positive constant value. Under this assumption, from the inequality (7.105) results that the difference of Lyapunov function is negative-definite. The designed control signal guarantees a reduction of filtered tracking error $\mathbf{s}_{\{k\}}$ when $\left|s_{j\{k\}}\right| \geq \rho_j$. In the case of initial condition $\left|s_{j\{k=0\}}\right| < \rho_j$ the filtered tracking error remains limited $\left|s_{j\{k\}}\right| < \rho_j, \forall k \geq 0, j = 1, 2, 3$.

The synthesis and the results of numerical tests of tracking control system of the RM with GDHP algorithm are presented in [87].

## 7.6.2  Simulation Tests

Simulation tests were carried out on tracking control systems of dynamic objects with the use of NDP algorithm in GDHP configuration. The first subchapter presents parameters of the WMR's control system with the GDHP structure, the second results obtained by the WMR motion simulation. The third subchapter presents parameters of the RM's control system with the GDHP structure, the fourth results obtained during the simulation of the RM motion.

### 7.6.2.1 Numerical Test of the WMR Motion Control with the Use of GDHP Algorithm

In a virtual computing environment a numerical tests were carried out on WMR's tracking control with a control algorithm with NDP structure in GDHP configuration. From among carried out numerical tests chosen was the sequence obtained after implementation of the trajectory with path in the shape of digit 8, as presented in Sect. 2.1.1.2. The numerical test was carried out with parametric disturbance in the form of changes in the parameters of model that corresponds to loading of the WMR with additional mass.

The following parameters of the control system with GDHP structure were applied in the numerical test:

- PD controller parameters:

  - $\mathbf{K}_D = 3.6h\mathbf{I}$ – fixed diagonal matrix of the PD controller gains,
  - $\mathbf{\Lambda} = 0.5\,\mathbf{I}$ – fixed diagonal matrix,

- the type of actor's NNs applied: RVFL,

  - $m_A = 8$ – number of actor's NN neurons,
  - $\mathbf{\Gamma}_{Aj} = \mathrm{diag}\,\{0.1\}$ – diagonal matrix of gain coefficients for the adaptation of weights $\mathbf{W}_{Aj}, j = 1, 2$,
  - $\mathbf{D}_A$ – NN input layer weights randomly selected from the interval $D_{A[i,g]} \in \langle -0.1, 0.1\rangle$,
  - $\mathbf{W}_{Aj\{k=0\}} = \mathbf{0}$ – zero initial values of the output layer weights,

- type of critic's NN applied: RVFL,

  - $m_C = 8$ – number of critic's NN neurons,
  - $\mathbf{\Gamma}_C = \mathrm{diag}\,\{0.9\}$ – diagonal matrix of gain coefficients for the adaptation of weights $\mathbf{W}_{Cj}$,
  - $\mathbf{D}_C$ – NN input layer weights randomly selected from the interval $D_{C[i,g]} \in \langle -0.2, 0.2\rangle$,
  - $\mathbf{W}_{Cj\{k=0\}} = \mathbf{0}$ – zero initial values of output layer weights,
  - $\eta_1 = \eta_2 = 1$ – positive constant, influence of individual performance indexes in the critic's weights adaptation process,

- $\beta_N = 2$ – slope coefficient of bipolar sigmoid neuron activation functions,
- $\mathbf{R} = \mathrm{diag}\,\{1\}$ – matrix of scaling coefficients in the cost function,
- $l_{AC} = 3$ – number of inner loop iteration steps for the actor's and critic's structure adaptation in the $k$th step of the iterative process,
- $\rho_j = 3$ – supervisory algorithm parameter,
- $\mathbf{F} = [0.08, 0.08]^T$ – vector of constant values of assumed limitations of $Y_{fj}\left(\mathbf{z}_{2\{k\}}\right)$ function,
- $\sigma_j = 0.01$ – supervisory algorithm parameter,
- $h = 0.01\,[\mathrm{s}]$ – numerical test calculation step.

In analyses of designed control systems with algorithms including NNs, elements of input vectors were scaled to the interval $\langle -1, 1 \rangle$.

The same parameters of PD controller and performance indexes, as in Sect. 7.4.2, were applied. The number of discrete measurement data of the WMR's trajectory with path in the shape of digit 8 amounts $n = 4500$.

In the description of numerical tests of algorithms operating in the discrete time domain the notation of variables was simplified by resigning from the $k$th index of the process step. In order to facilitate the analysis of results and the ability to compare continuous sequences of previously presented tracking control systems of point $A$ of the WMR the abscissa axis of discrete charts of sequences were calibrated in continuous time $t$.

The following are results of conducted numerical test of tracking execution with trajectory in the shape of digit 8 and parametric disturbance.

### 7.6.2.2  Examination of GDHP-Type Control, 8-Shaped Path, Disturbance Trial

Numerical test was carried out on tracking control algorithm of point $A$ of the WMR with NDP structure in GDHP configuration with zero initial values of actor's and critic's NNs output layer weights. The task of the control system was to generate control signals ensuring the execution of the desired trajectory of point $A$ of the WMR with path in the shape of digit 8 with double parametric disturbance in the form of parameters change in the WMR model corresponding to the robot's frame load with $m_{RL} = 4.0$ [kg] mass in time $t_{d1} = 12.5$ [s] and the removal of the additional mass in time $t_{d2} = 32.5$ [s].

Values of tracking control signals of individual structures included in the control system are shown in Fig. 7.50. The overall control signals $u_1$ and $u_2$ of wheel 1 and 2 are shown in Fig. 7.50a, respectively. They consist of control signals generated by the GDHP structure $u_{A1}$ and $u_{A2}$ (Fig. 7.50b), control signals of the PD controller $u_{PD1}$ and $u_{PD2}$ (Fig. 7.50c), supervisory control signals $u_{S1}$ and $u_{S2}$, and additional control signals $u_{E1}$ and $u_{E2}$ (Fig. 7.50d).

The adaptation process of the GDHP structure parameters in the initial phase of the first stage of motion runs faster when using the HDP structures with zero initial values of NNs output layers weights, but slower when using DHP algorithm. The GDHP algorithm generates control signals that have a large percentage in overall control signals from the beginning of the WMR point $A$ motion. The control signals of the PD controller have small values in the initial phase of motion and then are reduced to values close to zero. During the occurrence of disturbance at $t_{d1}$ and $t_{d2}$ the control signals generated by the GDHP structure compensate for the effects of changes in the parameters of the controlled object. In the second stage of motion, at $t = 22.5$ [s], compensation for the WMR nonlinearity, implemented by the GDHP structure, provides good tracking quality along the desired trajectory due to information on dynamics of the controlled object included in the values of NNs output layer weights. Values of the control signals $\mathbf{u}_E$ depend mainly on $\mathbf{z}_{d3}$; changes in the value of tracking

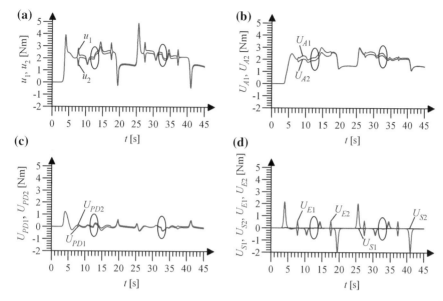

**Fig. 7.50** **a** Overall control signals $u_1$ and $u_2$ for wheels 1 and 2, **b** actor's NNs control signals $U_{A1}$ and $U_{A2}$, $\mathbf{U}_A = -\frac{1}{h}\mathbf{Mu}_A$, **c** PD control signals $U_{PD1}$ and $U_{PD2}$, $\mathbf{U}_{PD} = -\frac{1}{h}\mathbf{Mu}_{PD}$, **d** supervisory term's control signals $U_{S1}$ and $U_{S2}$, $\mathbf{U}_S = -\frac{1}{h}\mathbf{Mu}_S$, and additional control signals $u_{E1}$ and $u_{E2}$, $\mathbf{U}_E = -\frac{1}{h}\mathbf{Mu}_E$

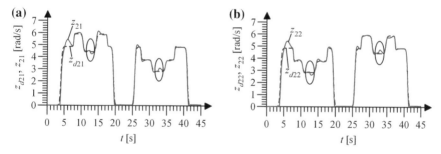

**Fig. 7.51** **a** Desired ($z_{d21}$) and executed ($z_{21}$) angular velocity of the WMR's wheel 1, **b** desired ($z_{d22}$) and executed ($z_{22}$) angular velocity of the WMR's wheel 2

errors have minor influence on sequences of these signals. Values of the supervisory control signals $\mathbf{u}_S$ remain equal to zero because the filtered tracking errors $\mathbf{s}$ does not exceed the assumed acceptable level.

The values of executed angular velocities of WMR drive wheels rotation are shown in Fig. 7.51 and are similar to the desired ones. Only in the initial phase of the first stage of motion a clear difference between the sequences of $z_{d21}$, $z_{d22}$ and $z_{21}$, $z_{22}$ resulting from the application of zero initial values of NNs output layer weights of GDHP structure can be noticed. During the disturbance occurrence, at time

$t_{d1}$, simulating the loading of the WMR with additional mass the values of angular velocities of motion execution are decreasing because the generated control signals are insufficient to execute the WMR motion at a desired velocity due to the increased resistance to motion. Increase in the values of nonlinearity compensation signals generated by the GDHP structure causes increase in the values of angular velocities and further execution of trajectory with desired motion parameters. Similarly, during the second parametric disturbance, at the time $t_{d2}$, simulating the removal of WMR additional mass, that is the return of the model's parameters to nominal values, the values of generated control signals are too large in relation to the required ones due to less resistance to motion of the object not loaded with additional mass. That is why the value of executed velocity of WMR point $A$ is greater than the desired value. As a result of NNs output layer weights adaptation process of GDHP structure the values of control signals $u_{A1}$ and $u_{A2}$ are reduced, velocity of motion execution decreases and reaches the desired value.

Tracking error values of the WMR wheel 1 and 2 are presented in Fig. 7.52a, b, respectively. Figure 7.52c presents the values of filtered tracking error $s_1$; the sequence of filtered tracking error $s_2$ is similar. Figure 7.52d shows a sliding manifold of drive wheel 1. The largest values of tracking errors occur in the initial phase of motion. They are minimized due to the NNs output layer weights adaptation process of GDHP structure. The influence of parametric disturbance on the tracking quality is small and the increase in the absolute value of tracking errors caused by changes

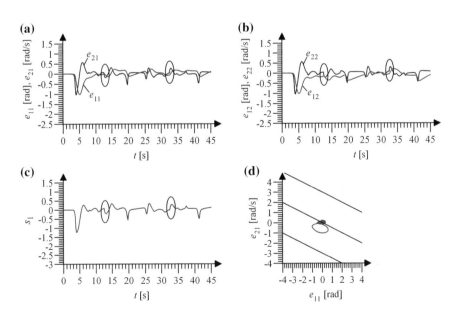

**Fig. 7.52**  **a** Tracking errors $e_{11}$ and $e_{21}$, **b** tracking errors $e_{12}$ and $e_{22}$, **c** filtered tracking error $s_1$, **d** sliding manifold of wheel 1

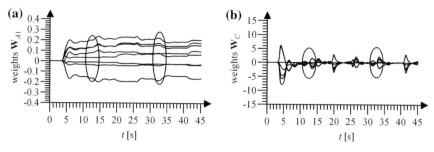

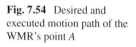

**Fig. 7.53**  **a** Actor's 1 NN output layer weights $\mathbf{W}_{A1}$, **b** critic's NN output layer weights $\mathbf{W}_C$

**Fig. 7.54**  Desired and
executed motion path of the
WMR's point $A$

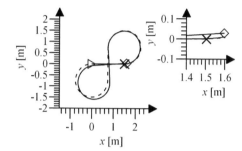

in the dynamics of the controlled object is reduced by adopting NNs output layer
weights of GDHP algorithm.

Values of actor's NN output layer weights generating signal controlling the motion
of wheel 1 and values of critic's NN output layer weights of GDHP algorithm are
shown in Fig. 7.53. Values of actor's NN output layer weights generating signal
controlling the motion of wheel 2 are similar to those presented in Fig. 7.53a. Zero
initial values of actor's and critic's NN output layer weights were assumed. Weights
values are stabilizing around certain values and remain limited during the numerical
test. During the parametric disturbance it is possible to notice changes in the values of
actor's NNs output layer weights resulting from the adjustment of generated signal
compensating for the nonlinearity of the WMR to the changed parameters of the
controlled object.

The desired (dashed line) and executed (continuous line) trajectory of point $A$ of
the WMR is shown in Fig. 7.54. The initial position of point $A$ of the WMR was
marked by a triangle and the desired final position was marked by "**X**" while the
final position of the point $A$ was marked by a rhombus.

Tables 7.26 and 7.27 show the values of individual performance indexes for the
execution of the tracking motion with the application of a control system with ACD
in GDHP configuration.

**Table 7.26**  Values of selected performance indexes

| Index | $e_{max1j}$ [rad] | $\varepsilon_{1j}$ [rad] | $e_{max2j}$ [rad/s] | $\varepsilon_{2j}$ [rad/s] | $s_{maxj}$ [rad/s] | $\sigma_j$ [rad/s] |
|---|---|---|---|---|---|---|
| Wheel 1, $j = 1$ | 1.01 | 0.22 | 1.03 | 0.19 | 1.24 | 0.22 |
| Wheel 2, $j = 2$ | 1.01 | 0.26 | 1.03 | 0.19 | 1.24 | 0.23 |

**Table 7.27**  Values of selected performance indexes of path execution

| Index | $d_{max}$ [m] | $d_n$ [m] | $\rho_d$ [m] |
|---|---|---|---|
| Value | 0.158 | 0.096 | 0.096 |

**Table 7.28**  Values of selected performance indexes in I and II stage of motion

| Index | $e_{max1j}$ [rad] | $\varepsilon_{1j}$ [rad] | $e_{max2j}$ [rad/s] | $\varepsilon_{2j}$ [rad/s] | $s_{maxj}$ [rad/s] | $\sigma_j$ [rad/s] |
|---|---|---|---|---|---|---|
| Stage I, $k = 1, \ldots, 2250$ | | | | | | |
| Wheel 1, $j = 1$ | 1.01 | 0.29 | 1.03 | 0.25 | 1.24 | 0.28 |
| Wheel 2, $j = 2$ | 1.01 | 0.33 | 1.03 | 0.24 | 1.24 | 0.29 |
| Stage II, $k = 2251, \ldots, 4500$ | | | | | | |
| Wheel 1, $j = 1$ | 0.24 | 0.11 | 0.46 | 0.11 | 0.49 | 0.13 |
| Wheel 2, $j = 2$ | 0.39 | 0.15 | 0.42 | 0.12 | 0.57 | 0.14 |

As a result of the carried out numerical test the obtained values of performance indexes were significantly lower than when applying in the algorithm of tracking control only a PD controller. Obtained values of performance indexes are lower than in the case of applying the adaptive, neural control system or control algorithm with HDP structure and zero initial values of actor's and critic's NN output layer weights presented in Sect. 7.4.2.1, and higher than those obtained in the numerical test Sect. 7.4.2.2 where assumed were non-zero initial values of critic's NN output layer weights in HDP algorithm, or obtained in the numerical test of control algorithm with DHP structure Sect. 7.5.2.2.

Table 7.28 presents the values of selected performance indexes of first and second stage of motion along the path in the shape of digit 8 calculated according to the methodology described in the numerical test Sect. 7.1.2.2.

Obtained values of applied performance indexes were lower in the second stage of motion than in the first stage. This results from the application of non-zero values of NN output layer weights in the initial phase of motion of the second stage of numerical test. Knowledge of the dynamics of controlled object included in the NN output layer weights of GDHP structure enables generation of signals compensating for nonlinearities of the WMR, which improves the quality of the generated tracking control.

### 7.6.2.3 Numerical Test of the RM Motion Control with the Use of GDHP Algorithm

In a virtual computing environment a simulation tests were carried out on tracking control of RM with the use of a control algorithm with NDP structure in GDHP configuration. From the performed numerical tests selected was the sequence obtained after implementation of the trajectory with path in the shape of a semicircle, as described in the Sect. 2.2.1.2.

The following parameters of the control system with GDHP structure were assumed in the numerical test:

- PD controller parameters:

  - $\mathbf{K}_D = \text{diag}\{0.4, 0.7, 0.7\}$ – constant diagonal matrix of PD controller gains,
  - $\mathbf{\Lambda} = \mathbf{I}$ – constant diagonal matrix,

- type of actor's NNs applied: RVFL,

  - $m_A = 8$ – number of the actor's NN neurons,
  - $\mathbf{\Gamma}_{A1} = \text{diag}\{0.2\}$ – diagonal matrix of gain coefficients for the adaptation of weights $\mathbf{W}_{A1}$,
  - $\mathbf{\Gamma}_{A2} = \mathbf{\Gamma}_{A3} = \text{diag}\{0.125\}$ – diagonal matrix of gain coefficients for the adaptation of weights $\mathbf{W}_{A2}$ and $\mathbf{W}_{A3}$,
  - $\mathbf{D}_A$ – NN input layer weights randomly selected from the interval $D_{A[i,g]} \in \langle -0.2, 0.2 \rangle$,
  - $\mathbf{W}_{Aj\{k=0\}} = \mathbf{0}$ – zero initial values of the output layer weights, $j = 1, 2, 3$,
  - $k_A = 0.1$ – positive constant,

- type of critic's NNs applied: RBF,

  - $m_C = 3^3 = 27$ – number of the critic's NN neurons,
  - $\mathbf{\Gamma}_C = \text{diag}\{0.0005\}$ – diagonal matrix of gain coefficients for the adaptation of weights $\mathbf{W}_C$,
  - $\mathbf{W}_{C\{k=0\}} = \mathbf{0}$ – zero initial values of output layer weights,
  - $k_{C1} = k_{C2} = 0.1$ – positive constant,
  - $\eta_1 = \eta_2 = 1$ – positive constant, influence of individual performance indexes in the critic's weights adaptation process,

- $\beta_N = 2$ – slope coefficient of bipolar sigmoid neuron activation functions,
- $\mathbf{R} = \text{diag}\{1\}$ – matrix of scaling coefficient in the cost function,
- $l_{AC} = 5$ – number of inner loop iteration steps for the actor's and critic's structure adaptation in the $k$th step of the iterative process,
- $\rho_j = 0.2$ – supervisory algorithm parameter,
- $\mathbf{F} = [0.25, 0.25, 0.35]^T$ – vector of constant values of assumed restrictions of function $Y_{fj}(\mathbf{z}_{1\{k\}}, \mathbf{z}_{2\{k\}})$,
- $\sigma_j = 0.01$ – supervisory algorithm parameter,
- $h = 0.01$ [s] – numerical test calculation step.

The same performance indexes, as described in Sect. 7.1.2, were assumed. The number of data samples of trajectory of RM's end-effector with path in the shape of a semi-circle amounts to $n = 4000$.

In the description of the numerical tests of algorithms operating in discrete time domain the notation of variables was simplified by resigning from the index of $k$th step of the process. In order to facilitate the analysis of results and comparison of presented earlier tracking control systems of point $C$ of the RM to continuous sequences, the abscissa axis of charts of discrete sequences was scaled in time $t$.

The following are some results of carried out numerical tests on tracking execution with the use of trajectory with path in the shape of a semicircle.

### 7.6.2.4 Examination of GDHP-Type Control, Semicircle Path

In the numerical test of the RM control with NDP algorithm in GDHP configuration the desired trajectory with path in the shape of a semicircle, discussed in Sect. 2.2.1.2, was applied.

In Fig. 7.55 green line represents the desired and the red line executed motion path of point $C$ of the RM. The initial position (point $S$) on the graph was marked by a triangle. "**X**" represents the return position of trajectory (point $G$) and the reached end-position of point $C$ of the RM was marked by a rhombus.

Sequences of signals controlling the actuator systems of kinematic chain of the RM are presented in Fig. 7.56. Overall control signals $u_1$, $u_2$ and $u_3$ as shown in Fig. 7.56a in accordance with equation (7.123) consist of control signals generated by the GDHP structure $u_{A1}$, $u_{A2}$ and $u_{A3}$ (Fig. 7.56b), control signals of the PD controller $u_{PD1}$, $u_{PD2}$ and $u_{PD3}$ (Fig. 7.56c), supervisory control signals $u_{S1}$, $u_{S2}$ and $u_{S3}$ as well as additional control signals $u_{E1}$, $u_{E2}$ and $u_{E3}$ (Fig. 7.56d). Supervisory control signals are not shown on the diagrams, because their values during the numerical test amount to zero, which results from the assumed values of the parameter $\rho_j$ of the supervisory control algorithm.

In the initial position of the RM's arm, when $t = 0$ [s], the rotation angles of individual links amount to $z_{11} = 0$, $z_{12} = 0.16533$, $z_{13} = -0.16533$ [rad]. In such configuration the point $C$ of the robotic arm end-effector is at the maximal distance

**Fig. 7.55** Desired and executed path of point $C$ of the RM's end-effector

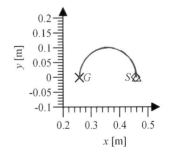

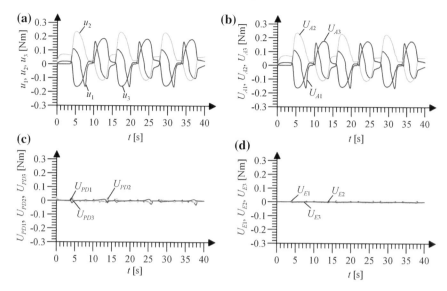

**Fig. 7.56** **a** The overall signals $u_1$, $u_2$ and $u_3$ controlling the motion of the RM's links, **b** the control signals generated by the actor NNs $U_{A1}$, $U_{A2}$ and $U_{A3}$, $\mathbf{U}_A = -\frac{1}{h}\mathbf{M}\mathbf{u}_A$, **c** control signals of PD controller $U_{PD1}$, $U_{PD2}$ and $U_{PD3}$, $\mathbf{U}_{PD} = -\frac{1}{h}\mathbf{M}\mathbf{u}_{PD}$, **d** control signals $U_{E1}$, $U_{E2}$ and $U_{E3}$, $\mathbf{U}_E = -\frac{1}{h}\mathbf{M}\mathbf{u}_E$

from the base of the robot (point $S$). In the initial stage of the numerical test, when $t \in \langle 0, 3.5 \rangle$, the manipulator is not moving, however, on the diagrams of control signals occur non-zero values of signals controlling the motion of link 2 and 3. It results from the adaptation of the NNs of GDHP structure associated with the prevention of unintended motion of the robot's arm under the influence of moments from the gravity forces acting on the individual robot's links. The dominant percentage in the overall control signals have control signals generated by the actor's structure of GDHP algorithm. The control signals of other components of the control system have values close to zero.

The value of desired (red line) and implemented (green line) angular velocity of link 2 rotation is shown in Fig. 7.57a as an example of the desired trajectory execution in the configuration space of the RM. The sequence of the absolute desired ($|v_{Cd}|$) and the absolute executed ($|v_C|$) value of velocity of point $C$ of the manipulator's end-effector is shown in Fig. 7.57b.

Tracking error values of the RM's link 1, 2 and 3 are presented in Fig. 7.58a, b. Figure 7.58c presents the values of filtered tracking errors $s_1$, $s_2$ and $s_3$. Figure 7.58d presents sliding manifold of link 1. The largest values of tracking errors occur during the first motion cycle of the RM's arm. However, because of the approach to the adaptation method of the NN output layer weights of GDHP structure the differences between the tracking errors in individual operating cycles are not marked.

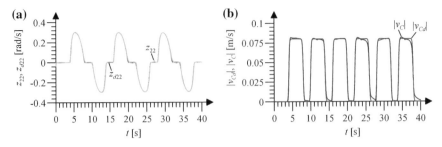

**Fig. 7.57**  **a** Desired ($z_{d22}$) and executed ($z_{22}$) angular velocity of the RM's link 2 rotation, **b** desired ($|v_{Cd}|$) and executed ($|v_C|$) absolute value of velocity of point $C$ of the manipulator's end-effector

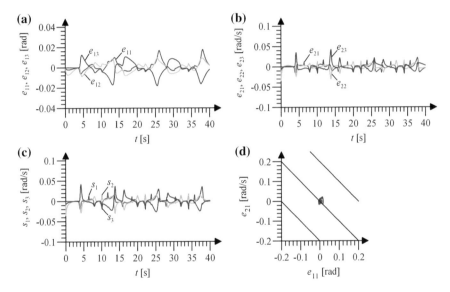

**Fig. 7.58**  **a** Tracking errors of rotation angles of links, $e_{11}$, $e_{12}$ and $e_{13}$ **b** tracking errors of angular velocities, $e_{21}$, $e_{22}$ and $e_{23}$, **c** filtered tracking errors $s_1$, $s_2$ and $s_3$, **d** sliding manifold of link 1

The values of actor's RVFL NN output layer weights and selected output layer weights of critic's RBF NN of GDHP structure are shown in Fig. 7.59. Zero initial values of actor's and critic's NNs output layer weights were assumed. Values of weights stabilize around a certain values and remain limited during numerical test. Output layer weights of actor's NNs generating control signals of link 2 and 3 motion, are adopted from the beginning of the numerical test, which is associated with the need to generate control signals to resist the motion of the robot's arm under the influence of moments from gravity forces. Values of output layer weights of actor's NN controlling the motion of link 1 change since $t = 3$ [s], due to the beginning of desired trajectory execution.

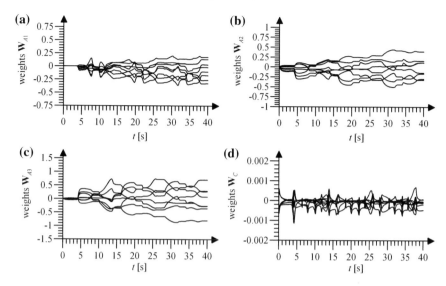

**Fig. 7.59** **a** Values of actor's 1 NN output layer weights $\mathbf{W}_{A1}$, **b** values of actor's 2 NN output layer weights $\mathbf{W}_{A2}$, **c** values of actor's 3 NN output layer weights $\mathbf{W}_{A3}$, **d** values of critic's NN output layer weights $\mathbf{W}_C$

**Table 7.29** Values of selected performance indexes

| Index | $e_{max1j}$ [rad] | $\varepsilon_{1j}$ [rad] | $e_{max2j}$ [rad/s] | $\varepsilon_{2j}$ [rad/s] | $s_{maxj}$ [rad/s] | $\sigma_j$ [rad/s] |
|---|---|---|---|---|---|---|
| Link 1, $j = 1$ | 0.0177 | 0.0047 | 0.0219 | 0.0059 | 0.0353 | 0.0074 |
| Link 2, $j = 2$ | 0.0095 | 0.0045 | 0.0343 | 0.0061 | 0.0345 | 0.0075 |
| Link 3, $j = 3$ | 0.018 | 0.0068 | 0.0428 | 0.0078 | 0.0418 | 0.0102 |

**Table 7.30** Values of selected performance indexes of path execution

| Index | $d_{max}$ [m] | $d_n$ [m] | $\rho_d$ [m] |
|---|---|---|---|
| Value | 0.0084 | 0.001 | 0.0028 |

The values of individual performance indexes of tracking motion with the use of control system with the NDP structure in GDHP configuration are presented in Tables 7.29 and 7.30.

As a result of the carried out numerical test the obtained values of performance indexes were lower than those obtained with the use in the tracking control algorithm of the RM only a PD controller.

### 7.6.3  Conclusions

Application of the NDP structure in GDHP configuration in the control system guarantees high quality of tracking motion expressed by low values of performance indexes. In the case of assuming the worst case scenario of zero initial values of NN output layer weights of actor's and critic's structures in GDHP algorithm the obtained values of performance indexes are lower than those obtained during the execution of motion with the use of a PD controller, adaptive control system or neural control system. The quality of tracking motion execution expressed by the values of assumed performance indexes is higher than the one obtained with the use of control system with HDP algorithm (in the case of assuming the zero initial values of NN output layer weights), but lower in the case of applying the control system with DHP algorithm. Better quality of motion execution, in relation to the control system with HDP algorithm, results from the use of more complex adaptation law of critic's NN output layer weights, which is a combination of critic's NN output layer weights adaptation laws typical for HDP and DHP algorithms. At the same time, the lower quality of motion, in relation to the control system with DHP algorithm, can be caused by a different way of determining the derivative of the value function with respect to the state of the object in the individual algorithms. In the DHP algorithm the derivative of the value function with respect to the state of the controlled object, is approximated by the critic's NNs, while in the applied version of the GDHP algorithm is calculated analytically on the basis of the output from the critic's NN approximating the value function. That is why it can be mapped with larger error. The influence of design parameters values of the GDHP algorithm on the process of actor's and critic's NN output layer weights adaptation is similar to those in the others NDP algorithms and was discussed earlier. The knowledge of mathematical model of the controlled object is required in order to carry out the synthesis of GDHP algorithm.

## 7.7  Action Dependent Heuristic Dynamic Programming in Tracking Control of a WMR

The structure of NDP algorithm in ADHDP configuration [26, 43, 77, 79] is similar to the HDP algorithm where the critic approximates the value function, while the actor generates control law. The difference in construction results from the inclusion of control signals generated by the actor in the input vector of critic's NN, which enables the analytical determination of prediction of the value function derivative with respect to the control signal, as a result of critic's NN backpropagation of the control signal. This quantity is necessary to determine the actor's NN weights adaptation law. Its calculation, in the HDP algorithm, requires knowledge of mathematical model of the controlled object, while in the ADHDP structure control signals generated by the actor are included in the input vector of critic's NN, and because of that the knowledge of the controlled object model is not necessary. Critic in ADHDP algorithm approximates the value function and that is why it is built with the use of only one adaptive structure, e.g., NN. Thus, its structure is simplified in comparison to the structure of the critic in DHP algorithm. The structure of actor's depends on the number of independent signals controlling the dynamic object. In this chapter, the synthesis of the discrete tracking control system of WMR with the use of NDP algorithm in ADHDP configuration was carried out. The presented algorithm operates in discrete time domain. This chapter is based on the information on the ADHDP algorithm, which description is provided in Sect. 6.3.1.

## 7.7.1 Synthesis of ADHDP-type Control

Synthesis of the WMR tracking control algorithm with the use of NDP structure in ADHDP configuration was carried out similarly as the implementation of the control system with HDP algorithm described in Sect. 7.4.1.

In the synthesis of the WMR discrete tracking control algorithm the following was applied: the definition of tracking errors $\mathbf{e}_{1\{k\}}$ and $\mathbf{e}_{2\{k\}}$ according to the relation (7.53), the filtered tracking errors $\mathbf{s}_{\{k\}}$ according to the relation (7.54) and $\mathbf{s}_{\{k+1\}}$ according to the relation (7.55) formulated in Sect. 7.4.1.

Two NDP algorithms in ADHDP configuration were applied in the designed control algorithm. In every ADHDP algorithm the actor's structure was implemented in the form of RVFL NN, in contrast, the critic's structure was built with the use of NN with neurons activation functions of Gaussian curve type. A characteristic feature of ADHDP algorithm is that it does not require knowledge of mathematical model of the controlled object in the adaptation process of actor's and critic's structure parameters.

The task of the ADHDP structures is to generate a suboptimal control law $\mathbf{u}_{A\{k\}}$ that minimizes the value function expressed by the following relation

$$V_{j\{k\}}\left(\mathbf{s}_{\{k\}}, \mathbf{u}_{\{k\}}\right) = \sum_{k=0}^{n} \gamma^k L_{Cj\{k\}}\left(\mathbf{s}_{\{k\}}, \mathbf{u}_{\{k\}}\right) , \qquad (7.126)$$

with the cost function

$$L_{Cj\{k\}}\left(\mathbf{s}_{\{k\}}, \mathbf{u}_{\{k\}}\right) = \frac{1}{2}\left(R_j s_{j\{k\}}^2\right) + \frac{1}{2}\left(Q_j u_{j\{k\}}^2\right) , \qquad (7.127)$$

where $j = 1, 2, R_1, R_2, Q_1, Q_2$ – positive design coefficients. The task of the NDP structures in ADHDP configuration, when applying the value functions (7.126) at the local cost (7.127), is to generate a control law that minimizes the filtered tracking errors $\mathbf{s}_{\{k\}}$ and control signals $\mathbf{u}_{\{k\}}$.

The general scheme of the NDP structure in ADHDP configuration is shown in Fig. 7.60.

**Fig. 7.60** Scheme of the NDP structure in ADHDP configuration

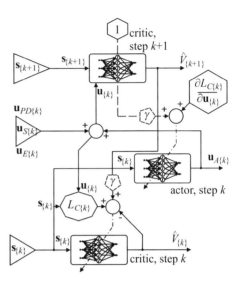

Each ADHDP structure included in the tracking controls system of WMR consists in:
- **Actor**, implemented in the form of RVFL NN that generates suboptimal control law $u_{Aj\{k\}}$ in $k$th step according to the relation

$$u_{Aj\{k\}}\left(\mathbf{x}_{Aj\{k\}}, \mathbf{W}_{Aj\{k\}}\right) = \mathbf{W}_{Aj\{k\}}^{T}\mathbf{S}\left(\mathbf{D}_{A}^{T}\mathbf{x}_{Aj\{k\}}\right), \qquad (7.128)$$

where $j = 1, 2$, actor's NNs of the ADHDP algorithm are built analogously as in the HDP structure shown in Sect. 7.4.1. The procedure of actor's NN output layer weights adaptation of ADHDP algorithm of designed control system is to minimize the performance index

$$e_{Aj\{k\}} = \frac{\partial L_{Cj\{k\}}\left(\mathbf{s}_{\{k\}}, \mathbf{u}_{\{k\}}\right)}{\partial u_{j\{k\}}} + \gamma\frac{\partial \hat{V}_{j\{k+1\}}\left(\mathbf{x}_{Cj\{k+1\}}, \mathbf{W}_{Cj\{k\}}\right)}{\partial u_{j\{k\}}}, \qquad (7.129)$$

where $\hat{V}_{j\{k+1\}}\left(\mathbf{x}_{Cj\{k+1\}}, \mathbf{W}_{Cj\{k\}}\right)$ – the value of output from the $j$th critic's NN determined in the next step [77]. The update of the actor's NN output layer weights is carried out according to the relation

$$\mathbf{W}_{Aj\{k+1\}} = \mathbf{W}_{Aj\{k\}} - e_{Aj\{k\}}\mathbf{\Gamma}_{Aj}\mathbf{S}\left(\mathbf{D}_{A}^{T}\mathbf{x}_{Aj\{k\}}\right), \qquad (7.130)$$

where $\mathbf{\Gamma}_{Aj}$ – fixed diagonal matrix of positive gain coefficients of the output layer weights adaptation of the actor's $j$th NN.
- **Critic**, estimating the value function (7.126) was implemented in the form of the NN with neurons activation functions of Gaussian curve type and generates signal

$$\hat{V}_{j\{k\}}\left(\mathbf{x}_{Cj\{k\}}, \mathbf{W}_{Cj\{k\}}\right) = \mathbf{W}_{Cj\{k\}}^{T}\mathbf{S}\left(\mathbf{x}_{Cj\{k\}}\right) , \tag{7.131}$$

where $j = 1, 2$, $\mathbf{W}_{Cj\{k\}}$ – vector of output layer weights of the $j$th critic's NN, $\mathbf{x}_{Cj\{k\}} = \boldsymbol{\kappa}_C \left[s_{j\{k\}}, u_{j\{k\}}\right]^T$ – input vector of the $j$th critic's NN, $\boldsymbol{\kappa}_C$ – fixed diagonal matrix of positive scaling coefficients of input values to the critic's NN, $\mathbf{S}\left(\mathbf{x}_{Cj\{k\}}\right)$ – vector of the neurons activation functions of Gaussian type. The input vector of the critic's NN, in the ADHDP algorithm, includes generated control signal $\mathbf{u}_{\{k\}}$. This results from assuming the value function (7.126) where the cost function (7.127) depends explicitly on the control signals.

The output layer weights of the critic's NN structure, in the ADHDP algorithm, were adapted using the backpropagation method of the temporal difference error defined as

$$e_{Cj\{k\}} = L_{Cj\{k\}}\left(\mathbf{s}_{\{k\}}, \mathbf{u}_{\{k\}}\right) + \gamma \hat{V}_{j\{k+1\}}\left(\mathbf{x}_{Cj\{k+1\}}, \mathbf{W}_{Cj\{k\}}\right) - \hat{V}_{j\{k\}}\left(\mathbf{x}_{Cj\{k\}}, \mathbf{W}_{Cj\{k\}}\right) . \tag{7.132}$$

Weights adaptation law was assumed in the following form

$$\mathbf{W}_{Cj\{k+1\}} = \mathbf{W}_{Cj\{k\}} - e_{Cj\{k\}}\boldsymbol{\Gamma}_{Cj}\mathbf{S}\left(\mathbf{x}_{Cj\{k\}}\right) , \tag{7.133}$$

where $\boldsymbol{\Gamma}_{Cj}$ – fixed diagonal matrix of positive gain coefficients of the output layer weights adaptation of the $j$th critic's NN.

The overall control signals $\mathbf{u}_{\{k\}}$, in the designed tracking control system of the WMR with NDP structure in ADHDP configuration, comprise of control signals generated by the actor's NNs $\mathbf{u}_{A\{k\}}$, signals generated by the PD controller $\mathbf{u}_{PD\{k\}}$, signals $\mathbf{u}_{E\{k\}}$, and additional supervisory control signals $\mathbf{u}_{S\{k\}}$ with structure resulting from the application of the Lyapunov stability theorem during the stability analysis of the designed control system. The structure of the tracking control system of point $A$ of the WMR with ADHDP algorithm is schematically presented in Fig. 7.61.

The analysis of the closed-loop control system stability was carried out as described in Sect. 7.4.1, assuming overall control signals in the form of relation (7.66) and the supervisory control signals $\mathbf{u}_{S\{k\}}$ in the form of relation (7.79).

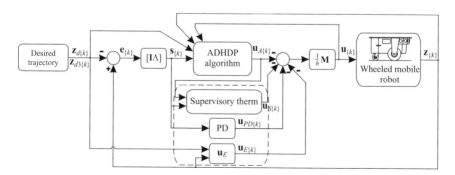

**Fig. 7.61** Scheme of the WMR tracking control system with ADHDP algorithm

The synthesis and results of experimental tests of tracking control system of the WMR with ADHDP algorithms are presented in [43].

## 7.7.2  Simulation Tests

The simulation tests were carried out on tracking control of the WMR with a control algorithm with NDP structure in ADHDP configuration in the virtual computing environment. From the performed numerical tests the sequence resulting from executed trajectory with path in the shape of digit 8, as described in Sect. 2.1.1.2, was selected. The numerical test was carried out with parametric disturbance in the form of changes in the model's parameters corresponding to the loading of the WMR with additional mass.

The following parameters of the control system with the ADHDP structure were assumed in the numerical test:

- PD controller parameters:

  - $\mathbf{K}_D = 3.6h\mathbf{I}$ – fixed diagonal matrix of PD controller gain,
  - $\mathbf{\Lambda} = 0.5\,\mathbf{I}$ – fixed diagonal matrix,

- type of actor's NNs applied: RVFL,

  - $m_A = 8$ – number of the actor's NN neurons,
  - $\mathbf{\Gamma}_{Aj} = \mathrm{diag}\,\{0.1\}$ – diagonal matrix of gain coefficients for the adaptation of weights $\mathbf{W}_{Aj}, j = 1, 2$,
  - $\mathbf{D}_A$ – NN input layer weights randomly selected from the interval $D_{A[i,g]} \in \langle -0.1, 0.1 \rangle$,
  - $\mathbf{W}_{Aj\{k=0\}} = \mathbf{0}$ – zero initial values of the output layer weights,

- type of critic's NNs applied: RBF,

  - $m_C = 19$ – number of the critic's NN neurons,
  - $\mathbf{\Gamma}_{Cj} = \mathrm{diag}\,\{0.01\}$ – diagonal matrix of gain coefficients for the adaptation of weights $\mathbf{W}_{Cj}$,
  - $\mathbf{c}_C$ – position of the centers of Gaussian curves in the two-dimensional input space of the critic's 1 NN is shown in Fig. 7.62; distribution of function centers is analogous in the case of critic's 2 NN,
  - $\mathbf{r}_C = [0.3, 0.2, 0.1, 0.08, 0.08, 0.1, 0.2, 0.3, 0.08, 0.1, 0.1, 0.2, 0.2, 0.2, 0.2, 0.3, 0.3, 0.3, 0.3]^T$ – widths of the Gaussian curves,
  - $\mathbf{W}_{Cj\{k=0\}} = \mathbf{0}$ – zero initial values of output layer weights of the critic,

- $\beta_N = 2$ – slope coefficient of bipolar sigmoid neuron activation functions,
- $R_j = 1, Q_j = 0.01$ – scaling coefficients in the cost function,
- $l_{AC} = 3$ – number of inner loop iteration steps for the actor's and critic's structure adaptation in the $k$th step of the iterative process,
- $\rho_j = 3$ – supervisory algorithm parameter,

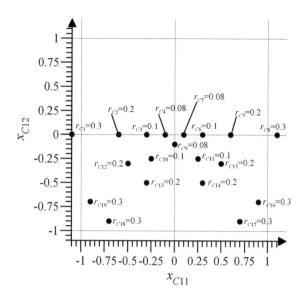

**Fig. 7.62** Position of critic's NN Gaussian curve type neurons centers

- $\sigma_j = 0.01$ – supervisory algorithm parameter,
- $h = 0.01$ [s] – numerical test calculation step.

The performance indexes from point Sect. 7.4.2 were assumed in order to evaluate the quality of tracking motion.

The following are presented selected results of numerical tests of tracking motion with path in the shape of a digit 8 with a parametric disturbance.

### 7.7.2.1 Examination of ADHDP-Type Control, 8-Shaped Path, Disturbance Trial

Numerical test on tracking control algorithm of point $A$ of the WMR with NDP structures in ADHDP configuration with zero initial values of the output layer weights of actor's and critic's NNs was carried out. The task of the control system was to generate control signals in order to ensure the execution of the desired trajectory of point $A$ of the WMR with path in the shape of a digit 8 with double parametric disturbance in the form of changes in the parameters of the WMR model corresponding to the loading of robot's frame with $m_{RL} = 4.0$ [kg] mass in time $t_{d1} = 12.5$ [s] and the removal of the additional mass in time $t_{d2} = 32.5$ [s].

Values of the tracking control signals of individual structures included in the control system are shown in Fig. 7.63. Overall control signals $u_1$ and $u_2$ of wheel 1 and 2 are shown in Fig. 7.63a and they consist of control signals generated by the ADHDP structures $u_{A1}$ and $u_{A2}$ (Fig. 7.63b), control signals of the PD controller $u_{PD1}$ and $u_{PD2}$ (Fig. 7.63c), supervisory control signals $u_{S1}$ and $u_{S2}$, and additional control signals $u_{E1}$ and $u_{E2}$ (Fig. 7.63d).

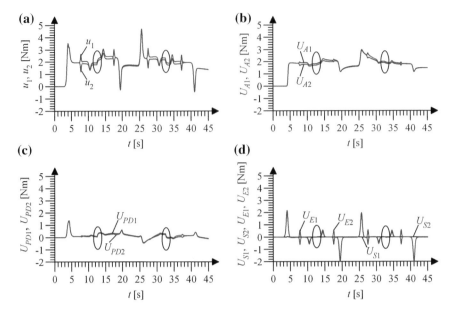

**Fig. 7.63** **a** Overall control signals $u_1$ and $u_2$ for wheels 1 and 2, **b** actor's NNs control signals $U_{A1}$ and $U_{A2}$, $\mathbf{U}_A = -\frac{1}{h}\mathbf{M}\mathbf{u}_A$, **c** PD control signals $U_{PD1}$ and $U_{PD2}$, $\mathbf{U}_{PD} = -\frac{1}{h}\mathbf{M}\mathbf{u}_{PD}$, **d** supervisory term's control signals $U_{S1}$ and $U_{S2}$, $\mathbf{U}_S = -\frac{1}{h}\mathbf{M}\mathbf{u}_S$, and additional control signals $u_{E1}$ and $u_{E2}$, $\mathbf{U}_E = -\frac{1}{h}\mathbf{M}\mathbf{u}_E$

   In the initial phase of the first stage of motion, acceleration, a small role in the overall control signals $\mathbf{u}$ perform signals generated by the PD controller $\mathbf{u}_{PD}$ and $\mathbf{u}_E$ signals. Small values of the control signals of ADHDP structures result from the application of zero initial values of actor's NNs output layer weights whose adaptation process begins when filtered tracking errors, different from zero, occur. In the next stages of motion in the NNs output layer weights of the NDP structures a certain knowledge of the controlled object dynamics is preserved, while the control signals of actor's NNs are dominant in the overall tracking control signals of point $A$ of the WMR. The occurrence of the first parametric disturbance in time $t_{d1}$ increases the values of control signals generated by the structure of the adaptive critic, which compensates for the changes of the controlled object dynamics. Similarly, the occurrence of the second disturbance in time $t_{d2}$ decreases the value of control signals of ADHDP structures. Also, it is possible to notice significant changes in the values of control signals of the PD controller. The values of the $\mathbf{u}_E$ control signals depend mainly on the $\mathbf{z}_{d3}$. The change in the value of tracking errors have a small influence on the sequences of these signals. The values of the supervisory control signals $\mathbf{u}_S$ remain equal to zero because the filtered tracking errors $\mathbf{s}$ do not exceed the acceptable level.

   The desired and executed angular velocities of WMR's drive wheels 1 and 2 are shown in Fig. 7.64a, b. The biggest differences between desired angular velocities of WMR's drive wheels and executed values occur in the initial phase of the first stage

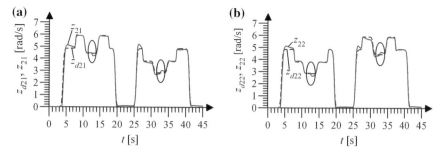

**Fig. 7.64** **a** Desired ($z_{d21}$) and executed ($z_{21}$) angular velocity of the WMR's wheel 1, **b** desired ($z_{d22}$) and executed ($z_{22}$) angular velocity of the WMR's wheel 2

of motion. During $t_{d1}$, the occurrence of disturbance simulating the loading of WMR with additional mass, the values of the angular velocities of motion execution are decreasing, because the generated control signals are insufficient in relation to the WMR motion with desired velocity due to the increased resistance to motion. The increase in the value of signals compensating for the nonlinearities generated by the ADHDP structures and control signals of the PD controller increases the values of the angular velocities and results in the further execution of the trajectory with desired parameters of motion. Analogously, during $t_{d2}$ of the second parametric disturbance occurrence simulating the removal of the additional mass of the WMR, and returning of the model's parameters to the nominal values. The values of generated control signals are too large in relation to the required ones due to the smaller resistances to motion of the object and that is why the value of the executed velocity of point $A$ of the WMR is larger than the desired value. As a result of the NNs output layer weights adaptation process of the ADHDP structures the values of the control signals $u_{A1}$ and $u_{A2}$ are reduced. Similarly, reduced are values of control signals of the PD controller, the velocity of tracking motion decreases and reaches the desired values.

Tracking errors of WMR wheel 1 and 2 are shown in Fig. 7.65a, b. Figure 7.65c presents the sequences of filtered tracking error $s_1$ and Fig. 7.65d sliding manifold of wheel 1. The values of tracking errors reach a larger absolute values than when using the previously tested algorithms with NDP structures in DHP configuration and HDP configurations with non-zero initial values of critic's NN output layer weights. The largest values of errors occur in the initial phase of the first stage of motion, then are reduced due to the realization of the NN output layer weights adaptation process of ADHDP structures. During the parametric disturbance occurrences the absolute values of tracking errors are increasing. Due to the slow process of ADHDP structures adaptation, errors arising from changes in the dynamics of the controlled object are reduced more slowly than in the case of previously mentioned algorithms. Tracking errors in the initial phase of the second stage of motion, along the trajectory in the shape of curved path in the rightward direction, reach lower values than at the beginning of the first stage of motion due to the information contained in the NN output layer weights of actor's and critic's structures.

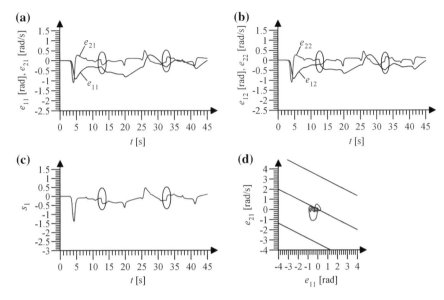

**Fig. 7.65** **a** Tracking errors $e_{11}$ and $e_{21}$, **b** tracking errors $e_{12}$ and $e_{22}$, **c** filtered tracking error $s_1$, **d** sliding manifold of wheel 1

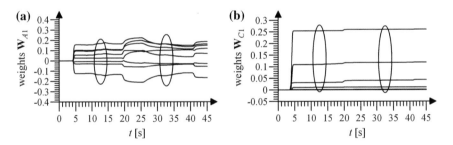

**Fig. 7.66** **a** Actor's 1 NN output layer weights $\mathbf{W}_{A1}$, **b** critic's 1 NN output layer weights $\mathbf{W}_{C1}$

The values of NN output layer weights of actor's and critic's first ADHDP structure are shown in Fig. 7.66a, b. The values of NN output layer weights of the second ADHDP structure, which generates a control signal of wheel 2, are similar. The structures of the critic include NNs with neurons activation functions of Gaussian curve type. The actor's structures were implemented as RVFL NNs. Zero initial values of actor's and critic's NNs output layer weights were applied. Values of weights stabilize around certain values and remain limited. During the parametric disturbance occurrence it is possible to notice small changes in the values of actor's NN output layer weights. This results from the slow process of ADHDP structures adaptation in relations to the changes of the WMR dynamics.

**Fig. 7.67** Desired and
executed motion path of the
WMR's point $A$

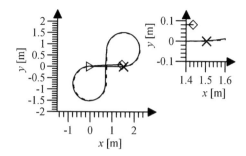

**Table 7.31** Values of selected performance indexes

| Index | $e_{max1j}$ [rad] | $\varepsilon_{1j}$ [rad] | $e_{max2j}$ [rad/s] | $\varepsilon_{2j}$ [rad/s] | $s_{maxj}$ [rad/s] | $\sigma_j$ [rad/s] |
|---|---|---|---|---|---|---|
| Wheel 1, $j = 1$ | 0.91 | 0.41 | 1.1 | 0.18 | 1.38 | 0.27 |
| Wheel 2, $j = 2$ | 0.91 | 0.38 | 1.1 | 0.18 | 1.38 | 0.26 |

**Table 7.32** Values of selected performance indexes of path execution

| Index | $d_{max}$ [m] | $d_n$ [m] | $\rho_d$ [m] |
|---|---|---|---|
| Value | 0.131 | 0.107 | 0.063 |

Desired (dashed line) and executed (continuous line) path of point $A$ of the WMR is shown in Fig. 7.67. The initial position of point $A$ of the WMR is marked by a triangle, the desired final position by "**X**", while the final position of point $A$ by a rhombus.

Tables 7.31 and 7.32 show the values of individual performance indexes for the execution of the tracking motion with the application of a control system with ACD in ADHDP configuration.

The obtained values of performance indexes are larger than values obtained during the numerical tests of control algorithms with NDP structures in DHP, GDHP and HDP configuration with non-zero initial values of critic's NN output layer weights.

Table 7.33 presents values of selected performance indexes of first and second stage of motion along the path in the shape of a digit 8, calculated in accordance with the methodology described in the numerical test Sect. 7.1.2.2.

Lower values of assumed performance indexes were obtained in the second stage of motion.

## 7.7.3  Conclusions

In the algorithm of tracking control of point $A$ of the WMR was applied compensation for the nonlinearity of the controlled object implemented in the form of NDP

**Table 7.33** Values of selected performance indexes in I and II stage of motion

| Index | $e_{max1j}$ [rad] | $\varepsilon_{1j}$ [rad] | $e_{max2j}$ [rad/s] | $\varepsilon_{2j}$ [rad/s] | $s_{maxj}$ [rad/s] | $\sigma_j$ [rad/s] |
|---|---|---|---|---|---|---|
| Stage I, $k = 1, \ldots, 2250$ | | | | | | |
| Wheel 1, $j = 1$ | 0.91 | 0.52 | 1.1 | 0.21 | 1.38 | 0.35 |
| Wheel 2, $j = 2$ | 0.91 | 0.44 | 1.1 | 0.21 | 1.38 | 0.31 |
| Stage II, $k = 2251, \ldots, 4500$ | | | | | | |
| Wheel 1, $j = 1$ | 0.39 | 0.18 | 0.39 | 0.13 | 0.43 | 0.16 |
| Wheel 2, $j = 2$ | 0.73 | 0.45 | 0.35 | 0.14 | 0.53 | 0.27 |

structure in ADHDP configuration. The control system provides good quality of tracking motion execution expressed by a low values of implemented performance indexes. In the numerical tests was assumed the least advantageous case of zero initial values of NNs output layer weights of NDP structures. The lack of knowledge of the mathematical model of the controlled object in the synthesis process of the control system is the characteristic feature of ADHDP algorithm, distinguishing this algorithm from the other NDP structures. The lack of knowledge of the controlled object model, as results from the numerical tests, is associated with lower quality of tracking control. In this case, the critic's NN is not only approximating the dependence of the value function in relation to the state of the object, but also the dependence of the value function in relation to the generated control signal. For this reason, the critic's NN structure of the ADHDP algorithm is being developed. The presented control system with ADHDP algorithms is characterized by the quality of tracking motion implementation comparable to the adaptive and neural control algorithm.

## 7.8  Behavioural Control of WMR's Motion

In the recent years, we saw a dynamic development of mobile robotics, which considerably broadened its focus, increasing at the same time the level of complexity of the tasks implemented in this field. This resulted in a need for developing relevant algorithms controlling the execution of the motion of mobile platforms. The growth in WMR design complexity, the amount of available information about the surrounding environment, and the level of advancement of microprocessors have made it possible to design control systems generating the WMR's trajectory in real time, with a possibility of its modification in response to environmental conditions, e.g. obstacles. The most frequent task of control of an autonomous WMR is implemented in the form of a hierarchical control system, in which the layer planning the trajectory of the motion of a selected WMR point generates a collision-free trajectory executed by a tracking control system. There are many methods of planning the trajectory of a WMR. They can be divided into two groups, assuming as the classification criterion the amount of available information about the environment, in the form of a map of the surroundings, fed to the control system.

The first group is constituted by algorithms making it possible to define a collision-free trajectory of the WMR in a known environment. They are called global methods. It is assumed that an exact map of the surrounding environment is known, which makes it possible to generate a collision-free trajectory before the motion is started, and then to execute this trajectory by the tracking control layer of the hierarchical control system. These methods are characterised by high computation costs and little resistance to changes in initial conditions of the task (change in the configuration of obstacles, emergence of a new or moving obstacle). Moreover, they require full knowledge of the environment, which excludes their application in open environments or environments equipped with sensors in an insufficient manner, e.g. due to economic reasons. An example of global motion planning methods is the artificial potential field method [2, 7, 23]. As part of this method, an artificial potential field is assigned to a known environment map, where a goal "attracts" the robot, while obstacles generate a repulsive potential. The motion path of a given WMR point is determined on the basis of the gradient of the artificial potential field. One of the disadvantages of the artificial potential field method is the possibility of occurrence of local minimums of the potential function where the robot can get stuck. Other examples of global methods are Voronoi diagram [6, 28] and visibility graph method [55, 57].

The second group are algorithms making it possible to define a collision-free trajectories in an unknown environment, called local methods. In these tasks, the map of the environment is not known, and the information about the distance to obstacles is obtained on an ongoing basis by the sensory system of the robot [19, 23, 37–39, 71, 72], which makes it possible to determine only the nearest section of the trajectory around the current state of the robot. The final trajectory is created by combining together local sections generated when the robot is in motion. These algorithms are called reactive algorithms, or sensor signal-based algorithms. Examples of such algorithms include systems generating behavioural (reactive) control, inspired by the animal world [9, 10, 38, 39, 47, 71, 72]. These are simple behaviours which can be observed, e.g. in insects. Among them, there is goal-seeking (GS) task (e.g. seeking for a food source), and "keep to the centre of the open space" tasks which can be viewed as an obstacle avoiding (OA) task.

The implementation schemes for individual elementary behaviours are presented in Fig. 7.68a, b. No behavioural control system, whether of OA or GS type, enables the implementation of a complex task such as "combination of behaviours OA and GS" (CB), shown schematically in Fig. 7.68c. The implementation of such a task is possible by way of combination of GS and OA behavioural controls, along with a relevant switching function. Another examples of local methods of generation of robots' motion trajectories are e.g. potential fields method [1, 60, 62] or elastic band approach [8].

**(a)**              **(b)**                          **(c)**

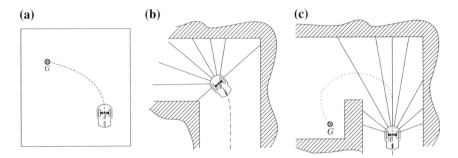

**Fig. 7.68**  **a** Execution scheme of an elementary behaviour of GS type, **b** execution scheme of an elementary behaviour of OA type, **c** execution scheme of a complex CB behaviour

**Fig. 7.69**  Schematic of
WMR Pioneer 2-DX with
laser rangefinder mounted

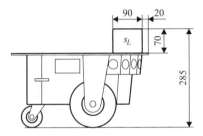

The sensory system of WMR Pioneer 2-DX, whose design is discussed in Sect. 2.1, was equipped with an additional Banner LT3PU laser rangefinder, installed on the upper panel of the WMR frame and marked as $s_L$ in Fig. 7.69.

The axis of the laser rangefinder beam is parallel to the symmetry axis of the WMR frame, $\omega_{Lr} = 0°$, and its dimensions are, respectively, $90 \times 70 \times 35$ [mm]. The laser rangefinder was calibrated to measure distance within the limits of $d_{Lr} \in \langle 0.4, 4.0 \rangle$ [m]. It is powered with 12 [V], the output signal $U_L \in \langle 0, 10 \rangle$ [V] is converted by a 12-bit A/D converter of the digital signal processing board into adequately scaled information about the distance of the selected WMR point to the obstacle.

Distance to obstacles is measured by the sensory system of WMR Pioneer 2-DX and consists in cyclical activation of measurements carried out by individual ultrasonic sensors $s_{u1}, \ldots, s_{u8}$ in a pre-defined sequence. The maximal measured distance to an obstacle is $d_{max} = 6$ [m]. Distance to obstacles located in front of the WMR was measured by the laser rangefinder irrespectively of measurements conducted with the use of ultrasonic sensors.

A schematic of WMR Pioneer 2-DX in the measurement environment is shown in Fig. 7.70, where the following notations were used: $l, l_1$ – dimensions resulting from WMR's geometry, $\beta$ – frame rotation angle, $A(x_A, y_A)$ – point located at the intersection of the WMR frame symmetry plane and the axis of driving wheels, $G(x_G, y_G)$ – selected goal of the WMR's motion, $x_1, y_1$ – axes of the movable system of coordinates connected with point $A$ in such a way so that axis $x_1$ is identical to the symmetry axis of the WMR frame, $p_G$ – straight line passing through the points

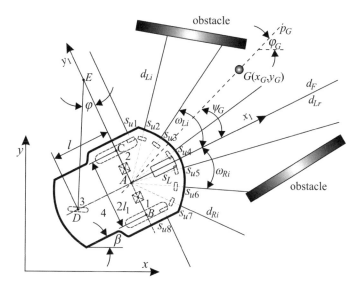

**Fig. 7.70** Scheme of WMR Pioneer 2-DX in the measurement environment

$A$ and $G$, $\psi_G$ – angle between the line $p_G$ and axis $x_1$, $\varphi_G$ – angle between the line $p_G$ and axis $x$ of the immovable system of coordinates, $d_{Li}$, $d_{Ri}$ – measurements of distance to obstacles, executed by the sensory system of the WMR, selected from all measurements and assigned to distance measurement groups for the left and right side of the WMR frame, $i = 1, 2$, $d_F$ – measurement of distance to an obstacle in front of the WMR frame, executed by the laser rangefinder, $d_F = d_{Lr}$, $\omega_{Li}$, $\omega_{Ri}$ – angles between axes of individual measurements and axis $x_1$ of the movable system of coordinates. The distance of WMR to the goal $d_G$ is determined as the length of the line segment $d_G = |AG|$.

Thus, an expanded sensory system of WMR Pioneer 2-DX is comprised of 8 ultrasonic sonars $s_{u1}, \ldots s_{u8}$ and a laser rangefinder $s_L$. In order to execute the task consisting in generating behavioural controls, selected sensors were assigned to three groups. The task of the first sensor group is to measure the distance to the obstacle on the left of the WMR frame; it includes sensors $s_{u2}$ and $s_{u3}$ whose measurement values are denoted, correspondingly, by $d_{L1\{k\}}$ ($s_{u2}$) and $d_{L2\{k\}}$ ($s_{u3}$). The distance to the obstacles located in front of the WMR is measured by the laser rangefinder $s_L$, whereas the measurement value is denoted by $d_{Lr\{k\}}$. The task of the third group of sensors is to measure the distance to the obstacle on the right side of the WMR frame. The third group of sensors includes sonars $s_{u6}$ and $s_{u7}$ whose measurement values are denoted, respectively, by $d_{R1\{k\}}$ ($s_{u6}$), $d_{R2\{k\}}$ ($s_{u7}$).

The algorithms generating behavioural control signals are based on feedback from the environment, fed by the sensory system of the WMR in the form of values of distance to obstacles, measured by individual sensor groups. Input signals to the structures of the trajectory generating layer of the hierarchical control system were scaled down to the interval $\langle 0, 1 \rangle$, by way of an adequate transformation of the signals. The normalised distance to the obstacle located in front of the WMR is determined on the basis of the scaled signal from the laser rangefinder, in accordance with the relation $d^*_{F\{k\}} = d_{Lr\{k\}}/d_{Fmax}$, where $d_{Fmax} = 4$ [m] is the range of the laser rangefinder. From among the signals of the sensor groups on the right and left side of the frame of WMR Pioneer 2-DX, in order to avoid a collision with an obstacle, minimal values are selected in accordance with the relations $d_{Lm\{k\}} = \min\left(d_{L1\{k\}}\left(s_{u2}\right), d_{L2\{k\}}\left(s_{u3}\right)\right), d_{Rm\{k\}} = \min\left(d_{R1\{k\}}\left(s_{u6}\right), d_{R2\{k\}}\left(s_{u7}\right)\right)$. Then, the values are scaled. The value $d^*_{L\{k\}} = 2\left[\left(d_{Lm\{k\}}/\left(d_{Lm\{k\}} + d_{Rm\{k\}}\right)\right) - 0.5\right]$ is the normalised distance to the obstacles on the left side of the WMR frame, $d^*_{R\{k\}} = 2\left[\left(d_{Rm\{k\}}/\left(d_{Lm\{k\}} + d_{Rm\{k\}}\right)\right) - 0.5\right]$ is the normalised distance to obstacles on the right side of the WMR frame.

### 7.8.1 Behavioural Control Synthesis

Difficulties in the execution of the WMR's autonomous motion result from the necessity of obtaining and processing information from the environment, enabling to detect obstacles in the vicinity of the robot, or to identify the goal of motion. The acquisition of information about the environment can be implemented in various ways. One of them is to apply ultrasonic distance sensors which - apart from the advantages in the form of low cost of a sensor and easiness of measurement processing - have also a number of disadvantages such as the range of measurement, usually limited to a few metres, susceptibility to disturbance, and if many sensors are used - cross-disturbance of measurements and a longer measurement time. Another type of measurement systems, more and more commonly used in WMRs in the recent years are laser rangefinders, offering higher range, measurement accuracy, shorter measurement time, and lower susceptibility to disturbance when compared to ultrasonic sensors. Moreover, acquisition of information using a group of such sensors does not result in cross-disturbance of measurements from individual sensors. Their disadvantage is a relatively high price and, in the context of application in sensory systems of robots, a point-to-point distance measurement. If we want to acquire data about the distance to obstacles around the robot, we have to use many sensors. An interesting solution without such a disadvantage is laser scanner, where the beam sweeps the space around the sensor with the use of a rotating mirror, which enables acquisition of information from a wide range of angles in relation to the sensor axis. However, such a method of measurement results in longer acquisition times of data relating to distances to obstacles around the robot. Currently, in the sensory systems of robots, vision systems are more and more commonly used. Such systems enable acquisition of large quantities of data, distance measurement and navigation in an unknown environment, but they require large computing capacity of the control unit to process the data. Another problem in the implementation of the WMR's autonomous motion is processing the information acquired from the sensory system, generating a collision-free trajectory in real time, and its execution. The methods of WMR's motion execution were discussed in the earlier subchapters. In this subchapter, we discuss elementary behaviours of GS and OA type, and present a synthesis of a behavioural control algorithm in a complex CB task, with the use of NDP algorithms.

### 7.8.1.1 "Goal-Seeking" Task

Behavioural control in an elementary GS behaviour consists in generation of control signals based on which a desired trajectory is calculated that guarantees reaching the selected target position in point $G$ by a selected point $A$ of the WMR. The orientation of the robot frame in the target position is irrelevant. This task can be viewed as an issue of minimisation of distance $d_{G\{k\}} = |AG|$ and angle $\psi_{G\{k\}}$ defined as deviation of the WMR frame's axis from the straight line $p_G$ passing through the points $A$ and $G$.

In this task, the velocity error $e_{Gv\{k\}}$ is defined, as well as the error of deviation of the axis of WMR frame from the line $p_G$, passing through points $A$ i $G$, $\psi_{G\{k\}}$, in the form of the relation

$$
\begin{aligned}
e_{Gv\{k\}} &= f\left(d_{G\{k\}}\right) - \frac{v_{A\{k\}}}{v_A^*} , \\
\psi_{G\{k\}} &= \varphi_{G\{k\}} - \beta_{\{k\}} ,
\end{aligned}
\tag{7.134}
$$

where $f(.)$ – scaling bipolar sigmoid function, $v_{A\{k\}}$ – executed velocity of motion of the point $A$ of the WMR, $v_A^*$ – maximal desired velocity of the point $A$ of the WMR. The angle $\varphi_{G\{k\}}$ is determined in accordance with the relation

$$
\varphi_{G\{k\}} = \arctan\left(\frac{y_{G\{k\}} - y_{A\{k\}}}{x_{G\{k\}} - x_{A\{k\}}}\right) .
\tag{7.135}
$$

The task of the behavioural control algorithm is to determine control signals $\mathbf{u}_{G\{k\}} = \left[u_{Gv\{k\}}, u_{G\dot{\beta}\{k\}}\right]^T$ minimising the values of the indicators (7.134) on the basis of information on the current location of the point $A$ of the WMR and the goal $G$. Control signals are as follows: $u_{Gv\{k\}} \in \langle 0, 1 \rangle$ – normalised signal controlling the desired velocity of the point $A$, $u_{G\dot{\beta}\{k\}} \in \langle -1, 1 \rangle$ – normalised signal controlling the desired angular velocity of the WMR frame's own rotation. The generated control signals are converted into the desired trajectory of the point $A$ of the WMR from the current position of the point $A$ to the point $G$.

### 7.8.1.2 "Obstacle Avoidance" Task

Behavioural control in a "keep to the centre of the open space" or OA task consists in generation of control signals ensuring that a desired trajectory of motion of the point $A$ of the WMR is determined that maximises distances to obstacles, which as a results - leads to generation of a desired trajectory in which the selected WMR's point occupies the centre of the open space. The control signals are generated on the basis of information from the sensory system of the robot.

In a task of OA type, velocity error $e_{Ov\{k\}}$ and error of keeping to the centre of an open space $e_{O\dot{\beta}\{k\}}$ are defined in the form of the relation

$$
\begin{aligned}
e_{Ov\{k\}} &= f\left(d^*_{F\{k\}}\right) - \frac{v_{A\{k\}}}{v^*_A} \ , \\
e_{O\dot{\beta}\{k\}} &= d^*_{R\{k\}} - d^*_{L\{k\}} \ .
\end{aligned}
\tag{7.136}
$$

In a task of OA type, the behavioural control algorithm generates control signals $\mathbf{u}_{O\{k\}} = \left[u_{Ov\{k\}}, u_{O\dot{\beta}\{k\}}\right]^T$ minimising the values of the errors (7.136), where $u_{Ov\{k\}} \in \langle 0, 1\rangle$ – normalised signal controlling desired velocity of the point $A$ of the WMR, $u_{O\dot{\beta}\{k\}} \in \langle 0, 1\rangle$ – normalised signal controlling the desired angular velocity of the WMR frame's own rotation. The generated control signals are converted into a desired collision-free trajectory of motion of the point $A$ of the WMR, whereas the motion goal is not defined.

### 7.8.1.3  "Goal-Seeking with Obstacle Avoidance" Task

The task of the layer generating a collision-free trajectory of the hierarchical control system in "goal-seeking with obstacle avoidance" tasks (CB) is to generate a desired trajectory of the point $A$ of the WMR that guarantees the execution of a complex task consisting in goal seeking with avoidance of static obstacles in a two-dimensional environment. Thus, this task is a combination of two tasks - of GS and OA type. It can be implemented in various ways, for example using control signals being a combination of control signals of behavioural control algorithms in tasks of GS and OA type, by assigning a fixed weight to control signals of individual tasks [11, 52], or making the influence of the control signals of individual tasks dependent on the information from the environment. The second solution requires an additional algorithm to be applied, generating a signal scaling the share of control signals of individual behavioural controls in the overall control signal of the trajectory generation layer. Such an algorithm can be implemented, for example, with the use of a fuzzy logic controller (FLC) [45, 52]. A typical structure of the trajectory generation layer in a CB task, implemented with the use of a FLC to scale behavioural controls, is presented in Fig. 7.71.

In this chapter, a CB task is implemented with the use of two NDP structures in the ADHDP configuration, and a proportional controller, simplifying thereby the structure of the trajectory generation layer. The overall control signal of the trajectory generator $\mathbf{u}_{B\{k\}} = \left[u_{Bv\{k\}}, u_{B\dot{\beta}\{k\}}\right]^T$ is comprised of control signals generated by two ADHDP algorithms, $\mathbf{u}_{BA\{k\}} = \left[u_{BAv\{k\}}, u_{BA\dot{\beta}\{k\}}\right]^T$, and control signals of the proportional controller P $\mathbf{u}_{BP\{k\}}$, in accordance with the relation

$$
\mathbf{u}_{B\{k\}} = \mathbf{u}_{BA\{k\}} + \mathbf{u}_{BP\{k\}} \ ,
\tag{7.137}
$$

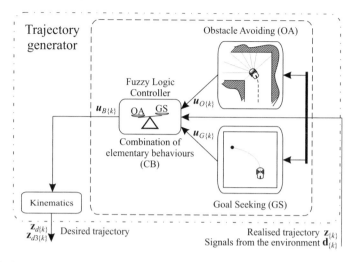

**Fig. 7.71** Scheme of the trajectory generation layer implemented with a fuzzy logic controller to combine behavioural controls

where the control signal of the controller P is expressed by the relation

$$\mathbf{u}_{BP\{k\}} = \mathbf{K}_{BP}\, \mathbf{e}_{B\{k\}}\,, \tag{7.138}$$

where $\mathbf{e}_{B\{k\}} = \left[e_{Bv\{k\}}, e_{O\dot{\beta}\{k\}}, \psi_{G\{k\}}\right]^T$ – error vector, defined as

$$
\begin{aligned}
e_{Bv\{k\}} &= \min\left(f\left(d^*_{F\{k\}}\right), f\left(d_{G\{k\}}\right)\right) - \frac{v_{A\{k\}}}{v^*_A}\,, \\
e_{O\dot{\beta}\{k\}} &= d^*_{R\{k\}} - d^*_{L\{k\}}\,, \\
\psi_{G\{k\}} &= \varphi_{G\{k\}} - \beta_{\{k\}}\,.
\end{aligned}
\tag{7.139}
$$

The notations occurring in the above relations are as follows: $f\,(.)$ – scaling bipolar sigmoid function, $\mathbf{K}_{BP}$ – constant matrix of positive coefficients of gains of the proportional controller,

$$
\mathbf{K}_{BP} = \begin{bmatrix} k_v & 0 & 0 \\ 0 & k_O & k_\psi \end{bmatrix}\,, \tag{7.140}
$$

where $k_v$, $k_O$, $k_\psi$ – constant, positive design coefficients.

The control signal $u_{Bv\{k\}} \in \langle 0, 1\rangle$ is a normalised signal of the layer generating a collision-free trajectory of motion, controlling the desired velocity of the point $A$ of the WMR, $u_{B\dot{\beta}\{k\}} \in \langle -1, 1\rangle$ is a normalised signal controlling the desired angular velocity of the WMR frame's own rotation.

A hierarchical control system with a layer planning the trajectory implemented with the use of a proportional controller and ADHDP structures, generating behavioural control signals in a CB task, is presented in a scheme in Fig. 7.72. The motion execution layer is constituted by the tracking control system in which a NDP algorithm

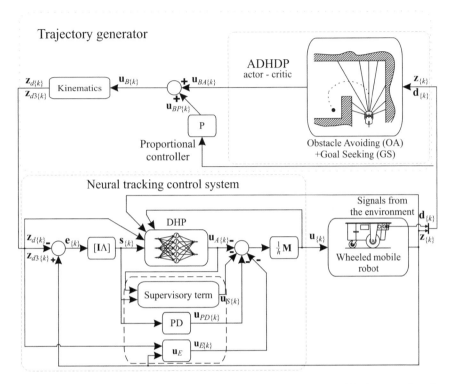

**Fig. 7.72** Scheme of a hierarchical WMR's motion control system applied in the execution of a CB task

is applied in DHP configuration, described in detail in Sect. 7.5.1. In the trajectory generation layer, an ADHDP algorithm was applied, since as the only one from the family of NDP structures it does not require the knowledge of the controled object model in the process of the actor's and critic's NNs weights adaptation. This makes it possible to apply it in the task of planning the trajectory of the point A of the WMR in an unknown environment. The control algorithm presented is discussed in [48, 49, 51, 54].

In the motion planning layer, two ADHDP algorithms were applied. Their task is to generate suboptimal control laws minimising the value functions

$$
\begin{aligned}
V_{Bv\{k\}}\left(e_{v\{k\}}, u_{Bv\{k\}}\right) &= \sum_{k=0}^{n} \gamma^k L_{Bv\{k\}}\left(e_{Bv\{k\}}, u_{Bv\{k\}}\right), \\
V_{B\dot{\beta}\{k\}}\left(e_{O\dot{\beta}\{k\}}, \psi_{G\{k\}}, u_{O\dot{\beta}\{k\}}\right) &= \sum_{k=0}^{n} \gamma^k L_{B\dot{\beta}\{k\}}\left(e_{O\dot{\beta}\{k\}}, \psi_{G\{k\}}, u_{\dot{\beta}\{k\}}\right).
\end{aligned}
\tag{7.141}
$$

The following cost functions were assumed

$$L_{Bv\{k\}}\left(e_{Bv\{k\}}, u_{Bv\{k\}}\right) = \tfrac{1}{2}R_v f^2\left(e_{Bv\{k\}}\right) + \tfrac{1}{2}Q_v u_{Bv\{k\}}^2 ,$$
$$L_{B\dot\beta\{k\}}\left(e_{O\dot\beta}, \psi_{G\{k\}}, u_{B\dot\beta\{k\}}\right) = \tfrac{1}{2}R_{O\dot\beta} e_{O\dot\beta\{k\}}^2 + \tfrac{1}{2}R_{G\dot\beta} f^2\left(\psi_{G\{k\}}\right) + \tfrac{1}{2}Q_{\dot\beta} u_{B\dot\beta\{k\}}^2 ,$$

$$(7.142)$$

where $R_v, Q_v, R_{O\dot\beta}, R_{G\dot\beta}, Q_{\dot\beta}$ – positive design coefficients, determining the influence of individual errors and control signals on the cost function value.

The task of ADHDP algorithms, taking into account the value functions (7.141) calculated on the basis of the cost functions (7.142) is to generate a control law minimising the values of individual errors and control signals.

In the task consisting in generation of control signals $u_{BAv\{k\}}$ and $u_{BA\dot\beta\{k\}}$ two ADHDP structures were applied, comprised of:

- **Actor**, implemented in the form of a RVFL NN which generates a suboptimal control law

$$u_{BAv\{k\}}\left(\mathbf{x}_{BAv\{k\}}, \mathbf{W}_{BAv\{k\}}\right) = \mathbf{W}_{BAv\{k\}}^T \mathbf{S}\left(\mathbf{D}_{BAv}^T \mathbf{x}_{BAv\{k\}}\right) ,$$
$$u_{BA\dot\beta\{k\}}\left(\mathbf{x}_{BA\dot\beta\{k\}}, \mathbf{W}_{BA\dot\beta\{k\}}\right) = \mathbf{W}_{BA\dot\beta\{k\}}^T \mathbf{S}\left(\mathbf{D}_{BA\dot\beta}^T \mathbf{x}_{BA\dot\beta\{k\}}\right) ,$$

$$(7.143)$$

where $\mathbf{x}_{BAv\{k\}} = \left[1, f\left(e_{Bv\{k\}}\right)\right]^T$, $\mathbf{x}_{BA\dot\beta\{k\}} = \left[1, e_{O\dot\beta\{k\}}, f\left(\psi_{G\{k\}}\right)\right]^T$ – vectors of inputs to individual actor's NNs, $\mathbf{D}_{BAv}, \mathbf{D}_{BA\dot\beta}$ – matrices of constant weights of the input layer of the actor's NN, selected randomly in the process of NN initialisation, $f(.)$ – scaling bipolar sigmoid function.

Weights of the actor's NN are adapted with the use of backpropagation method of performance indexes

$$e_{BAv\{k\}} = \frac{\partial L_{Bv\{k\}}\left(e_{Bv\{k\}}, u_{Bv\{k\}}\right)}{\partial u_{Bv\{k\}}} + \gamma \frac{\partial \hat V_{Bv\{k+1\}}\left(\mathbf{x}_{BCv\{k+1\}}, \mathbf{W}_{BCv\{k\}}\right)}{\partial u_{Bv\{k\}}} ,$$
$$e_{BA\dot\beta\{k\}} = \frac{\partial L_{B\dot\beta\{k\}}\left(e_{O\dot\beta}, \psi_{G\{k\}}, u_{B\dot\beta\{k\}}\right)}{\partial u_{B\dot\beta\{k\}}} + \gamma \frac{\partial \hat V_{B\dot\beta\{k+1\}}\left(\mathbf{x}_{BC\dot\beta\{k+1\}}, \mathbf{W}_{BC\dot\beta\{k\}}\right)}{\partial u_{B\dot\beta\{k\}}} .$$

$$(7.144)$$

- **Critic** approximating the value functions (7.141) in which a RVFL NNs were applied. The critic generates signals in the form of

$$\hat V_{Bv\{k\}}\left(\mathbf{x}_{BCv\{k\}}, \mathbf{W}_{BCv\{k\}}\right) = \mathbf{W}_{BCv\{k\}}^T \mathbf{S}\left(\mathbf{D}_{BCv}^T \mathbf{x}_{BCv\{k\}}\right) ,$$
$$\hat V_{B\dot\beta\{k\}}\left(\mathbf{x}_{BC\dot\beta\{k\}}, \mathbf{W}_{BC\dot\beta\{k\}}\right) = \mathbf{W}_{BC\dot\beta\{k\}}^T \mathbf{S}\left(\mathbf{D}_{BC\dot\beta}^T \mathbf{x}_{BC\dot\beta\{k\}}\right) ,$$

$$(7.145)$$

where $\mathbf{W}_{BCv\{k\}}, \mathbf{W}_{BC\dot\beta\{k\}}$ – vectors of weights of the output layer of the critic's NNs, $\mathbf{x}_{BCv\{k\}} = \left[f\left(e_{Bv\{k\}}\right), u_{Bv\{k\}}\right]^T$, $\mathbf{x}_{BC\dot\beta\{k\}} = \left[e_{O\dot\beta\{k\}}, f\left(\psi_{G\{k\}}\right), u_{B\dot\beta\{k\}}\right]^T$ – vectors of inputs to the critic's NNs, $\mathbf{D}_{BCv}, \mathbf{D}_{BC\dot\beta}$ – matrices of constant weights of the critic NNs' input layer, selected randomly in the process of NNs initiation.

Weights of the critic's NNs are adapted with the use of backpropagation method of the temporal difference errors

$$
\begin{aligned}
e_{BCv\{k\}} &= L_{Bv\{k\}}\left(e_{Bv\{k\}}, u_{Bv\{k\}}\right) + \gamma \hat{V}_{Bv\{k+1\}}\left(\mathbf{x}_{BCv\{k+1\}}, \mathbf{W}_{BCv\{k\}}\right) + \\
&\quad - \hat{V}_{Bv\{k\}}\left(\mathbf{x}_{BCv\{k\}}, \mathbf{W}_{BCv\{k\}}\right) , \\
e_{BC\dot{\beta}\{k\}} &= L_{B\dot{\beta}\{k\}}\left(e_{O\dot{\beta}}, \psi_{G\{k\}}, u_{B\dot{\beta}\{k\}}\right) + \gamma \hat{V}_{B\dot{\beta}\{k+1\}}\left(\mathbf{x}_{BC\dot{\beta}\{k+1\}}, \mathbf{W}_{BC\dot{\beta}\{k\}}\right) + \\
&\quad - \hat{V}_{B\dot{\beta}\{k\}}\left(\mathbf{x}_{BC\dot{\beta}\{k\}}, \mathbf{W}_{BC\dot{\beta}\{k\}}\right) .
\end{aligned}
\tag{7.146}
$$

The generated control signals $u_{Bv\{k\}}$, $u_{B\dot{\beta}\{k\}}$ make it possible to determine the desired angular velocities of rotation of the WMR's driving wheels.

In the global immovable system of coordinates $xy$, the position of the WMR is described by coordinates defined as $\left[x_{A\{k\}}, y_{A\{k\}}, \beta_{\{k\}}\right]^T$, where $x_{A\{k\}}, y_{A\{k\}}$ – coordinates of the point $A$ of the WMR, $\beta_{\{k\}}$ – the angle of the WMR frame's own rotation.

In order to generate the trajectory of the point $A$ of the WMR, a modified description of the inverse kinematics problem, presented in Sect. 2.1.1, was applied. Changes of the parameters defining the position of the point $A$ and orientation of the WMR's frame in the global system of coordinates $xy$, in the task consisting in generation of the desired trajectory with the use of behavioural controls are defined by the relation

$$
\begin{bmatrix} \dot{x}_{Ad\{k\}} \\ \dot{y}_{Ad\{k\}} \\ \dot{\beta}_{d\{k\}} \end{bmatrix} = \begin{bmatrix} v_A^* \cos(\beta) & 0 \\ v_A^* \sin(\beta) & 0 \\ 0 & \dot{\beta}^* \end{bmatrix} \begin{bmatrix} u_{Bv\{k\}} \\ u_{B\dot{\beta}\{k\}} \end{bmatrix} ,
\tag{7.147}
$$

where $v_A^*$ – desired maximal velocity of the point $A$ of the WMR, $\dot{\beta}^*$ – desired maximal angular velocity of the WMR frame's own rotation. Taking account of the relation (7.147), the Eq. (2.17) has the following form

$$
\begin{bmatrix} z_{d21\{k\}} \\ z_{d22\{k\}} \end{bmatrix} = \frac{1}{r} \begin{bmatrix} v_A^* & \dot{\beta}^* l_1 \\ v_A^* & -\dot{\beta}^* l_1 \end{bmatrix} \begin{bmatrix} u_{Bv\{k\}} \\ u_{B\dot{\beta}\{k\}} \end{bmatrix} ,
\tag{7.148}
$$

where desired discreet values of angular velocities of the WMR driving wheels' own rotation are generated on the basis of behavioural control signals $u_{Bv\{k\}}$, $u_{B\dot{\beta}\{k\}}$.

The generated desired values of angular velocities are converted into desired angles of driving wheels' own rotation $z_{d11\{k\}}$, $z_{d12\{k\}}$, and desired angular accelerations $z_{d31\{k\}}$, $z_{d32\{k\}}$. The generated desired trajectory is executed by the tracking control system.

### 7.8.2 Simulation Tests

Numerical tests of the trajectory planning layer in a CB task were carried out with the use of an algorithm implemented using two NDP structures in the ADHDP configuration, hereinafter referred to as neural trajectory generator. The task of the trajectory generation layer of the hierarchical control system is to plan a collision-free trajectory of the WMR, whose execution will result in moving the point $A$ of the WMR from the initial position to the goal $G$ with selected coordinates. The WMR frame's orientation as at the moment of reaching the goal is irrelevant. In order to execute a CB task, a virtual measurement environment of a complex shape was programmed, in which the initial position of the point $A$ of the WMR was determined in the point $S$ (0.5, 1.0). The positions of the goals of motion of the WMR were determined in such a way so that reaching them would be impossible with the use of behavioural control of OA or GS type. The application of an OA behavioural control does not guarantee reaching the goal $G$, while using a GS control can result in a collision with an obstacle. The designed WMR's trajectory planning algorithm implements the function of soft switching of behavioural controls in order to generate a trajectory within a complex task of goal seeking with avoidance of obstacles. The generated desired trajectory is executed by a neural tracking control system with the use of NDP structures in the DHP configuration, discussed in detail in Sect. 7.5.1.

The task of the hierarchical control system is to generate and execute a desired trajectory of the point $A$ of the WMR to the goal located in the points $G_A$ (4.9, 3.5), $G_B$ (10.3, 3.0) and $G_C$ (7.6, 1.5) of the virtual measurement environment. The paths of motion of the point $A$ of the WMR to the goal $G$, situated in individual locations, are presented in Fig. 7.73.

The triangle marks the initial position of the point $A$ of the WMR in the point $S$ (0.5, 1.0), "**X**" marks the location of the goal, and the rhombus – final position of the point $A$ of the WMR. The continuous line shows the motion path of the point $A$, while the positions of the WMR frame are marked every 2 [s] of motion.

In the tests of the hierarchical control system with a layer generating collision-free motion trajectories of the point $A$ of the WMR, implemented with the use of a neural trajectory generator with ADHDP structures, and a layer of motion execution with DHP structures, the following parameter values were used:

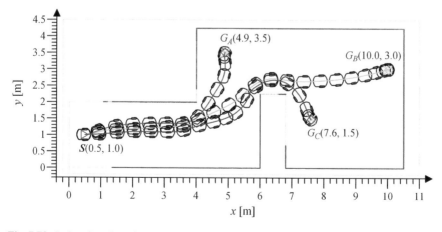

**Fig. 7.73** Paths of motion of the point $A$ of the WMR to individual locations of the goal $G$

- parameters of ADHDP structure generating control signal $u_{BAv\{k\}}$:

  - the type of actor's NN applied: RVFL,
    $m_A = 8$ – number of neurons of actor's NN,
    $\Gamma_{BAv} = \text{diag}\{0.5\}$ – weight adaptation gain matrix $\mathbf{W}_{BAv}$,
    $\mathbf{D}_{BAv}$ – NN input layer weights randomly selected from the interval $D_{BAvi,i} \in \langle -0.5, 0.5 \rangle$,
    $\mathbf{W}_{BAv\{k=0\}} = \mathbf{0}$ – zero initial values of the output layer weights,
  - the type of critic's NN applied: RVFL,
    $m_C = 8$ – number of neurons of critic's NN,
    $\Gamma_{BCv} = \text{diag}\{0.9\}$ – weight adaptation gain matrix $\mathbf{W}_{BCv}$,
    $\mathbf{D}_{BCv}$ – NN input layer weights randomly selected from the interval $D_{BCvi,i} \in \langle -1, 1 \rangle$,
    $\mathbf{W}_{BCv\{k=0\}} = \mathbf{0}$ – zero initial values of the output layer weights,
  - $R_v = 1$, $Q_v = 0.0001$ – scale coefficients in the cost function,

- parameters of ADHDP structure generating control signal $u_{BA\dot\beta}$:

  - the type of actor's NN applied: RVFL,
    $m_A = 8$ – number of neurons of actor's NN,
    $\Gamma_{BA\dot\beta} = \text{diag}\{0.02\}$ – weight adaptation gain matrix $\mathbf{W}_{BA\dot\beta}$,
    $\mathbf{D}_{BA\dot\beta}$ – NN input layer weights randomly selected from the interval $D_{BA\dot\beta i,i} \in \langle -1, 1 \rangle$,
    $\mathbf{W}_{BA\dot\beta\{k=0\}} = \mathbf{0}$ – zero initial values of the output layer weights,
  - the type of critic's NN applied: RVFL,
    $m_C = 8$ – number of neurons of critic's NN,
    $\Gamma_{BC\dot\beta} = \text{diag}\{0.02\}$ – weight adaptation gain matrix $\mathbf{W}_{BC\dot\beta}$,
    $\mathbf{D}_{BC\dot\beta}$ – NN input layer weights randomly selected from the interval $D_{BC\dot\beta i,i} \in \langle -1, 1 \rangle$,
    $\mathbf{W}_{BC\dot\beta\{k=0\}} = \mathbf{0}$ – zero initial values of the output layer weights,
  - $R_{O\dot\beta} = 1$, $R_{G\dot\beta} = 1$, $Q_{\dot\beta} = 1.5$ – scale coefficients in the cost function,

- $k_v = 0.1$, $k_O = 0.05$, $k_\psi = 0.1/\pi$ – gains of the proportional controller,
- $l_{AC} = 3$ – number of inner loop iteration steps for the actor's and critic's structure adaptation in the $k$th step of the iterative process,
- $h = 0.01$ [s] – value of the computation step in the numerical test.

In all numerical tests of the hierarchical control system in the motion execution layer, a tracking control system was applied with DHP structures, with settings presented in Sect. 7.5.2.1. Only the following parameters of the tracking control system were changed:

- $\mathbf{K}_D = 1.8h\mathbf{I}$ – PD controller gain matrix,
- $\Gamma_{Aj} = \text{diag}\{0.15\}$ – weights adaptation gain matrix of the actor's NN $\mathbf{W}_{Aj}, j = 1, 2$,
- $\Gamma_{Cj} = \text{diag}\{0.25\}$ – weights adaptation gain matrix of the critic's NN $\mathbf{W}_{Cj}$.

In numerical and verification tests of the hierarchical control system, the values of gain coefficients of weight adaptation of the actor's and critic's NN's output layer weights of DHP algorithms and parameters of the PD controller of the motion execution layer were lowered.

In the description of numerical tests of the algorithm generating the trajectory of the point $A$ of the WMR, operating in a discreet time domain, the notation of variables was simplified by resigning from the $k$th index step of the process. In order to facilitate the analysis of the results, the abscissa axis of charts of discreet sequences were calibrated in time $t$.

### 7.8.2.1 Behavioural Control Research

Numerical tests of the algorithm generating collision-free motion trajectories of the point $A$ of the WMR in an unknown environment were carried out with the use of NDP algorithms in the ADHDP configuration, in the motion planning layer. In this section, a numerical test was discussed in detail, in which the task of the hierarchical control system is to generate online a desired trajectory of the point $A$ of the WMR, and to execute it, from the selected initial position in the point $S$ (0.5, 1.0), to the goal $G_A$ (4.9, 3.5). The trajectory is planned on the basis of feedback from the environment in the form of the measurements executed by emulated distances sensors. The executed motion path of the point $A$ of the WMR is marked with a red line in Fig. 7.74.

The triangle marks the initial position of the point $A$ of the WMR in the point $S$, the rhombus - final position of the point $A$, while "X" marks the desired final position $G_A$ (4.9, 3.5). The positions of the WMR frame were presented in discreet time moments, in intervals of 2 [s]. Red dots mark positions of obstacles detected by the emulated distance sensors. Frequency of the sensors ($f_S = 100$ [Hz]) was equal to the frequency of the simulation iteration steps. A large number of measurements

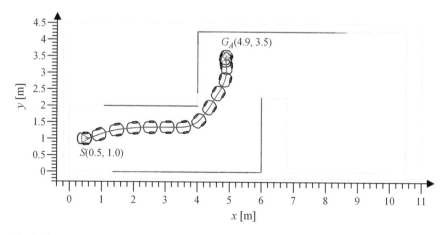

**Fig. 7.74** Motion path of the point $A$ of the WMR to the goal $G_A$ (4.9, 3.5)

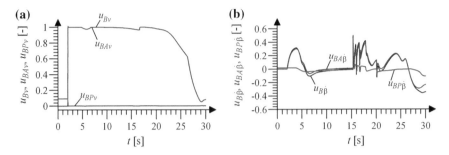

**Fig. 7.75** **a** Control signals $u_{Bv}$, $u_{BAv}$, $u_{BPv}$, **b** control signals $u_{B\dot{\beta}}$, $u_{BA\dot{\beta}}$, $u_{BP\dot{\beta}}$

was obtained. The measurements, marked in Fig. 7.74, formed a continuous red line on the circumference of the walls of the virtual measurement track.

The motion path of the WMR Pioneer 2-DX model in the virtual simulation environment results from the execution of the trajectory determined on the basis of control signals of the neural trajectory generator. The overall control signals of the neural trajectory generator ($u_{Bv}$ i $u_{B\dot{\beta}}$) are comprised of control signals generated by the actor's NNs ($u_{BAv}$ i $u_{BA\dot{\beta}}$) and by the proportional controller ($u_{BPv}$ i $u_{BP\dot{\beta}}$). The values of these signals are shown in Fig. 7.75.

The application of a proportional controller in the layer of the neural trajectory generator makes it possible to generate control signals executing the assumed task also at the initial stage of the motion, when weights of the NN's output layer of ADHDP structures are subject to adaptation from the initial (zero) values selected during the initiation. Such an approach makes it possible to limit the exploration executed by an NDP structure by way of a "hint" as to the correct direction of seeking of the suboptimal control. As a result of the adaptation of weights of the NN's output layer of ADHDP structures, control signals $u_{BAv}$ and $u_{BA\dot{\beta}}$ begin to dominate in the overall control, while control signals of the proportional controller assume values near to zero. Such an approach makes it possible to avoid time-consuming trial and error learning.

In the process of initialization of the actor-critic structures, the most unfavourable case was assumed in which the values of weights of the NN's output layer are equal to zero, which corresponds to the situation with no initial knowledge. The process of adaptation of the weights of the NN's output layer takes place online in accordance with the assumed adaptation algorithm; the values of weights of the NN's output layer of ADHDP structures are presented in Fig. 7.76. The weight values are limited.

In Fig. 7.77a the values of the distance between the point $A$ of the WMR and the goal, $d_G$, is presented; in Fig. 7.77b - the values of the angle $\psi_G$. The distance to goal $d_G$ is reduced during the motion to values close to zero. In the final phase of the motion, we can observe an increase of the value of the angle $\psi_G$, resulting from the WMR stopping in a place where the point $A$ is located farther than the point $G$, if we assume that the axis of the robot's frame determines the direction. In such a case, the value of the angle $\psi_G$ is $\psi_G = \pi$ or $\psi_G = -\pi$ [rad].

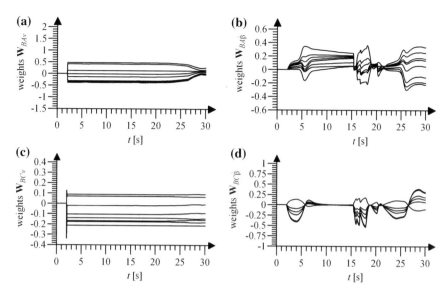

**Fig. 7.76** **a** Values of the NN's output layer weights of the actor $\mathbf{W}_{BAv}$, **b** values of the NN's output layer weights of the actor $\mathbf{W}_{BA\dot{\beta}}$, **c** values of the NN's output layer weights of the critic $\mathbf{W}_{BCv}$, **d** values of the NN's output layer weights of the critic $\mathbf{W}_{BC\dot{\beta}}$

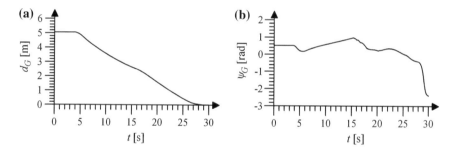

**Fig. 7.77** **a** Values of the distance between the point $A$ of the WMR and the target $G$, $d_G$, **b** values of the angle $\psi_G$

On the basis of the values of control signals of the trajectory generation layer, in accordance with the relation (7.148), values of motion parameters of the WMR's driving wheels are determined. The generated values of desired angular velocities of the WMR driving wheels' own rotation $z_{d21}$ and $z_{d22}$ are presented in Fig. 7.78a. The desired trajectory is executed by the tracking control system in which DHP algorithm was applied to compensate for the nonlinearities of the controlled object. The values of the overall signal controlling the motion of wheel 1, $u_1$, with scaled values of the actor NN's control signal $U_{A1}$, and PD control signal $U_{PD1}$, are shown in Fig. 7.78b. The trajectory is executed with tracking errors which in the case of

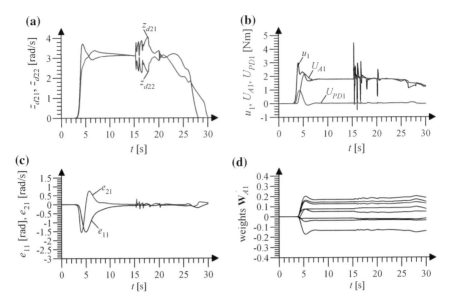

**Fig. 7.78** **a** Values of desired angular velocities of the driving wheels' own rotation, $z_{d21}$ and $z_{d22}$, **b** control signal $u_1$, scaled control signal generated by the actor's NN $U_{A1}$, $\mathbf{U}_A = -\frac{1}{h}\mathbf{M}\mathbf{u}_A$, scaled control signal generated by PD controller $U_{PD1}$, $\mathbf{U}_{PD} = -\frac{1}{h}\mathbf{M}\mathbf{u}_{PD}$, **c** values of tracking errors $e_{11}$ and $e_{21}$, **d** values of the weights of the actor NN's output layer $\mathbf{W}_{A1}$

wheel 1 of the WMR are presented in Fig. 7.78c. Weights of the first NN's output layer of the structure of actor $\mathbf{W}_{A1}$ of the DHP algorithm are presented in Fig. 7.78d.

The application of zero initial values of the weights of the NN's output layer of the DHP actor-critic structure in the motion execution layer results in the highest values of tracking errors at the initial stage of the motion, when the main role in generation of overall control signals is played by control signals of the PD controller $\mathbf{u}_{PD}$. The continuation of the weights adaptation process results in an increase of values of the control signals generated by the actor's NNs, which begin to dominate in the overall control signals, and - as a result - tracking errors are reduced.

In addition, numerical tests were carried out of the presented control system in a complex virtual measurement environment [48]. The paths of motion of the point $A$ of the WMR to individual positions of the goal are presented in Fig. 7.79.

Due to the complexity and size of the virtual measurement environment from Fig. 7.79, it was impossible to carry out verification tests with regard to the proposed algorithm of the neural trajectory generator for the presented environment config-uration in laboratory conditions. Verification tests of the proposed algorithm were carried out within the measurement environment presented in Fig. 7.73.

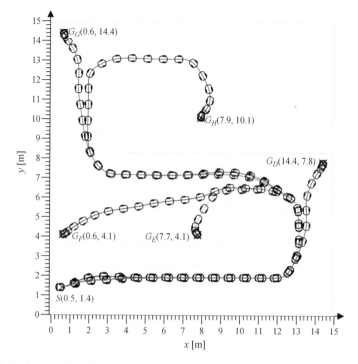

**Fig. 7.79** Paths of motion of the point $A$ of the WMR to individual locations of the goal $G$

## 7.8.3 Conclusions

Behavioural control of the WMR's motion in an unknown environment is a complex task requiring the acquisition of information from the environment about distances to obstacles with the use of the sensory system of the robot, the determination of position and orientation of the robot on the basis of available information, and control of actuating systems in order to execute the trajectory generated in real time.

The presented algorithm generating collision-free motion trajectories of the WMR in an unknown environment with static obstacles executes the task of planning the trajectory. The execution of the generated trajectory results in displacement of the point $A$ of the robot from the selected initial position to the goal located in a specific, accessible point of the test path.

The simulation tests checked the operation of the designed algorithm generating motion trajectories to a goal located in selected points of the test environment, which would not be accessible to point $A$ of the WMR with the use of one of the behavioural controls of OA or GS type, due to the distribution of obstacles. In this case the implementation of the motion towards the goal is guaranteed by the control of the trajectory planning layer, generated as a combination of behavioural controls of OA and GS type, with soft switching between individual tasks. This problem was solved by applying adaptive critic structures in the ADHDP configuration, which generate control signals enabling the execution of a complex task.

## 7.9  Summary

This subchapter contains a summary of the numerical tests conducted on the algorithms of tracking control of the point $A$ of the WMR. The quality of motion execution achieved with the use of individual algorithms is compared on the basis of the obtained values of selected performance indexes. Numerical tests of tracking control systems were carried out in the same conditions, applying the same trajectory with a path shaped as "8".

The values of averaged performance indexes obtained in individual numerical tests were determined on the basis of the following relations:
– average maximal error value for the execution of the desired angle of driving wheel 1 and 2 own rotation [rad], for continuous systems

$$e_{maxa} = \frac{1}{2}\left(e_{max1} + e_{max2}\right) , \tag{7.149}$$

or discreet systems

$$e_{maxa} = \frac{1}{2}\left(e_{max11} + e_{max12}\right) , \tag{7.150}$$

– average root mean square error of the execution of the desired rotation angle [rad] for continuous systems

$$\varepsilon_{av} = \frac{1}{2}\left(\varepsilon_1 + \varepsilon_2\right) , \tag{7.151}$$

or discreet systems

$$\varepsilon_{av} = \frac{1}{2}\left(\varepsilon_{11} + \varepsilon_{12}\right) , \tag{7.152}$$

– average maximal error value for the execution of the desired angular velocity [rad/s], for continuous systems

$$\dot{e}_{maxa} = \frac{1}{2}\left(\dot{e}_{max1} + \dot{e}_{max2}\right) , \tag{7.153}$$

or discreet systems

$$\dot{e}_{maxa} = \frac{1}{2}\left(e_{max21} + e_{max22}\right) , \tag{7.154}$$

– average root mean square error for the execution of the desired angular velocity [rad/s], for continuous systems

$$\dot{\varepsilon}_{av} = \frac{1}{2} \left( \dot{\varepsilon}_1 + \dot{\varepsilon}_2 \right) , \tag{7.155}$$

or discreet systems

$$\dot{\varepsilon}_{av} = \frac{1}{2} \left( \varepsilon_{21} + \varepsilon_{22} \right) , \tag{7.156}$$

– average maximal value of filered tracking error [rad/s],

$$s_{maxa} = \frac{1}{2} \left( s_{max1} + s_{max2} \right) , \tag{7.157}$$

– average root mean square filtered tracking error [rad/s],

$$\sigma_{av} = 0.5 \left( \sigma_1 + \sigma_2 \right) . \tag{7.158}$$

Values of performance indexes of the following tracking control algorithms were compared:

1. PD controller,
2. adaptive control system,
3. neural control system,
4. control system with an ADHDP algorithm,
5. control system with an HDP algorithm with zero initial values of the weights of NN's output layer of the critic $\mathbf{W}_{C\{0\}} = \mathbf{0}$ (HDP$_{W0}$),
6. control system with an HDP algorithm with nonzero initial values of the weights of NN's output layer of the critic $\mathbf{W}_{C\{0\}} = \mathbf{1}$ (HDP$_{W1}$),
7. control system with a GDHP algorithm,
8. control system with a DHP algorithm.

The values of selected performance indexes of the execution of the tracking motion of the point $A$ of the WMR were compared with the use of a trajectory with a path shaped as "8", without parametric disturbance. The values of performance indexes are presented in Table 7.34 with relevant numbering ordering the sequence of the control systems in the tables and on comparative charts, which is applied further in this subchapter.

Graphical interpretation of the values of performance indexes $s_{maxa}$ and $\sigma_{av}$ is presented on bar charts in Fig. 7.80a, b. In white (1), values of performance indexes of the PD controller are marked, while in grey (2) - values of performance indexes of the adaptive control system, and in green (3) values of performance indexes of the neural control algorithm. Performance indexes of the control system with NDP structures in the ADHDP configuration are marked in purple (4), performance indexes of the control system with HDP algorithms with zero initial values of the weights of the critic NN's output layer - in dark blue (5), performance indexes of the control system with HDP structures and nonzero initial values of the weights of the critic NN's output layer - in light blue (6); the values obtained with the use of a tracking

**Table 7.34** Values of selected performance indexes of the execution of a trajectory with a motion path shaped as "8", numerical tests without disturbance

| No. | Algorithm | $e_{maxa}$ [rad] | $\varepsilon_{av}$ [rad] | $\dot{e}_{maxa}$ [rad/s] | $\dot{\varepsilon}_{av}$ [rad/s] | $s_{maxa}$ [rad/s] | $\sigma_{av}$ [rad/s] |
|---|---|---|---|---|---|---|---|
| 1 | PD | 4.16 | 3.31 | 2.56 | 0.47 | 3.25 | 1.74 |
| 2 | Adaptive | 1.3 | 0.34 | 1.76 | 0.21 | 2.04 | 0.27 |
| 3 | Neural | 1.13 | 0.2 | 1.87 | 0.28 | 2.19 | 0.29 |
| 4 | ADHDP | 0.91 | 0.32 | 1.1 | 0.18 | 1.38 | 0.24 |
| 5 | $HDP_{W0}$ | 1.9 | 0.43 | 1.66 | 0.34 | 2.22 | 0.41 |
| 6 | $HDP_{W1}$ | 0.75 | 0.2 | 0.81 | 0.13 | 0.96 | 0.17 |
| 7 | GDHP | 1.01 | 0.22 | 1.03 | 0.18 | 1.24 | 0.21 |
| 8 | DHP | 0.83 | 0.18 | 0.98 | 0.15 | 1.18 | 0.18 |

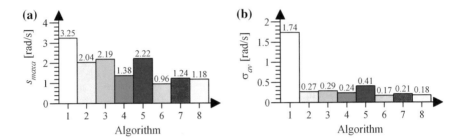

**Fig. 7.80** Values of selected performance indexes of the execution of a trajectory with a motion path shaped as "8", numerical tests without disturbance

control system with the algorithm of adaptive critic of GDHP type are marked in red (7), and the values obtained with the use of a DHP algorithm - in yellow (8).

From the point of view of the analysis of tracking control quality of the WMR's motion, the value of the performance index $\sigma_{av}$, which depends on the values of filtered tracking errors (being a function of the rotation angle error and angular velocity error of the own rotation of WMR's driving wheels, determined in each of discreet steps of the numerical test), is significant. Important information is presented by the maximum values of trajectory execution errors $e_{maxa}$, $\dot{e}_{maxa}$, and $s_{maxa}$, which - if adaptive algorithms with zero initial values of the adapted parameters (e.g. NN weights) are used - occur at the initial stage of the motion, in most unfavourable conditions, when in the adapted parameters of the WMR nonlinearity compesators no information about the dynamics of the controlled object is encoded. On the basis of maximal tracking error values, it is possible to make a conclusion about the speed of adaptation of individual algorithm, since faster adaptation of the control system parameters ensures more precise execution of the tracking motion in unfavourable conditions.

The reference point in the comparative analysis of the quality of execution of the tracking motion of the point $A$ of the WMR in individual control systems are values of selected performance indexes obtained in numerical tests of the PD controller, marked in tables as 1. Other control algorithms include the PD controller and additional modules, e.g. a module approximating nonlinearities of the controlled object, which improves the quality of execution of the WMR's tracking motion. When we take the values of the performance index $\sigma_{av}$ (Fig. 7.80b) as a comparison criterion, the tested tracking control algorithms applied to the motion of the point $A$ can be divided into four groups: A, B, C, D, where algorithms ensuring the highest quality of the tracking, expressed with the lowest values of the assumed performance indexes, are included into group A.

The lowest quality of motion execution, expressed with the highest values of performance indexes, occurs for the PD controller (1). This results from the principle of operation of the PD controller, in which the occurrence of trajectory execution errors is necessary to generate control signals. The PD controller is less accurate when it comes to tracking control of the motion of objects described with nonlinear dynamics equations in comparison to control systems in which there is a PD controller and an adaptive structure compensating for the nonlinearities of the controlled object. The PD controller was classified to group D, the control system with $HDP_{W0}$ algorithms was included to group C. The value $\sigma_{av}$ of the control system using HDP algorithms and zero initial values of the weights of the critic NN's output layer results from large tracking error values at the initial stage of weight adaptation. The application of the values of initial weights of critic's NN output layer equal to 1 would result in the improvement of the quality of generated control. Group B includes the adaptive control algorithm (2), the neural control algorithm (3), and the control system using NDP structures in the ADHDP configuration (4), the values $\sigma_{av}$ of these control systems are 0.24-0.27. The tracking control systems using $HDP_{W1}$ (6), GDHP (7) and DHP (8) algorithms were included to group A, ensuring the highest quality of execution of the tracking motion of point $A$ of the WMR. However, one should remember that in the case of $HDP_{W1}$ algorithm (6), nonzero initial values of the weights of critic NN's output layer were used, which significantly improved the quality of control, especially at the initial stage of the robot's motion. In the other control algorithms, the most unfavourable case of zero initial values of the adapted parameters was applied. Therefore, the results of the $HDP_{W1}$ algorithm (6) shall be used for information purposes only, in comparison to the results of $HDP_{W0}$ algorithm (5), in the context of differences in the control quality resulting from a change in the initial conditions of the adapted parameters (weights of the NN's output layer). The presented classification of control systems is valid also for other performance indexes. The highest quality of execution of the tracking motion was achieved using control systems with NDP algorithms.

Table 7.35 presents values of selected performance indexes of motion execution with the use of tracking control systems for the point $A$ of the WMR described earlier, at individual stages of numerical tests of the execution of a trajectory with a path shaped as "8".

Charts comparing the values of selected performance indexes obtained at individual stages of the motion are presented in Fig. 7.81, with the notation and colours as defined earlier. The bar charts corresponding to the values of performance indexes

**Table 7.35** Values of selected performance indexes at individual stages of the execution of a trajectory with a motion path shaped as "8", in numerical tests without parametric disturbance

| No. | Algorithm | Stage I | | | | Stage II | | | |
| | | $k = 1, \ldots, 2250$ | | | | | | $k = 2251, \ldots, 4500$ | |
| | | $e_{maxa}$ [rad] | $\varepsilon_{av}$ [rad] | $s_{maxa}$ [rad/s] | $\sigma_{av}$ [rad/s] | $e_{maxa}$ [rad] | $\varepsilon_{av}$ [rad] | $s_{maxa}$ [rad/s] | $\sigma_{av}$ [rad/s] |
|---|---|---|---|---|---|---|---|---|---|
| 1 | PD | 4.04 | 3.13 | 3.11 | 1.7 | 4.04 | 3.13 | 3.11 | 1.7 |
| 2 | Adaptive | 1.3 | 0.48 | 2.04 | 0.37 | 0.13 | 0.06 | 0.42 | 0.09 |
| 3 | Neural | 1.13 | 0.25 | 2.19 | 0.36 | 0.49 | 0.12 | 0.93 | 0.19 |
| 4 | ADHDP | 0.91 | 0.33 | 1.38 | 0.27 | 0.74 | 0.33 | 0.75 | 0.22 |
| 5 | $\text{HDP}_{W0}$ | 1.9 | 0.56 | 2.22 | 0.55 | 0.3 | 0.19 | 0.56 | 0.15 |
| 6 | $\text{HDP}_{W1}$ | 0.75 | 0.27 | 0.96 | 0.22 | 0.29 | 0.1 | 0.49 | 0.1 |
| 7 | GDHP | 1.01 | 0.3 | 1.24 | 0.28 | 0.27 | 0.1 | 0.53 | 0.11 |
| 8 | DHP | 0.83 | 0.25 | 1.18 | 0.23 | 0.24 | 0.08 | 0.44 | 0.1 |

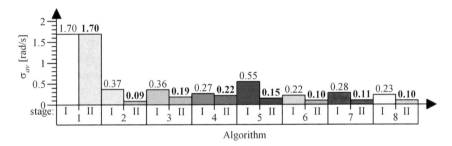

**Fig. 7.81** Values of selected performance indexes at individual stages of the execution of a trajectory with a motion path shaped as "8", in numerical tests without parametric disturbance

of the 2nd stage of motion are highlighted with diagonal fill, and their values are written in bold.

The continuation of the process of parameters adaptation positively influences the quality of the control generated, as it can be seen from the comparison of the values of selected performance indexes of the 1st and the 2nd stage of motion. The large difference in values $\sigma_{av}$ at the 1st and 2nd stage of tracking motion of the point $A$ of the WMR, implemented with the use of an adaptive control algorithm and a control system with $\text{HDP}_{W0}$ structures, draws particular attention. The lowest value of the performance index $\sigma_{av} = 0.09$ at the 2nd stage of the motion was obtained with the use of an adaptive control system.

The values of selected performance indexes obtained in numerical tests of tracking control algorithms for the motion of the point $A$ of the WMR with the use of a trajectory with a motion path shaped as "8", with parametric disturbance, are presented in Table 7.36.

**Table 7.36** Values of selected performance indexes of the execution of a trajectory with a motion path shaped as "8", version with parametric disturbance

| No. | Algorithm | $e_{maxa}$ [rad] | $\varepsilon_{av}$ [rad] | $\dot{e}_{maxa}$ [rad/s] | $\dot{\varepsilon}_{av}$ [rad/s] | $s_{maxa}$ [rad/s] | $\sigma_{av}$ [rad/s] |
|-----|-----------|------------------|--------------------------|--------------------------|----------------------------------|--------------------|------------------------|
| 1 | PD | 4.94 | 3.7 | 2.56 | 0.49 | 3.76 | 1.93 |
| 2 | Adaptive | 1.3 | 0.35 | 1.76 | 0.21 | 2.04 | 0.28 |
| 3 | Neural | 1.13 | 0.21 | 1.87 | 0.29 | 2.19 | 0.31 |
| 4 | ADHDP | 0.91 | 0.4 | 1.1 | 0.18 | 1.38 | 0.27 |
| 5 | HDP$_{W0}$ | 1.9 | 0.45 | 1.66 | 0.35 | 2.22 | 0.42 |
| 6 | HDP$_{W1}$ | 0.75 | 0.21 | 0.81 | 0.14 | 0.96 | 0.18 |
| 7 | GDHP | 1.01 | 0.24 | 1.03 | 0.19 | 1.24 | 0.23 |
| 8 | DHP | 0.83 | 0.19 | 0.98 | 0.16 | 1.18 | 0.18 |

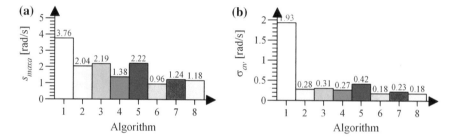

**Fig. 7.82** Values of selected performance indexes of the execution of a trajectory with a motion path shaped as "8", version with parametric disturbance

Charts comparing the values of selected performance indexes are presented in Fig. 7.82, with the notation and colours as defined earlier.

Introduction of double parametric disturbance resulted in a slight increase of the performance index $\sigma_{av}$ in numerical tests of all tracking control algorithms, apart from the tracking control system using NDP structure in the DHP configuration, in the case of which the value $\sigma_{av}$ did not change.

Table 7.37 contains the values of selected performance indexes used in numerical tests of tracking control systems of the point $A$ of the WMR, at individual stages of execution of the trajectory with a path shaped as "8", with parametric disturbance.

Charts comparing the values of the performance index $\sigma_{av}$ obtained at individual stages of the motion are presented in Fig. 7.83.

Introducing double parametric disturbance resulted in deterioration of the quality of tracking motion execution with the use of majority of the systems tested.

The lowest values of the assumed performance indexes for tracking control of the motion of the point $A$ of the WMR occur in numerical tests of the control systems using NDP algorithms in DHP, GDHP, and HDP$_{W1}$ configuration (where nonzero initial values of the weights of the critic NN's output layer are set). The values of performance indexes of the control system using ADHDP algorithms are comparrable to the values obtained in numerical tests of the adaptive and neural control system.

**Table 7.37** Values of selected performance indexes at individual stages of the execution of a trajectory with a motion path shaped as "8", in numerical tests with parametric disturbance

| No. | Algorithm | Stage I $k = 1, \ldots, 2250$ | | | | Stage II $k = 2251, \ldots, 4500$ | | | |
|---|---|---|---|---|---|---|---|---|---|
| | | $e_{maxa}$ [rad] | $\varepsilon_{av}$ [rad] | $s_{maxa}$ [rad/s] | $\sigma_{av}$ [rad/s] | $e_{maxa}$ [rad] | $\varepsilon_{av}$ [rad] | $s_{maxa}$ [rad/s] | $\sigma_{av}$ [rad/s] |
| 1 | PD | 4.81 | 3.45 | 3.11 | 1.88 | 4.46 | 3.43 | 3.51 | 1.86 |
| 2 | Adaptive | 1.3 | 0.49 | 2.04 | 0.37 | 0.31 | 0.12 | 0.47 | 0.12 |
| 3 | Neural | 1.13 | 0.25 | 2.19 | 0.36 | 0.36 | 0.12 | 0.68 | 0.18 |
| 4 | ADHDP | 0.91 | 0.48 | 1.38 | 0.33 | 0.56 | 0.32 | 0.48 | 0.22 |
| 5 | HDP$_{W0}$ | 1.9 | 0.55 | 2.22 | 0.55 | 0.47 | 0.3 | 0.62 | 0.21 |
| 6 | HDP$_{W1}$ | 0.75 | 0.27 | 0.96 | 0.22 | 0.24 | 0.11 | 0.47 | 0.12 |
| 7 | GDHP | 1.01 | 0.31 | 1.24 | 0.29 | 0.27 | 0.13 | 0.53 | 0.14 |
| 8 | DHP | 0.83 | 0.25 | 1.18 | 0.24 | 0.26 | 0.08 | 0.45 | 0.11 |

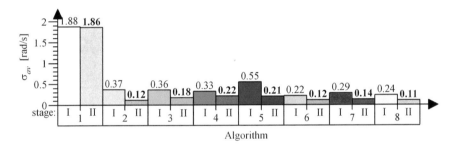

**Fig. 7.83** Values of selected performance indexes at individual stages of the execution of a trajectory with a motion path shaped as "8", in numerical tests with parametric disturbance

The lowest value of the performance index $\sigma_{av} = 0.11$ at the 2nd stage of the motion was obtained with the use of a tracking control system with a DHP algorithm.

## 7.9.1  Selection of Value of the Future Reward Discount Factor $\gamma$

Numerical tests of control systems using NDP structures were carried out with various values of the future reward/penalty discount factor, $\gamma$, assumed. The applied parameters of control systems were identical as in other numerical tests, and a trajectory with a motion path shaped as "8" was selected. During the numerical tests, parametric disturbance was simulated. The obtained values of the performance index $\sigma_{av}$ are presented in Table 7.38.

**Table 7.38** Values of the performance index $\sigma_{av}$ depending on the selection of $\gamma$; trajectory with a path shaped as "8", numerical test with parametric disturbance

| No. | Algorithm | $\gamma = 0.1$ | $\gamma = 0.3$ | $\gamma = 0.5$ | $\gamma = 0.9$ |
|---|---|---|---|---|---|
| 4 | ADHDP | 0.4 | 0.3 | 0.27 | 0.24 |
| 6 | HDP$_{W1}$ | – | 0.35 | 0.18 | 0.27 |
| 7 | GDHP | 0.55 | 0.31 | 0.23 | 0.98[a] |
| 8 | DHP | 0.31 | 0.21 | 0.18 | 0.16 |

[a]In the case of a tracking control system of the point $A$ of the WMR with a GDHP algorithm, with other design coefficients of the algorithm remaining unchanged, for $\gamma > 0.6$ induction of oscillation was observed for the signal of the GDHP control algorithm, which suggests that too high settings of other design parameters for this value of the coefficient $\gamma$ were selected (for $\gamma = 0.7$ $\sigma_{av} = 0.23$, while for $\gamma = 0.8$ $\sigma_{av} = 0.49$)

Where ADHDP or DHP algorithms were used, with other parameters of the control systems remaining unchanged, the growth of value of the factor $\gamma$ leads to the improvement of the tracking control quality for the point $A$ of the WMR. More specifically, increasing the value of the factor $\gamma$ from $\gamma = 0.5$ to $\gamma = 0.9$ results in a slight increase of the control quality expressed by the value of the performance index lower by approx. 10%. Where an HDP algorithm is used, the lowest values of the performance index $\sigma_{av}$ were obtained using $\gamma = 0.5$, while when $\gamma = 0.1$, the adaptation process was incorrect. This can result from the assumed values of the control system parameters or the selection of used NNs structures. In the case of a control algorithm with a GDHP structure, the quality of tracking motion improves along with the growth of value of the factor $\gamma$, however for $\gamma > 0.6$, oscillation begin to occur in the control signal which, for $\gamma > 0.7$, result in control deterioration. Therefore, to facilitate the comparison of the WMR tracking motion execution with the use of control systems with various adaptive critic structures, the value of factor $\gamma = 0.5$ was used in all NDP algorithms. We should remember that in the case of such complex structures as NDP algorithms, there are many design parameters influencing the process of adaptation of the weights of actor's and critic's NNs, which is reflected in the quality of generated control.

# References

1. Adeli, H., Tabrizi, M.H.N., Mazloomian, A., Hajipour, E., Jahed, M.: Path planning for mobile robots using iterative artificial potential field method. IJCSI **8**, 28–32 (2011)
2. Arkin, R.C.: Behavior-Based Robotics. MIT Press, Cambridge (1998)
3. Astrom, K.J., Wittenmark, B.: Adaptive Control. Addison-Wesley, New York (1979)
4. Balaji, P.G., German, X., Srinivasan, D.: Urban traffic signal control using reinforcement learning agents. IET Intell. Transp. Syst. **4**, 177–188 (2010)
5. Barto, A., Sutton, R., Anderson, C.: Neuronlike adaptive elements that can solve difficult learning problems. IEEE Trans. Syst. Man Cybern. Syst. **13**, 834–846 (1983)
6. Bhattacharya, P., Gavrilova, M.L.: Voronoi diagram in optimal path planning. In: Proceedings of 4th International Symposium on Voronoi Diagrams in Science and Engineering, Glamorgan, pp. 38–47 (2007)

7. Borenstain, J., Koren, J.: Real-time obstacle avoidance for fast mobile robots. IEEE Trans. Syst. Man Cybern. Syst. **19**, 1179–1186 (1989)

8. Brock, O., Khatib, O.: Elastic strips: a framework for motion generation in human environments. Int. J. Robot. Res. **21**, 1031–1052 (2002)

9. Burghardt, A.: Behavioral control of a mobile mini robot (in Polish). Meas. Autom. Monit. **11**, 26–29 (2004)

10. Burghardt, A.: Proposal for a rapid prototyping environment for algorithms intended for autonomous mobile robot control. Mech. Mech. Eng. **12**, 5–16 (2008)

11. Burghardt, A.: Accomplishing tasks in reaching the goal of robot formation. Solid State Phenom. **220–221**, 27–32 (2015)

12. Burghardt, A., Hendzel, Z.: Adaptive neural network control of underactuated system, In: Proceedings of 4th International Conference on Neural Computation Theory and Applications, Barcelona, pp. 505–509 (2012)

13. Burghardt, A., Gierlak, P., Szuster, M.: Reinforcement learning in tracking control of 3DOF robotic manipulator. In: Awrejcewicz, J., Kamierczak, M., Olejnik, P., Mrozowski, J. (eds.) Dynamical Systems: Theory, pp. 163–174. WPL, Lodz (2013)

14. Burns, R.S.: Advanced Control Engineering. Butterworth-Heinemann, Oxford (2001)

15. Canudas de Wit, C., Siciliano, B., Bastin, G.: Theory of Robot Control. Springer, London (1996)

16. Chang, Y.-H., Chan, W.-S., Chang, C.-W., Tao, C.W.: Adaptive fuzzy dynamic surface control for ball and beam system. Int. J. Fuzzy Syst. **13**, 1–13 (2011)

17. Chang, Y.-H., Chang, C.-W., Tao, C.-W., Lin, H.-W., Taur, J.-S.: Fuzzy sliding-mode control for ball and beam system with fuzzy ant colony optimization. Expert Syst. Appl. **39**, 3624–3633 (2012)

18. Doya, K.: Reinforcement learning in continuous time and space. Neural Comput. **12**, 219–245 (2000)

19. Drainkov, D., Saffiotti, A.: Fuzzy Logic Techniques for Autonomous Vehicle Navigation. Springer, New York (2001)

20. Emhemed, A.A.: Fuzzy control for nonlinear ball and beam system. Int. J. Fuzzy Log. Intell. Syst. **3**, 25–32 (2013)

21. Ernst, D., Glavic, M., Wehenkel, L.: Power systems stability control: reinforcement learning framework. IEEE Trans. Power Syst. **19**, 427–435 (2004)

22. Fabri, S., Kadirkamanathan, V.: Dynamic structure neural networks for stable adaptive control of nonlinear systems. IEEE Trans. Neural Netw. **12**, 1151–1167 (1996)

23. Fahimi, F.: Autonomous Robots. Modeling, Path Planning, and Control. Springer, New York (2009)

24. Fairbank, M., Alonso, E., Prokhorov, D.: Simple and fast calculation of the second-order gradients for globalized dual heuristic dynamic programming in neural networks. IEEE Trans. Neural Netw. Learn. Syst. **23**, 1671–1676 (2012)

25. Ferrari, S., Stengel, R.F.: An adaptive critic global controller. In: Proceedings of American Control Conference, vol. 4, pp. 2665–2670. Anchorage, Alaska (2002)

26. Ferrari, S., Stengel, R.F.: Model-based adaptive critic designs in learning and approximate dynamic programming. In: Si, J., Barto, A., Powell, W., Wunsch, D.J. (eds.) Handbook of Learning and Approximate Dynamic Programming, pp. 64–94. Wiley, New York (2004)

27. Fierro, R., Lewis, F.L.: Control of a nonholonomic mobile robot using neural networks. IEEE Trans. Neural Netw. **9**, 589–600 (1998)

28. Garrido, S., Moreno, L., Abderrahim M., Martin, F.: Path planning for mobile robot navigation using Voronoi diagram and fast marching. In: Proceedings of IEEE International Conference on Intelligent Robots and Systems, Beijing, China, pp. 2376–2381 (2006)

29. Giergiel, J., Hendzel, Z., Zylski, W.: Kinematics, Dynamics and Control of Wheeled Mobile Robots in Mechatronic Aspect (in Polish). Faculty IMiR AGH, Krakow (2000)

30. Giergiel, J., Hendzel, Z., Zylski, W.: Modeling and Control of Wheeled Mobile Robots (in Polish). Scientific Publishing PWN, Warsaw (2002)

31. Gierlak, P., Szuster, M., Zylski, W.: Discrete dual-heuristic programming in 3DOF manipulator control. Lect. Notes Artif. Int. **6114**, 256–263 (2010)
32. Gierlak, P., Muszyska, M., Zylski, W.: Neuro-fuzzy control of robotic manipulator. Int. J. Appl. Mech. Eng. **19**, 575–584 (2014)
33. Glower, J.S., Munighan, J.: Designing fuzzy controllers from a variable structures standpoint. IEEE Trans. Fuzzy Syst. **5**, 138–144 (1997)
34. Godjevac, J., Steele, N.: Neuro-fuzzy control of a mobile robot. Neurocomputing **28**, 127–143 (1999)
35. Han, D., Balakrishnan, S.: Adaptive critic based neural networks for control-constrained agile missile control. Proc. Am. Control Conf. **4**, 2600–2605 (1999)
36. Hendzel, Z.: Tracking Control of Wheeled Mobile Robots (in Polisch). OWPRz, Rzeszw (1996)
37. Hendzel, Z.: Fuzzy reactive control of wheeled mobile robot. JTAM **42**, 503–517 (2004)
38. Hendzel, Z., Burghardt, A.: Behavioral fuzzy control of basic motion of a mobile robot (in Polish). Meas. Autom. Monit. **11**, 23–25 (2004)
39. Hendzel, Z., Burghardt, A.: Behaviouralc Control of Wheeled Mobile Robots (in Polish). Rzeszow University of Technology Publishing House, Rzeszw (2007)
40. Hendzel, Z., Szuster, M.: A dynamic structure neural network for motion control of a wheeled mobile robot. In: Rutkowski, L., Tadeusiewicz, R., Zadeh, L.A., Zurada, J. (eds.) Computational Intelligence: Methods and Applications, pp. 365–376. EXIT, Warszawa (2008)
41. Hendzel, Z., Szuster, M.: Discrete model-based dual heuristic programming in wheeled mobile robot control. In: Awrejcewicz, J., Kamierczak, M., Olejnik, P., Mrozowski, J. (eds.) Dynamical Systems - Theory and Applications, pp. 745–752. Left Grupa, Lodz (2009)
42. Hendzel, Z., Szuster, M.: Heuristic dynamic programming in wheeled mobile robot control. In: Kaszyski, R., Pietrusewicz, K. (eds.) Methods and Models in Automation and Robotics, pp. 513–518. IFAC, Poland (2009)
43. Hendzel, Z., Szuster, M.: Discrete action dependant heuristic dynamic programming in wheeled mobile robot control. Solid State Phenom. **164**, 419–424 (2010)
44. Hendzel, Z., Szuster, M.: Discrete model-based adaptive critic designs in wheeled mobile robot control. Lect. Notes Artif. Int. **6114**, 264–271 (2010)
45. Hendzel, Z., Szuster, M.: Adaptive critic designs in behavioural control of a wheeled mobile robot. In: Awrejcewicz, J., Kamierczak, M., Olejnik, P., Mrozowski J. (eds.) Dynamical Systems, Nonlinear Dynamics and Control, pp. 133–138, WPL, Łodz (2011)
46. Hendzel, Z., Szuster, M.: Discrete neural dynamic programming in wheeled mobile robot control. Commun. Nonlinear Sci. Numer. Simulat. **16**, 2355–2362 (2011)
47. Hendzel, Z., Szuster, M.: Neural dynamic programming in behavioural control of WMR (in Polish). Acta Mech. Autom. **5**, 28–36 (2011)
48. Hendzel, Z., Szuster, M.: Neural dynamic programming in reactive navigation of wheeled mobile robot. Lect. Notes Comput. Sc. **7268**, 450–457 (2012)
49. Hendzel, Z., Szuster, M.: Approximate dynamic programming in sensor-based navigation of wheeled mobile robot. Solid State Phenom. **220–221**, 60–66 (2015)
50. Hendzel, Z., Trojnacki, M.: Neural Network Control of Mobile Wheeled Robots (in Polish). Rzeszow University of Technology Publishing House, Rzeszow (2008)
51. Hendzel, Z., Burghardt, A., Szuster, M.: Artificial intelligence methods in reactive navigation of mobile robots formation. In: Proceedings of 4th International Conference on Neural Computation Theory and Applications, Barcelona, pp. 466–473 (2012)
52. Hendzel, Z., Burghardt, A., Szuster, M.: Adaptive critic designs in control of robots formation in unknown environment. Lect. Notes Artif. Int. **7895**, 351–362 (2013)
53. Hendzel, Z., Burghardt, A., Szuster, M.: Reinforcement learning in discrete neural control of the underactuated system. Lect. Notes Artif. Int. **7894**, 64–75 (2013)
54. Hendzel, Z., Burghardt, A., Szuster, M.: Artificial intelligence algorithms in behavioural control of wheeled mobile robots formation. SCI **577**, 263–277 (2015)
55. Huan, H.P., Chung, S.Y.: Dynamic visibility graph for path planning. In: Proceedings of IEEE/RSJ International Conference on Intelligent Robots and Systems, Sendai, Japan, vol. 3, pp. 2813–2818 (2004)

56. Jamshidi, M., Zilouchian, A.: Intelligent Control Systems Using Soft Computing Methodologies. CRC Press, London (2001)
57. Janet, J.A., Luo, R.C., Kay, M.G.: The essential visibility graph: an approach to global motion planning for autonomous mobile robots. In: Proceedings of IEEE International Conference on Robotics and Automation, Nagoya, Japan, vol. 2, pp. 1958–1963 (1995)
58. Kecman, V.: Learning and Soft Computing. MIT Press, Cambridge (2001)
59. Keshmiri, M., Jahromi, A.F., Mohebbi, A., Amoozgar, M.H., Xie, W.F: Modeling and control of ball and beam system using model based and non-model based control approaches. Int. J. Smart Sens. Intell. Syst. 5, 14–35 (2012)
60. Khatib, O.: Real-time obstacle avoidance for manipulators and mobile robots. Proc. IEEE Int. Conf. Robot. Autom. 2, 500–505 (1985)
61. Kim, Y.H., Lewis, F.L.: A dynamical recurrent neural-network-based adaptive observer for a class of nonlinear systems. Automatica 33, 1539–1543 (1997)
62. Koren, Y., Borenstein, J.: Potential field methods and their inherent limitations for mobile robot navigation. Proc. IEEE Int. Conf. Robot. Autom. 2, 1398–1404 (1991)
63. Koshkouei, A.J., Zinober, A.S.: Sliding mode control of discrete-time systems. J. Dyn. Syst. Meas. Control. 122, 793–802 (2000)
64. Kozowski, K., Dutkiewicz, P., Wrblewski, W.: Modeling and Control of Robots (in Polish). Scientific Publishing PWN, Warsaw (2003)
65. Lee, G.H., Jung, S.: Line tracking control of a two-wheeled mobile robot using visual feedback. Int. J. Adv. Robot. Syst. 10, 1–8 (2013)
66. Lendaris, G., Schultz, L., Shannon, T.: Adaptive critic design for intelligent steering and speed control of a 2-axle vehicle. Proc. IEEE INNS-ENNS Int. Jt. Conf. Neural Netw. 3, 73–78 (2000)
67. Lewis, F.L., Campos, J., Selmic, R.: Neuro-Fuzzy Control of Industrial Systems with Actuator Nonlinearities. Society for Industrial and Applied Mathematics, Philadelphia (2002)
68. Levis, F.L., Liu, K., Yesildirek, A.: Neural net robot controller with guaranted tracking performance. IEEE Trans. Neural Netw. 6, 703–715 (1995)
69. Liu, G.P.: Nonlinear Identification and Control. Springer, London (2001)
70. Liu, D., Wang, D., Yang, X.: An iterative adaptive dynamic programming algorithm for optimal control of unknown discrete-time nonlinear systems with constrained inputs. Inform. Sci. 220, 331–342 (2013)
71. Maaref, H., Barret, C.: Sensor-based navigation of a mobile robot in an indoor environment. Robot. Auton. Syst. 38, 1–18 (2002)
72. Millán, J., del, R.: Reinforcement learning of goal-directed obstacle-avoiding reaction strategies in an autonomous mobile robot. Robot. Auton. Syst. 15, 275–299 (1995)
73. Miller, W.T., Sutton, R.S., Werbos, P.J.: Neural Networks for Control. A Bradford Book, MIT Press, Cambridge (1990)
74. Mohagheghi, S., Venayagamoorthy, G.K., Harley, R.G.: Adaptive critic design based neuro-fuzzy controller for a static compensator in a multimachine power system. IEEE Trans. Power Syst. 21, 1744–1754 (2006)
75. Muskinja, N., Riznar, M.: Optimized PID position control of a nonlinear system based on correlating the velocity with position error. Math. Probl. Eng. 2015, 1–11 (2015)
76. Ng, K.C., Trivedi, M.T.: A neuro-fuzzy controller for mobile robot navigation and multirobot convoying. IEEE Trans. Syst. Man Cybern. B Cybern. 28, 829–840 (1998)
77. Prokhorov, D., Wunch, D.: Adaptive critic designs. IEEE Trans. Neural Netw. 8, 997–1007 (1997)
78. Samarjit, K., Sujit, D., Ghosh, P.K.: Applications of neuro fuzzy systems: a brief review and future online. Appl. Soft Comput. 15, 243–259 (2014)
79. Si, J., Barto, A.G., Powell, W.B., Wunsch, D.: Handbook of Learning and Approximate Dynamic Programming. IEEE Press, Wiley-Interscience, Hoboken (2004)
80. Sira-Ramirez, H.: Differential geometric methods in variable-structure control. Int. J. Control 48, 1359–1390 (1988)
81. Slotine, J.J., Li, W.: Applied Nonlinear Control. Prentice Hall, New Jersey (1991)
82. Spong, M.W., Vidyasagar, M.: Robot Dynamics and Control (in Polish). WNT, Warsaw (1997)

83. Spooner, J.T., Passio, K.M.: Stable adaptive control using fuzzy systems and neural networks. IEEE Trans. Fuzzy Syst. **4**, 339–359 (1996)
84. Syam, R., Watanabe, K., Izumi, K.: Adaptive actor-critic learning for the control of mobile robots by applying predictive models. Soft. Comput. **9**, 835–845 (2005)
85. Szuster, M.: Globalised dual heuristic dynamic programming in tracking control of the wheeled mobile robot. Lect. Notes Artif. Int. **8468**, 290–301 (2014)
86. Szuster, M., Gierlak, P.: Approximate dynamic programming in tracking control of a robotic manipulator. Int. J. Adv. Robot. Syst. **13**, 1–18 (2016)
87. Szuster, M., Gierlak, P.: Globalised dual heuristic dynamic programming in control of robotic manipulator. AMM **817**, 150–161 (2016)
88. Szuster, M., Hendzel, Z.: Discrete globalised dual heuristic dynamic programming in control of the two-wheeled mobile robot. Math. Probl. Eng. **2014**, 1–16 (2014)
89. Szuster, M., Hendzel, Z., Burghardt, A.: Fuzzy sensor-based navigation with neural tracking control of the wheeled mobile robot. Lect. Notes Artif. Int. **8468**, 302–313 (2014)
90. Tadeusiewicz, R.: Neural Networks (in Polish). AOWRM, Warsaw (1993)
91. Tcho, K., Mazur, A., Dulba, I., Hossa, R., Muszyski, R.: Mobile Manipulators and Robots (in Polish). Academic Publishing Company PLJ, Warsaw (2000)
92. Wang, L.: A Course in Fuzzy Systems and Control. Prentice-Hall, New York (1997)
93. Yu, W.: Nonlinear PD regulation for ball and beam system. Int. J. Elec. Eng. Educ. **46**, 59–73 (2009)
94. Zylski, W., Gierlak, P.: Verification of multilayer neuralnet controller in manipulator tracking control. Solid State Phenom. **164**, 99–104 (2010)
95. Zylski, W., Gierlak, P.: Tracking Control of Robotic Manipulator (in Polish). Rzeszow University of Technology Publishing House, Rzeszow (2014)

# Chapter 8
# Reinforcement Learning in the Control of Nonlinear Continuous Systems

This chapter deals with the implementation of reinforcement learning methods during the development of adaptive control systems of the nonlinear objects which are learning the optimal solutions in real time on the basis of the measurement data of kinematic parameters of the controlled object's motion. From the point of view of the control systems development, adaptive control and optimal control are based on different approaches. The main task of the optimal control is to determine the optimal control signal resulting from the solution of the Hamilton–Jacobi–Bellman (HJB) equation, which minimizes the assumed quality rating. It is an off-line type solution. Such an approach to solution requires the determination of an accurate mathematical model of the controlled object, which is a difficult task from the practical point of view. In addition, the determination of an optimal control within a nonlinear object requires off-line solution of the nonlinear HJB equation which may be insolvable. However, in the adaptive control, the adaptation process of the controlled object nonlinearity and its changeable working conditions is implemented in real time, based on the measured kinematic parameters of motion of the controlled object. Although, the adaptive control is not an optimal control, which means, that it does not minimize the optimization criterion assumed by the designer. Reinforcement learning is a part of the process of machine learning. This method is inspired by the biological processes occurring among humans and animals, in which the learning process is based on an interaction with environment and evaluation of responses of an environment to certain action with an assumed objective function. Each living organism interacts with environment and uses these interactions to improve its own operation, in order to survive and develop. Reinforcement learning is closely linked with adaptive and optimal control. It is included in a class of methods enabling the determination of an optimal adaptive control of nonlinear objects in real time. This way of machine learning was extensively implemented in Markov decision-making processes in discrete state spaces. This subject matter was widely described in Chap. 7.

© Springer International Publishing AG 2018
M. Szuster and Z. Hendzel, *Intelligent Optimal Adaptive Control for Mechatronic Systems*, Studies in Systems, Decision and Control 120, https://doi.org/10.1007/978-3-319-68826-8_8

Algorithms of adaptive dynamic programming are examples of control algorithms, in which the learning process includes reinforcement. These algorithms are an interesting alternative in practical applications. The implementation of an adaptive dynamic programming is an extension of the existing solutions in the theoretical area and in terms of application. Currently, implementation of reinforcement learning in the nonlinear continuous systems is intensively developed. The difficulties arise from the fact, that the Hamiltonian in discrete systems is not dependent from the dynamic properties of the controlled object, whereas for continuous systems the Hamiltonian includes dynamic properties of the object. Adaptive dynamic programming is based on an iterative determination of value function and control signals, and on the principle of optimality, time difference method. Also uses two parametric structures critic and actor (ASE-ACE) and is an approximate solution of the Hamilton–Jacobi–Bellman equation.

The proposed methods of motion synthesis of the WMRs are aimed to extend the existing solutions with emphasis on the application aspects regarding the control methods of the intelligent WMRs. The theoretical basis and the implementation of reinforcement learning, for five synthesis methods of WMR tracking control, will be discussed. The classical concept of application of reinforcement learning method in the control of WMR will be discussed in Sect. 8.1. The reinforcement signal in the binary form will be implemented and the simulations of the adopted solution will be carried out. The approximation of classical reinforcement learning in the WMR's control will be applied in Sect. 8.2. The subject matter of Sect. 8.3 will be the application of the actor-critic structure in the tracking control of the robot. Section 8.4 will present reinforcement learning of actor-critic type in optimal adaptive control. Approach, presented in this section, focuses on the application of actor-critic parametric structure (ASE-ACE) with learning with a critic. Section 8.5 presents the application of the adaptive critic structure in the optimal tracking control of the robot.

## 8.1 Classical Reinforcement Learning

The reinforcement signal, in the classical reinforcement learning method, adopts binary values $+1$ or $-1$ [2, 3, 13] generated by the critic. This means that the reinforcement learning method, when determining the adaptive control, is easy to calculate, as shown in the example of tracking control of the WMR.

### 8.1.1 Control Synthesis, Stability of a System, Reinforcement Learning Algorithm

The dynamic equations of WMR motion are written in the following form:

$$\mathbf{M\ddot{q}} + \mathbf{C}\left(\dot{\mathbf{q}}\right)\dot{\mathbf{q}} + \mathbf{F}\left(\dot{\mathbf{q}}\right) + \boldsymbol{\tau}_d\left(t\right) = \boldsymbol{\tau}\,, \tag{8.1}$$

where matrices $\mathbf{M}$, $\mathbf{C}\,(\dot{\mathbf{q}})$ and vectors $\mathbf{F}\,(\dot{\mathbf{q}})$, $\boldsymbol{\tau}$ result from the description of the wheeled mobile robot dynamics with Maggi's formalism introduced in Sect. 2.1.2, $\boldsymbol{\tau}_d\,(t)$ is the unknown disturbance vector. Let's write the Eq. (8.1) in the state space

$$\dot{\mathbf{x}}_1 = \mathbf{x}_2 \,,$$
$$\dot{\mathbf{x}}_2 = \mathbf{f}\,(\mathbf{x}) + \mathbf{u} + \mathbf{d} \,, \tag{8.2}$$

where $\mathbf{x}_1 = [\alpha_1, \alpha_2]^T$, $\mathbf{x}_2 = [\dot{\alpha}_1, \dot{\alpha}_2]^T$, $\mathbf{u} = \mathbf{M}^{-1}\boldsymbol{\tau} \in \Re^2$, $\mathbf{d} = -\mathbf{M}^{-1}\boldsymbol{\tau}_d\,(t) \in \Re^2$, and the nonlinear function is formulated as follows

$$\mathbf{f}\,(\mathbf{x}) = -\mathbf{M}^{-1}\,(\mathbf{C}\,(\mathbf{x}_2)\,\mathbf{x}_2 + \mathbf{F}\,(\mathbf{x}_2)) \in \Re^2 \,. \tag{8.3}$$

We assume that the inertia matrix $\mathbf{M}$ is known, its reciprocal exists and the moments of the drive wheels will be calculated from the equation $\boldsymbol{\tau} = \mathbf{Mu}$, moreover, the unknown disturbance vector satisfies the upper limit $\|\mathbf{d}\,(t)\| < d_{max}$, $d_{max}$=const. > 0. The desired trajectory of point $A$ of the robot is known, $\mathbf{x}_d\,(t) = [\mathbf{x}_d\,(t)\,, \dot{\mathbf{x}}_d\,(t)]^T$. The control algorithm synthesis is to determine such a control law $\boldsymbol{\tau}$ for which the trajectory of selected point $A$ of the robot will be convergent with a desired trajectory. The tracking error $\mathbf{e}$, filtered tacking error $\mathbf{s}$, interpreted as a measurement of the control quality (current cost), were defined in the form of

$$\mathbf{e}\,(t) = \mathbf{x}_d\,(t) - \mathbf{x}\,(t) \,, \tag{8.4}$$

$$\mathbf{s}\,(t) = \dot{\mathbf{e}}\,(t) + \boldsymbol{\Lambda}\mathbf{e}\,(t) \,, \tag{8.5}$$

where $\boldsymbol{\Lambda}$ – a positive-definite diagonal matrix. Then, Eq. (8.1) can be converted to the following form

$$\dot{\mathbf{s}} = \mathbf{f}\,(\mathbf{x}, \mathbf{x}_d) + \mathbf{u} + \mathbf{d} \,. \tag{8.6}$$

If we will select control signal with compensation of the object's nonlinearity

$$\mathbf{u} = -\hat{\mathbf{f}}\,(\mathbf{x}, \mathbf{x}_d) - \mathbf{K}_D\mathbf{s} + \mathbf{v}\,(t) \,, \tag{8.7}$$

where $\mathbf{K}_D = \mathbf{K}_D^T > 0$ is a design matrix and $\mathbf{K}_D\mathbf{s}$ term is an equation of the PD controller

$$\mathbf{K}_D\mathbf{s} = \mathbf{K}_D\dot{\mathbf{e}} + \mathbf{K}_D\boldsymbol{\Lambda}\mathbf{e} \,, \tag{8.8}$$

and $\mathbf{v}\,(t)$ is a vector of a robust control, which we will determine further on. The task of the robust control will be compensating for the inaccuracies resulting from the error of the robot's nonlinearity approximation $\boldsymbol{\varepsilon}$ by the NN and disturbance vector $\mathbf{d}$. Then the description of a closed control system will be as follows

$$\dot{\mathbf{s}} = -\mathbf{K}_D\mathbf{s} + \tilde{\mathbf{f}}\,(\mathbf{x}, \mathbf{x}_d) + \mathbf{d} + \mathbf{v}\,(t) \,, \tag{8.9}$$

where the error of nonlinearity estimation of the WMR is defined as $\tilde{\mathbf{f}}(\mathbf{x}, \mathbf{x}_d) = \mathbf{f}(\mathbf{x}, \mathbf{x}_d) - \hat{\mathbf{f}}(\mathbf{x}, \mathbf{x}_d)$. The linear NN with respect to weights of output layer was applied in the approximation of the nonlinear vector function $\mathbf{f}(\mathbf{x}, \mathbf{x}_d)$. Thus, the nonlinear function approximated by the network will be formulated as

$$\mathbf{f}(\mathbf{x}, \mathbf{x}_d) = \mathbf{W}^T \boldsymbol{\varphi}(\mathbf{x}) + \boldsymbol{\varepsilon} , \tag{8.10}$$

where $\boldsymbol{\varepsilon}$ – an approximation error which satisfies the constrain $\|\varepsilon\| \leq \varepsilon_N, \varepsilon_N$=const. > 0 and values of the ideal weights are constrained $\|\mathbf{W}\| \leq W_{max}$. Estimate of the function $\mathbf{f}(\mathbf{x}, \mathbf{x}_d)$ will be formulated as follows

$$\hat{\mathbf{f}}(\mathbf{x}, \mathbf{x}_d) = \hat{\mathbf{W}}^T \boldsymbol{\varphi}(\mathbf{x}) , \tag{8.11}$$

where $\hat{\mathbf{W}}$ – an estimate of weights of the ideal NN. By implementing (8.11) into the control law with nonlinearity compensation of the robot (8.7), the obtained control law was in the following form

$$\mathbf{u} = -\hat{\mathbf{W}}^T \boldsymbol{\varphi}(\mathbf{x}) - \mathbf{K}_D \mathbf{s} + \mathbf{v}(t) . \tag{8.12}$$

By substituting (8.10) and (8.11) into (8.9) the following was received

$$\dot{\mathbf{s}} = -\mathbf{K}_D \mathbf{s} + \tilde{\mathbf{W}}^T \boldsymbol{\varphi}(\mathbf{x}) + \boldsymbol{\varepsilon} + \mathbf{d} + \mathbf{v}(t) , \tag{8.13}$$

in which the following relation was used

$$\tilde{\mathbf{f}}(\mathbf{x}, \mathbf{x}_d) = \mathbf{f}(\mathbf{x}, \mathbf{x}_d) - \hat{\mathbf{f}}(\mathbf{x}, \mathbf{x}_d) = \mathbf{W}^T \boldsymbol{\varphi}(\mathbf{x}) - \hat{\mathbf{W}}^T \boldsymbol{\varphi}(\mathbf{x}) + \boldsymbol{\varepsilon} = \tilde{\mathbf{W}}^T \boldsymbol{\varphi}(\mathbf{x}) + \boldsymbol{\varepsilon} , \tag{8.14}$$

where $\tilde{\mathbf{W}} = \mathbf{W} - \hat{\mathbf{W}}$ – an estimation error of the NN weights. It was assumed, that the ideal weights of NN $\mathbf{W}$ are constrained. It was assumed, during the reinforcement control problem, that the reinforcement signal is a measure function of the control quality in the form of

$$\mathbf{R} = \text{sgn}(\mathbf{s}) , \tag{8.15}$$

where $\text{sgn}(\mathbf{s}) = \left[\text{sgn}(s_1), \text{sgn}(s_1)\right]^T$ assuming the $\pm 1$ values. The aim of the reinforcement control is to reduce the influence of nonlinearity and external disturbances on the control system quality. These effects have a significant influence on the quality of the nonlinear objects control systems [9, 11, 13, 16]. Figure 8.1 presents a scheme of a closed control system.

**Theorem 8.1** *If we will adopt a control* $\mathbf{u}$ *in the form of (8.12) with a robust control defined as*

$$\mathbf{v}(t) = -k_s \mathbf{R} \|\mathbf{R}\|^{-1} , \tag{8.16}$$

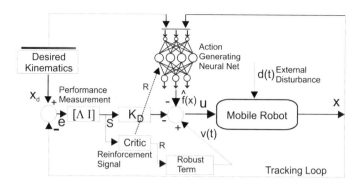

**Fig. 8.1** Architecture of reinforcement neural network control

*with a coefficient that satisfies the relation $k_s \geq d_{max}$, reinforcement signal (8.15) and the learning law of network weights*

$$\dot{\hat{\mathbf{W}}} = \mathbf{F}\boldsymbol{\varphi}\left(\mathbf{x}\right)\mathbf{R}^T - \gamma\mathbf{F}\hat{\mathbf{W}} , \qquad (8.17)$$

*where $\mathbf{F} = \mathbf{F}^T > 0$ – a design matrix, $\gamma > 0$ – a coefficient providing quick convergence of learning, then the filtered tracking error $\mathbf{s}$ (performance measure) and the estimation error of NN weights $\tilde{\mathbf{W}} = \mathbf{W} - \hat{\mathbf{W}}$ are uniformly ultimately bounded to the $\psi_s$ and $\psi_W$ sets.*

*Proof* Let's consider a Lyapunov function in the form

$$V = \sum_{i=1}^{2} |s_i| + \frac{1}{2}\mathrm{tr}\left(\tilde{\mathbf{W}}^T\mathbf{F}^{-1}\tilde{\mathbf{W}}\right) . \qquad (8.18)$$

The derivative of function $V$ with respect to time is in the form of

$$\dot{V} = \mathrm{sgn}\left(\mathbf{s}\right)^T\dot{\mathbf{s}} + \mathrm{tr}\left(\tilde{\mathbf{W}}^T\mathbf{F}^{-1}\dot{\tilde{\mathbf{W}}}\right) . \qquad (8.19)$$

By substituting into relation (8.19) the error dynamics Eq. (8.13) the following was obtained

$$\dot{V} = -\mathrm{sgn}\left(\mathbf{s}\right)^T\mathbf{K}_D\mathbf{s} + \mathrm{sgn}\left(\mathbf{s}\right)^T\left(\boldsymbol{\varepsilon} + \mathbf{d} + \mathbf{v}\left(t\right)\right) + \mathrm{tr}\left\{\tilde{\mathbf{W}}^T\left[\mathbf{F}^{-1}\dot{\tilde{\mathbf{W}}} + \boldsymbol{\varphi}\left(\mathbf{x}\right)\mathrm{sgn}\left(\mathbf{s}\right)^T\right]\right\} . \qquad (8.20)$$

By introducing into relation (8.20) the learning law of network weights (8.17) and the robust control (8.16) we will receive

$$\dot{V} \leq -\sqrt{2}\delta_{min}\left(\mathbf{K}_D\right)\|\mathbf{s}\| + \varepsilon_{max}\sqrt{2} + \gamma\left(\|\tilde{\mathbf{W}}\|W_{max} - \|\tilde{\mathbf{W}}\|^2\right) , \qquad (8.21)$$

where relations $\|\text{sgn}(\mathbf{s})\| \leq \sqrt{2}$ and $\text{tr}\left\{\tilde{\mathbf{W}}^T \hat{\mathbf{W}}\right\} = \text{tr}\left\{\tilde{\mathbf{W}}^T \left(\mathbf{W} - \tilde{\mathbf{W}}\right)\right\} \leq \|\tilde{\mathbf{W}}\|$ $\left(W_{max} - \|\tilde{\mathbf{W}}\|\right)$ were applied.

By squaring the binomial the following was obtained

$$\dot{V} \leq -\sqrt{2}\delta_{min}(\mathbf{K}_D)\|\mathbf{s}\| + \varepsilon_{max}\sqrt{2} - \gamma\left(\|\tilde{\mathbf{W}}\| - W_{max}/2\right)^2 + \gamma W_{max}^2/4 , \quad (8.22)$$

where $\delta_{min}(\mathbf{K}_D)$ – a smallest eigenvalue of the $\mathbf{K}_D$ matrix.

The derivative of Lyapunov function $\dot{V}$ is negative definite until the conditions are satisfied

$$\|\mathbf{s}\| > \frac{\sqrt{2}\varepsilon_{max} + \gamma W_{max}^2/4}{\sqrt{2}\delta_{min}(\mathbf{K}_D)} = \psi_s , \quad (8.23)$$

$$\|\tilde{\mathbf{W}}\| > \sqrt{W_{max}^2/4 + \varepsilon_{max}\sqrt{2}/\gamma} + W_{max}/2 = \psi_W . \quad (8.24)$$

It can be stated, that the derivative of Lyapunov function $\dot{V}$ is negative outside the set

$$\mathbf{\Omega} : \mathbf{\Omega} = \left\{\left(\|\mathbf{s}\|, \|\tilde{\mathbf{W}}\|\right), 0 \leq \|\mathbf{s}\| \leq \psi_s, 0 \leq \|\tilde{\mathbf{W}}\| \leq \psi_W\right\} . \quad (8.25)$$

This guarantees the error convergence of quality measure $\mathbf{s}$ of the designed control system and estimation of the neural network $\tilde{\mathbf{W}}$ weights to the $\mathbf{\Omega}$ set limited by $\psi_s$, $\psi_W$ constants. In summary, on the basis of the extended Lyapunov stability theory [15, 16] it was demonstrated that the errors are uniformly ultimately bounded to the set (8.25).                                                                              □

**Notice**:

1. From the relation (8.25) results, that the filtered tracking error $\mathbf{s}$ is uniformly ultimately bounded to the $\mathbf{\Omega}$ set with practical limit $\psi_s$. By increasing the coefficients of $\mathbf{K}_D$ matrix the tracking error $\mathbf{s}$ and errors $\mathbf{e}$ and $\dot{\mathbf{e}}$, which are also limited, can be reduced. The synthesis of neural adaptive control in this form enables proper operation of the control system with PD controller until the NN starts to adapt.
2. The proposed solution does not require a linear form with respect to unknown parameters of the object, which is required in the adaptive systems. The implementation of NN allows us to notate the nonlinearities of the controlled object in a linear form and at the same time, enables to introduce such a form when the linear parameterization of nonlinearity of the object is not possible.
3. In the proposed structure of reinforcement learning the current cost (8.15) can be interpreted as a reinforcement signal evaluating the quality of control resulting from the nonlinearity of the controlled object and disturbance.

**Table 8.1** Assumed
parameters of the WMR

| Parameter | Value | Unit |
|-----------|-------|------|
| $a_1$ | 0.265 | kgm$^2$ |
| $a_2$ | 0.666 | kgm$^2$ |
| $a_3$ | 0.007 | kgm$^2$ |
| $a_4$ | 0.002 | kgm$^2$ |
| $a_5$ | 2.411 | Nm |
| $a_6$ | 2.411 | Nm |

4. After assuming zero initial values of the network weights in the proposed control algorithm, the stability of the system will be provided by the PD controller until the network starts to learn. This means that the network does not require pre-learning phase or application of the "trial and error" method in the solutions of reinforcement learning.

## 8.1.2 Simulation Tests

Simulation of the proposed solution was implemented for the motion of a selected point $A$ of the WMR along the desired trajectory with path in the shape of loop consisting of five characteristic stages of motion described in Sect. 2.1.1. The form of matrices and vectors occurring in Eq. (8.1) is expressed by relation (2.35). Vector $\mathbf{a} = [a_1, \ldots, a_6]^T$ contains parameters resulting from the geometry, mass distribution, robot's resistance to motion. Their assumed values are provided in Table 8.1.

The assumed zero initial conditions of the controlled object, in the simulation, were as follows $\boldsymbol{\alpha}(0) = \mathbf{0}$, $\dot{\boldsymbol{\alpha}}(0) = \mathbf{0}$ and other data $\boldsymbol{\Lambda} = \text{diag}[1, 1]$, $\mathbf{K}_D = \text{diag}[2, 2]$, $\mathbf{F} = \text{diag}[10, \ldots, 10]$, $\gamma = 0.0001$, $k_s = 0.01$. The NN applied in the simulation, presented in Fig. 8.2, consists of six bipolar sigmoid neurons. Also, it is a network with functional extensions, whose first layer weights were generated by the random number generator from the following interval $[-0.1, 0.1]$. The separate NN was applied for each element of the nonlinear function $\mathbf{f}(\mathbf{x}) = [f_1(\mathbf{x}) \quad f_2(\mathbf{x})]^T$, in order to approximate the nonlinear function. Calculations were performed in Matlab/Simulink with Euler's integration method with time discretization step 0.01 [s].

Two tests were carried out in order to demonstrate the efficiency of the reinforcement learning algorithm.

The application of PD controller, without neural compensating for the nonlinearity of the robot, with the assumption that the parametric disturbance occurs, was analyzed in **the first test**. For time $t \geq 12$ [s] a parametric disturbance $\mathbf{a} + \Delta\mathbf{a}$ where $\Delta\mathbf{a} = [0\ 0\ 0\ 0\ 1\ 1]^T$ occurs, which is associated with the change of resistance to motion occurring, for example, when the WMR will drive on the path with a different type of surface. The obtained motion parameters and control signals are shown in Fig. 8.3.

**Fig. 8.2** Neural network of
RVFL type approximating
nonlinear functions $m = 2$

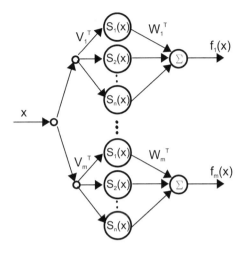

Figure 8.3a, b presents, respectively, the obtained tracking errors of angles of
rotation and angular velocities of wheels 1 and 2. During the transitional period the
control errors, resulting from the desired trajectory and parametric disturbance, occur.
As results from the sequences of errors they tend to zero. Sequences of measurement
function of quality control and the PD control signal are presented in Fig. 8.3c, d,
respectively. These sequences map the aforementioned changeable working conditions of the robot. These signals vary in value, which results from the description
of PD controller (8.8). The differences in the values of the angular velocities of
wheels are presented in Fig. 8.3e. The biggest errors occur during the accelerating
and braking of the robot and when the changeable resistance to motion occurs. These
errors are shown in Fig. 8.3b. Figure 8.3f presents desired trajectory (dashed line) and
the trajectory executed by point $A$ of the mobile robot. The executed trajectory is
delayed in relation to the desired trajectory and thus, an error of motion execution,
whose maximum value amounts to 0.103 [m], occurs. Also, it should be noted that
the solution to the tracking problem, in the analyzed case, is stable, which means that
all signals are constrained. This is due to the fact that the PD controller may be interpreted as robust to parameter disturbances when its properties are presented on the
error and derivative error phase plane. For a quantitative evaluation of the quality of
generated control and implementation of tracking motion, the performance indexes,
presented in Sect. 7.1.2, were assumed.

The values of individual performance indexes for execution of the tracking motion
with control signals generated by the PD controller are presented in the Tables 8.2
and 8.3.

Results of this test, in the further part of this chapter, will act as a reference point in
assessing the quality of tracking motion of the WMR by implementing other control
methods, in which the PD control is included.

Neural control system was implemented in **the second test** in order to carry
out the tracking motion of the robot. The PD controller was included in control

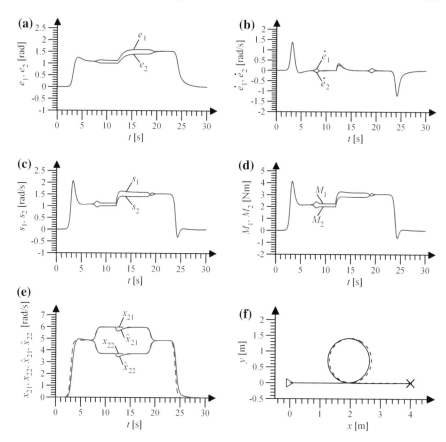

**Fig. 8.3** Results of the PD control system simulation: **a** values of tracking errors of angles of rotation of wheels 1 and 2, **b** values of tracking errors of angular velocities of wheels 1 and 2, **c** errors of quality control measure $s_1$, $s_2$, **d** PD control signals, **e** desired (dashed line) and executed angular velocities of rotation of wheels, **f** desired trajectory (dashed line) and trajectory executed by point $A$ of the WMR

**Table 8.2** Values of selected performance indexes for PD controller

| Index | $e_{maxj}$ [rad] | $\varepsilon_j$ [rad] | $\dot{e}_{maxj}$ [rad/s] | $\dot{\varepsilon}_j$ [rad/s] | $\sigma_j$ [rad/s] |
|---|---|---|---|---|---|
| wheel 1, $j = 1$ | 1.591 | 1.147 | 1.381 | 0.2825 | 1.18 |
| wheel 2, $j = 2$ | 1.525 | 1.072 | 1.381 | 0.2806 | 1.108 |

**Table 8.3** Values of selected performance indexes of path execution for PD controller

| Index | $d_{max}$ [m] | $\rho$ [m] |
|---|---|---|
| Value | 0.103 | 0.0642 |

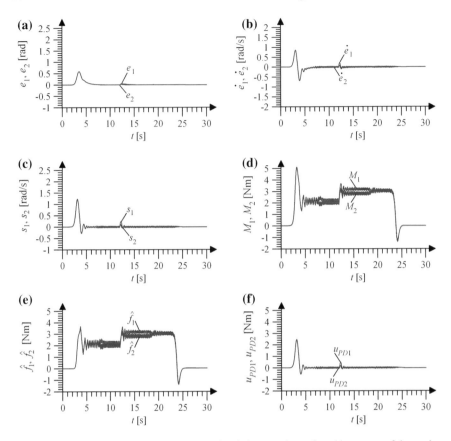

**Fig. 8.4** Results of the neural control system simulation: **a** values of tracking errors of the angles of rotation of wheels 1 and 2, **b** values of the tracking errors of angular velocities of wheels 1 and 2, **c** errors of quality control measure $s_1$, $s_2$, **d** overall control signals, **e** compensation control signals generated by the NN, **f** PD control signals

system. Additionally, the neural network, for each element of nonlinear function $\mathbf{f}(\mathbf{x}) = [f_1(\mathbf{x}) \ \ f_2(\mathbf{x})]^T$, was also included in the test in order to approximate the function (8.11). Zero initial value of the output layer weights were assumed in the process of NNs initialization. Other conditions were assumed as in the first test. The obtained values of control signals and tracking errors are shown in Fig. 8.4.

From the obtained values results the significant efficiency of implementation of neural compensation of the robot's nonlinearity in the learning structure of network weights based on the reinforcement signal. The obtained tracking errors presented in Fig. 8.4a–c are significantly lower than in the case of control with the use of a PD controller. Their values are provided in Tables 8.4 and 8.5. The PD controller is predominant in the initial phase of the motion execution, Fig. 8.4f, which results from the fact that the initial conditions of network weights were equal to zero, Fig. 8.5c, d. The signal of PD controller, when the neural network learns the nonlinearity, is close

**Table 8.4** Values of selected performance indexes for neural control system

| Index | $e_{maxj}$ [rad] | $\varepsilon_j$ [rad] | $\dot{e}_{maxj}$ [rad/s] | $\dot{\varepsilon}_j$ [rad/s] | $\sigma_j$ [rad/s] |
|---|---|---|---|---|---|
| wheel 1, $j = 1$ | 0.5891 | 0.1034 | 0.8583 | 0.1414 | 0.1746 |
| wheel 2, $j = 2$ | 0.5891 | 0.103 | 0.8583 | 0.1401 | 0.1733 |

**Table 8.5** Values of selected performance indexes of path execution for neural control system

| Index | $d_{max}$ [m] | $\rho$ [m] |
|---|---|---|
| Value | 0.0486 | 0.0092 |

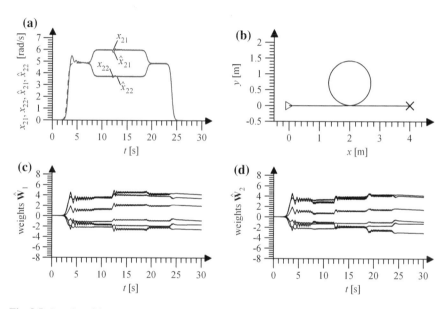

**Fig. 8.5** Results of the neural control system simulation: **a** desired and executed angular velocities of wheels rotation, **b** desired and executed trajectory by selected point $A$ of the mobile robot, **c** values of the NN weights approximating the nonlinearity $f_1$, **d** values of the NN weights approximating the nonlinearity $f_2$

to zero. In the overall control signal, shown in Fig. 8.4d, the compensating control, shown in Fig. 8.4e, is predominant. Figure 8.5 presents a comparison of desired $\alpha_{d1}$, $\alpha_{d2}$ and executed angular velocities $\dot{\alpha}_1$, $\dot{\alpha}_2$ by the WMR. The desired and executed motion path of point $A$ of the WMR and the values of output layer weights from the NNs for wheels 1 and 2.

The values provided in Fig. 8.5c, d confirm the conclusions resulting from the theoretical considerations regarding the limitations of the NN weights. Figure 8.5b presents desired trajectory (dashed line) and the trajectory executed by the point $A$ of the WMR. Because the error is small the difference is unnoticeable. In this case, the maximum value of the motion error amounts to 0.0486 [m], which is expressed by a better accuracy of executed motion in comparison to the used of only a PD controller. The values of individual performance indexes for the execution of tracking motion with control generated by the neural control system is presented in Tables 8.4 and 8.5.

### 8.1.3   Conclusions

This section presents the results of the implementation of classical reinforcement learning method in tracking control of the WMR. The reinforcement signal in binary form was applied in order to train the NN, which task was to compensate for the nonlinearity of the robot. Simulations of the adopted solution were performed. By analyzing the changes in the values of individual performance indexes in the control with PD controller and the compensation for the nonlinearity by the NN, it can be stated that the compensation for the robot's nonlinearity, with the use of the reinforcement learning algorithm, improves the quality of the desired trajectory execution by a selected point of the mobile robot. However, the implementation of reinforcement signal in accordance with Eq. (8.15) results in "vibrations" of the control signal, which adversely influences the control actuator. The modification of reinforcement control, in order to eliminate the adverse effects, was carried out in the next subchapter.

## 8.2   Approximation of Classical Reinforcement Learning

As demonstrated in Sect. 8.1 the implementation of classical reinforcement learning, in which the reinforcement signal assumes the binary values 1 or $-1$, is computationally simple and includes a small mapping error of desired trajectory of selected point of the WMR while the determination of adaptive control is in progress, as shown in the example of tracking control of the WMR. However, the implementation of the reinforcement signal in accordance with the Eq. (8.15) results in "vibrations" of the control signal which adversely influences the actuator system of the generated control. In order to avoid these adverse effects the continuous relation must approximate the discontinuous relation of the reinforcement signal. This issue will be the subject matter of this chapter.

**Fig. 8.6** Dependence of reinforcement signal from the signal of quality control measure $s_j$; $a_j = 3$, $r_j$ is a reinforcement signal, $o_g$ represents the upper limit and $o_d$ the lower limit

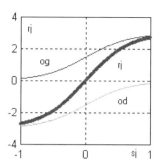

## 8.2.1 Control System Synthesis, Stability of the System, Reinforcement Learning Algorithm

Similarly to Sect. 8.1, in this section the task of the control system will be to bring the state vector of the WMR described in the form of (8.2), (8.3) to the desired trajectory of point $A$ of the robot discussed in Sect. 8.1.2. For this purpose, let's define the tracking error and quality control measure (current cost) in the form of

$$\mathbf{e}(t) = \mathbf{x}_d(t) - \mathbf{x}(t) , \tag{8.26}$$

$$\mathbf{s}(t) = \dot{\mathbf{e}}(t) + \mathbf{\Lambda}\mathbf{e}(t) , \tag{8.27}$$

where $\mathbf{\Lambda}$ – a positive-definite diagonal matrix. In the further discussion we will apply assumptions similar to those described in Sect. 8.1. Let's introduce critic's reinforcement signal approximation (8.15) as [13]

$$r_j(t) = \frac{a_j}{1 + e^{-a_j s_j(t)}} - \frac{a_j}{1 + e^{a_j s_j(t)}} , \tag{8.28}$$

where $j = 1, 2$, $a_j$ – a positive constant value, the reinforcement value $r_j$ is limited to the interval $\left[-a_j, a_j\right]$ when $a_j > 0$ as shown in Fig. 8.6.

The quality control measure $s_j(t)$ was assumed in Eq. (8.28) as an assessment of the value of generated reinforcement for the mobile robot control. When measure $s_j(t)$ is small, the quality of tracking execution is good.

**Theorem 8.2** *The control signal* $\mathbf{u}$ *will be in the form of (8.12) with robust control defined as*

$$\mathbf{v}(t) = -k_s \mathbf{R} \|\mathbf{R}\|^{-1} , \tag{8.29}$$

*with a coefficient that satisfies the relation* $k_s \geq d_{max}$, *reinforcement signal (8.28) and the learning law of network's weights*

$$\dot{\hat{\mathbf{W}}} = \mathbf{F}\boldsymbol{\varphi}(\mathbf{x})\mathbf{R}^T - \gamma\mathbf{F}\hat{\mathbf{W}} , \tag{8.30}$$

*where* $\mathbf{F} = \mathbf{F}^T > 0 - a\ design\ matrix,\ \gamma > 0 - a\ coefficient\ assuring\ fast\ convergence$
*of learning, then the quality control measure* $\mathbf{s}$ *and the estimation error of the NN*
*weights* $\tilde{\mathbf{W}} = \mathbf{W} - \hat{\mathbf{W}}$ *are respectively uniformly ultimately bounded to* $\psi_s$, $\psi_W$ *sets.*

*Proof* Let's consider a proposal of Lyapunov function in the form of

$$V = \sum_{j=1}^{2} \ln\left\{1 + \exp\left(-s_j\right)\right\} + \ln\left\{1 + \exp\left(-s_j\right)\right\} + \frac{1}{2}\mathrm{tr}\left(\tilde{\mathbf{W}}^T \mathbf{F}^{-1}\tilde{\mathbf{W}}\right) . \quad (8.31)$$

The derivative of the function $V$ with respect to time is in the form of

$$\dot{V} = \mathbf{R}^T\dot{\mathbf{s}} + \mathrm{tr}\left(\tilde{\mathbf{W}}^T \mathbf{F}^{-1}\dot{\tilde{\mathbf{W}}}\right) . \quad (8.32)$$

By substituting equation of error dynamics (8.13) into relation (8.32) the following
formula was obtained

$$\dot{V} = -\mathbf{R}^T\mathbf{K}_D\mathbf{s} + \mathbf{R}^T\left(\boldsymbol{\varepsilon} + \mathbf{d} + \mathbf{v}\left(t\right)\right) + \mathrm{tr}\left\{\tilde{\mathbf{W}}^T\left[\mathbf{F}^{-1}\dot{\tilde{\mathbf{W}}} + \boldsymbol{\varphi}\left(\mathbf{x}\right)\mathbf{R}^T\right]\right\} . \quad (8.33)$$

By introducing into relation (8.33) the learning law of network weights (8.30) and
robust control (8.29), we will receive

$$\dot{V} \leq -a\delta_{min}\left(\mathbf{K}_D\right)\|\mathbf{s}\| + \varepsilon_{max}a + \gamma\left(\|\tilde{\mathbf{W}}\|W_{max} - \|\tilde{\mathbf{W}}\|^2\right) , \quad (8.34)$$

where the following relations $\|\mathbf{R}\| \leq a$ and $\mathrm{tr}\left\{\tilde{\mathbf{W}}^T\hat{\mathbf{W}}\right\} = \mathrm{tr}\left\{\tilde{\mathbf{W}}^T\left(\mathbf{W} - \tilde{\mathbf{W}}\right)\right\} \leq$
$\|\tilde{\mathbf{W}}\|\left(W_{max} - \|\tilde{\mathbf{W}}\|\right)$ were implemented.

By squaring the binomial the following was received

$$\dot{V} \leq -a\delta_{min}\left(\mathbf{K}_D\right)\|\mathbf{s}\| + \varepsilon_{max}a - \gamma\left(\|\tilde{\mathbf{W}}\| - W_{max}/2\right)^2 + \gamma W_{max}^2/4 , \quad (8.35)$$

where $\delta_{min}\left(\mathbf{K}_D\right)$ – a smallest eigenvalue of $\mathbf{K}_D$ matrix.

The derivative of the Lyapunov function $\dot{V}$ is negative definite until the conditions
are satisfied

$$\|\mathbf{s}\| > \frac{a\varepsilon_{max} + \gamma W_{max}^2/4}{a\delta_{min}\left(\mathbf{K}_D\right)} = \psi_s , \quad (8.36)$$

$$\|\tilde{\mathbf{W}}\| > \sqrt{W_{max}^2/4 + \varepsilon_{max}a/\gamma} + W_{max}/2 = \psi_W . \quad (8.37)$$

It can be stated, that the derivative of Lyapunov function $\dot{V}$ is negative outside the set

$$\mathbf{\Omega} : \mathbf{\Omega} = \left\{ \left( \|\mathbf{s}\|, \|\tilde{\mathbf{W}}\| \right), 0 \leq \|\mathbf{s}\| \leq \psi_s, 0 \leq \|\tilde{\mathbf{W}}\| \leq \psi_W \right\}, \tag{8.38}$$

what guarantees the convergence of errors of quality measure $\mathbf{s}$ of designed control system and estimation of weights of the NN $\tilde{\mathbf{W}}$ into the $\mathbf{\Omega}$ set which is limited by $\psi_s$, $\psi_W$ constants. In summary, on the basis of the extended Lapunov stability theory it was proven that errors are uniformly ultimately bounded to the following set (8.38).

$\square$

### 8.2.2 Simulation Tests

It was assumed, in the simulation tests, that the obtained results, provided in the Sect. 8.1.2, after implementation of the PD controller in the control system are known. This subchapter presents the results of simulation tests with approximation of the critic of the reinforcement signal (8.28).

The neural control system was applied in the simulation test during execution of robot's tracking motion. In the initialization process of NNs the zero initial values of weights of the output layer were assumed. Other conditions were assumed analogously as in Sect. 8.1.2. The obtained values of control and tracking errors were presented in Fig. 8.7.

After comparison of obtained results with those presented in Fig. 8.4, we can notice the lack of oscillation in the obtained signals. Furthermore, the obtained tracking errors are much smaller in comparison to errors provided in Fig. 8.4a–c. The presence of signals in the overall control is similar to those presented in Fig. 8.7d–f. It can be noticed that the influence of neural compensation control of robot's nonlinearity was increased by reducing the influence of PD control, which justifies reduction of control errors.

Figure 8.8 presents a comparison of the desired $\dot{\alpha}_{d1}$, $\dot{\alpha}_{d2}$ and executed angular velocities $\dot{\alpha}_1$, $\dot{\alpha}_2$ by the WMR, desired and executed path of point $A$ of the WMR and values of weights of the output layers from the NNs for wheels 1 and 2.

On the basis of visual evaluation of results presented in Fig. 8.8, the differences in comparison to results shown in Fig. 8.5 are unnoticeable. The values of individual performance indexes for the implementation of tracking motion with control generated by the neural control system and approximation of critic of the reinforcement signal were presented in Tables 8.6 and 8.7.

Increase in accuracy of the desired trajectory execution in the analyzed case was more than 50% which results from the comparison of the maximum distance error of the desired and executed path. The analysis of the quantitative results provided in Tables 8.6 and 8.7 proves the validity of the assumed solution.

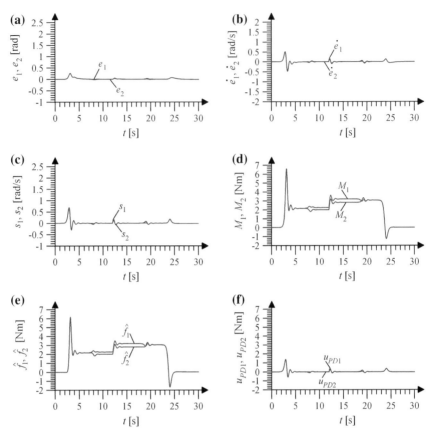

**Fig. 8.7** Results of neural control system simulation: **a** values of tracking errors of angles of rotation of wheels 1 and 2, **b** values of tracking errors of angular velocities of wheels 1 and 2, **c** errors of quality control measure $s_1$, $s_2$, **d** overall control signals, **e** compensation control signals generated by the neural network, **f** PD control signals

### 8.2.3  Conclusions

Approximation of classical reinforcement learning in control of the WMR was applied in Sect. 8.2. Synthesis of the tracking control algorithm of the mobile robot was performed on the basis of Lyapunov stability theory, demonstrating that the errors are uniformly ultimately bounded to the set (8.38). Theoretical considerations were confirmed by the results obtained within the simulation. The obtained simulation results of the assumed solution validate the occurrence of better accuracy of execution of the desired motion of robot's selected point.

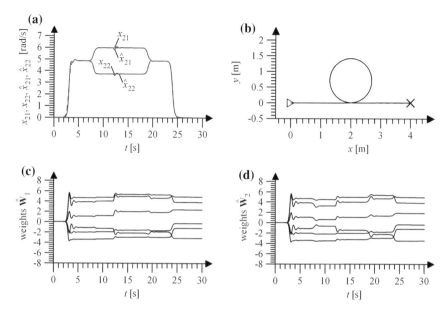

**Fig. 8.8** Results of the neural control system simulation: **a** desired and executed angular velocities of wheels rotation, **b** desired and executed path by selected point $A$ of the mobile robot; the difference is unnoticeable due to the small error, **c** values of neural network weights approximating nonlinearity $f_1$, **d** values of neural network weights approximating nonlinearity $f_2$

**Table 8.6** Values of selected performance indexes for neural control system

| Index | $e_{maxj}$ [rad] | $\varepsilon_j$ [rad] | $\dot{e}_{maxj}$ [rad/s] | $\dot{\varepsilon}_j$ [rad/s] | $\sigma_j$ [rad/s] |
|---|---|---|---|---|---|
| wheel 1, $j = 1$ | 0.262 | 0.04096 | 0.5172 | 0.07814 | 0.08791 |
| wheel 2, $j = 2$ | 0.262 | 0.04103 | 0.5172 | 0.07782 | 0.08766 |

**Table 8.7** Values of the selected performance indexes for neural control system

| Index | $d_{max}$ [m] | $\rho$ [m] |
|---|---|---|
| Value | 0.02161 | 0.0041 |

## 8.3 Reinforcement Learning in the Actor-Critic Structure

In this chapter, the algorithm based on the method of reinforcement learning with actor-critic structure for continuous time, in which the element generating the control (ASE-actor) and the element generating the reinforcement signal (ACE-critic) are implemented in the form of an artificial neural network, was proposed as an alternative. The system does not require the initial learning phase. It operates on-line without knowledge on the controlled object model. Stability of control system provides additional control signal designed on the basis of Lyapunov stability theory.

### 8.3.1  Synthesis of Control System, System Stability, Reinforcement Learning Algorithm

In this section, similarly to Sect. 8.1, the task of control system will be to bring the state vector of the WMR in the following form

$$\mathbf{M}\ddot{\mathbf{q}} + \mathbf{C}\left(\dot{\mathbf{q}}\right)\dot{\mathbf{q}} + \mathbf{F}\left(\dot{\mathbf{q}}\right) + \boldsymbol{\tau}_d\left(t\right) = \boldsymbol{\tau}\,, \tag{8.39}$$

to the desired trajectory of point $A$ of the robot, discussed in Sect. 8.1.2. The tracking error $\mathbf{e}$, the filtered tracking error $\mathbf{s}$ and the auxiliary quantity $\mathbf{v}$ were defined as

$$\mathbf{e} = \mathbf{q} - \mathbf{q}_d\,, \tag{8.40}$$

$$\mathbf{s} = \dot{\mathbf{e}} + \boldsymbol{\Lambda}\mathbf{e}\,, \tag{8.41}$$

$$\mathbf{v} = \dot{\mathbf{q}}_d - \boldsymbol{\Lambda}\mathbf{e}\,, \tag{8.42}$$

where $\boldsymbol{\Lambda}$ is a positive-definite diagonal matrix. Then the Eq. (8.39) can be expressed in the following form

$$\mathbf{M}\dot{\mathbf{s}} = -\mathbf{u} - \mathbf{C}\left(\dot{\mathbf{q}}\right)\mathbf{s} + \mathbf{M}\dot{\mathbf{v}} + \mathbf{C}\left(\dot{\mathbf{q}}\right)\mathbf{v} + \mathbf{F}\left(\dot{\mathbf{q}}\right) + \boldsymbol{\tau}_d\left(t\right)\,. \tag{8.43}$$

Adopting the nonlinear function as

$$\mathbf{f}\left(\mathbf{x}\right) = \mathbf{M}\dot{\mathbf{v}} + \mathbf{C}\left(\dot{\mathbf{q}}\right)\mathbf{v} + \mathbf{F}\left(\dot{\mathbf{q}}\right)\,, \tag{8.44}$$

where $\mathbf{x} = \left[\mathbf{v}^T, \dot{\mathbf{v}}^T, \dot{\mathbf{q}}^T\right]^T$, relation (8.43) was written in the form of

$$\mathbf{M}\dot{\mathbf{s}} = -\mathbf{u} - \mathbf{C}\left(\dot{\mathbf{q}}\right)\mathbf{s} + \mathbf{f}\left(\mathbf{x}\right) + \boldsymbol{\tau}_d\left(t\right)\,. \tag{8.45}$$

Figure 8.9 presents the scheme of the two-wheeled mobile robot control system. The control signal generated by the system is composed of a compensation control signal in the form of $\mathbf{u}_{RL}$ and the supervisory control signal $\mathbf{u}_S$, whose implementation ensures the stability of the closed control system which includes, among others, a PD controller.

In the control system which is based on the algorithm specified in [4], modification [9, 11, 12] of this algorithm, which includes the concept of the reinforcement learning in continuous time, was applied. In the above mentioned the two-layer NN was

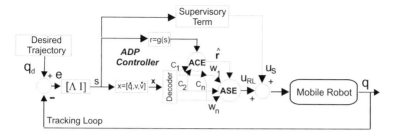

**Fig. 8.9** Scheme of the mobile robot tracking control system with ACE-ASE structure

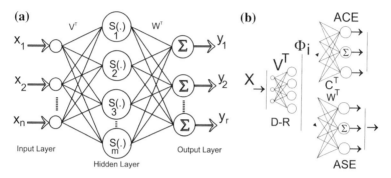

**Fig. 8.10**  **a** Scheme of the neural network, **b** implementation of the actor-critic structure

applied in order to implement the tasks of two parametric structures: actor and critic's. The output from the two-layer NN can be formulated as follows

$$\mathbf{y} = \mathbf{W}^T \mathbf{S}\left(\mathbf{V}^T \mathbf{x}\right) . \tag{8.46}$$

If the weights of input layer $\mathbf{V}$ are defined by a random selection and remain constant in the network's training process, than the value of the output from the network depends only on the output layer weights $\mathbf{W}$ adaptation. In the applied solution, the neural network shown in Fig. 8.10a was broken down into 3 elements as presented in Fig. 8.10b; (D-R) decoder, ACE and ASE structures.

Our goal is to determine the following compensation control

$$\mathbf{u}_{RL} = \hat{\mathbf{f}}\left(\mathbf{x}\right) . \tag{8.47}$$

for the dynamic system (8.45), which optimizes the value function in the form of

$$V_{uj}\left(\mathbf{s}\left(t\right)\right) = \int_t^\infty \frac{1}{\tau_c} e^{\frac{\tau-t}{\tau_c}} r_j\left(\mathbf{s}\left(\tau\right), \mathbf{u}\left(\tau\right)\right) d\tau , \tag{8.48}$$

where $j = 1, 2$, $\tau_c$ assumes a constant values, but $\mathbf{r}(t) = [r_1(t), r_2(t)]^T$ is a value of instantaneous reinforcement signal. Let's define the process of constrained reinforcement signal $\mathbf{r}(t)$ generation as presented in Sect. 8.2, according to the relation

$$r_j(t) = \frac{a_j}{1 + e^{-a_j s_j(t)}} - \frac{a_j}{1 + e^{a_j s_j(t)}} , \tag{8.49}$$

where $j = 1, 2$, $a_j$ is a positive constant value and the reinforcement value $r_j$ is limited to the interval $\left[-a_j, a_j\right]$ for $a_j > 0$, as shown in Fig. 8.6. The filtered tracking error $s_j(t)$ was applied, in the Eq. (8.49), as an assessment of value of the generated reinforcement for the mobile robot control. When the filtered tracking error $s_j(t)$ is small, the quality of the tracking motion execution is appropriate. Solutions which use learning with a critic are based on algorithms which operate on the basis of cyclical calculations related with adaptation of control law and the approximation of value function. Time differences error for the continuous system will be formulated as [3, 4, 7]

$$\tau_c \dot{V}_{uj}(\mathbf{s}(t)) = V_{uj}(\mathbf{s}(t)) - r_j(t) . \tag{8.50}$$

Let $P_j(t)$ be the prediction of the value function. If the prediction of the value function is correct, the following condition occurs

$$\tau_c \dot{P}_j(t) = P_j(t) - r_j(t) . \tag{8.51}$$

If the prediction is not appropriate, the prediction error of the value function is defined as [17]

$$\hat{r}_j(t) = r_j(t) - P_j(t) + \tau_c \dot{P}_j(t) . \tag{8.52}$$

Critic (ACE) is a parametric structure $\mathbf{P}(t)$ with parameters $\mathbf{C}$. The task of the above mentioned parametric structure is to approximate the prediction of the value function and generates signal in the form of

$$P_j(t) = \mathbf{C}^T \boldsymbol{\varphi} . \tag{8.53}$$

where on the basis of (8.46) the following occurs

$$\varphi_i = S_i\left(\mathbf{V}^T \mathbf{x}\right) , \tag{8.54}$$

where $i = 1, 2, \ldots, N$, $N$ is the number of neurons. Critic's weights are adapted according to the relation

$$\dot{C}_i(t) = \beta \hat{r}_j(t) \overline{e}_i(t) . \tag{8.55}$$

In contrast, the activity path $\overline{e}_i(t)$ is described by equation

$$\lambda_C \dot{\overline{e}}_i(t) = \varphi_i(t) - \overline{e}_i(t) , \tag{8.56}$$

where $\beta$ is a weights learning rate, while $\lambda_C$ is a design index. The $\mathbf{u}_{RL}$ control signal generated by the ASE structure assums the following form

$$\mathbf{u}_{RL} = \delta \mathbf{S} \left( \mathbf{W}^T \boldsymbol{\varphi} \right) \in \Re^2 , \tag{8.57}$$

where $\delta$ is a scaling index of actor's output. By analogy, the actor's (ASE) weights adaptation law is in the form of

$$\dot{W}_i (t) = \alpha \hat{r}_j (t) \, \varphi_i , \tag{8.58}$$

where $\alpha$ is a positive constant index. Our task is to provide the stability of the closed system without changing the designed control signal $\mathbf{u}_{RL}$. This means that the control signal $\mathbf{u}$ must be designed in such a way to ensure stability of the closed control system, which means that the filtered tracking error $\mathbf{s}$ must be limited, i.e.

$$|s_i| \leq \Phi_i , \quad \forall t > 0 , \tag{8.59}$$

where $\Phi_i$ assumes constant values, $i = 1, 2$. For this purpose, the existing control signal $\mathbf{u}_{RL}$ was enriched by a supervisory control $\mathbf{u}_S$ [23] generated by an additional supervisory system.

For this particular problem let's introduce the control law

$$\mathbf{u} = \mathbf{u}_{RL} + \mathbf{u}_S . \tag{8.60}$$

By substituting (8.60) into (8.45) we receive a description of the closed system in the form of

$$\mathbf{M}\dot{\mathbf{s}} = -\mathbf{C}\mathbf{s} + \mathbf{f} (\mathbf{x}) - \mathbf{u}_{RL} - \mathbf{u}_S + \boldsymbol{\tau}_d . \tag{8.61}$$

Assuming the additional signal

$$\mathbf{u}^* = \mathbf{f} (\mathbf{x}) + \mathbf{K}\mathbf{s} , \tag{8.62}$$

relation (8.61) will be in the following form

$$\mathbf{M}\dot{\mathbf{s}} = -\mathbf{C}\mathbf{s} + \mathbf{u}^* - \mathbf{K}\mathbf{s} - \mathbf{u}_{RL} - \mathbf{u}_S + \boldsymbol{\tau}_d . \tag{8.63}$$

Let's define positive-definite function

$$L = \frac{1}{2}\mathbf{s}^T \mathbf{M}\mathbf{s} . \tag{8.64}$$

By determining the derivative of function (8.64) and with the use of relation (8.63), we receive

$$\dot{L} = \mathbf{s}^T \mathbf{M}\dot{\mathbf{s}} + \frac{1}{2}\mathbf{s}^T \dot{\mathbf{M}}\mathbf{s} = -\mathbf{s}^T \mathbf{K}\mathbf{s} + \frac{1}{2}\mathbf{s}^T \left(\dot{\mathbf{M}} - 2\mathbf{C}\right)\mathbf{s} + \mathbf{s}^T \left[\mathbf{u}^* - \mathbf{u}_{RL} + \tau_d\right] - \mathbf{s}^T \mathbf{u}_S .$$

$$(8.65)$$

Assuming that $\dot{\mathbf{M}} - 2\mathbf{C}$ is a skew-symmetric matrix, the derivative (8.65) has the following form

$$\dot{L} \leq -\mathbf{s}^T \mathbf{K}\mathbf{s} + \sum_{i=1}^{2} |s_i| \left[|f_i(\mathbf{x}) - u_{RLi} + k_i s_i + \tau_{di}|\right] + \sum_{i=1}^{2} s_i u_{Si} .$$

$$(8.66)$$

If we will select a supervisory control signal in the form of

$$u_{Si} = -\left[F_i + |k_i s_i| + Z_i + \eta_i\right] \mathrm{sgn} s_i ,$$

$$(8.67)$$

assuming that

$$|f_i(\mathbf{x}) - \hat{f}_i(\mathbf{x})| \leq F_i ,$$

$$(8.68)$$

where $F_i > 0$, $|\tau_{di}| \leq Z_i$, $Z_i > 0$, $\eta_i > 0$, $i = 1, 2$, we will finally receive

$$\dot{L} \leq -\mathbf{s}^T \mathbf{K}\mathbf{s} - \sum_{i=1}^{2} \eta_i |s_i| .$$

$$(8.69)$$

The derivative of Lyapunov function is negative semidefinite, thus the closed system (8.63) is stable in relation to the state vector **s**. The control signal (8.67) includes $\mathrm{sgn}(s_i)$ function which activates a high switching frequency of the actuator. In order to eliminate the high switching frequency of the actuator, approximation of the $\mathrm{sgn}(s_i)$ function in its neighbourhood $\Phi_i$ is introduced by function [20]

$$\mathrm{sat}(s_i) = \begin{cases} \mathrm{sgn}(s_i) & for \ |s_i| > \Phi_i , \\ \frac{s_i}{\Phi_i} & for \ |s_i| \leq \Phi_i . \end{cases}$$

$$(8.70)$$

If $\mathbf{e}(0) = \mathbf{q}_d(0) - \mathbf{q}(0) = \mathbf{0}$, then the accuracy of control is provided in finite time [20]

$$\forall t \geq 0 , \ |s_i(t)| \leq \Phi_i \ \Rightarrow \ \forall t \geq 0 , \ |e_i^{(i)}(t)| \leq (2\lambda_i)^i \varepsilon_i ,$$

$$(8.71)$$

where $i = 0, 1, \ldots, n_i - 1$, $\varepsilon_i = \Phi_i / \lambda_i^{n_i - 1}$, $n_i = 2$ is the rank of the $i$-th system and the superscript $(i)$ is an error derivative.

## 8.3.2  Simulation Tests

Simulation of the designed control system was carried out for the motion of point $A$ of the WMR along the desired path in the shape of loop consisting of five characteristic stages of motion, as described in Sect. 8.1.2.

The simulation was carried out for zero initial conditions of the controlled object $\alpha\,(0) = \mathbf{0}$, $\dot{\alpha}\,(0) = \mathbf{0}$ and other data, $\mathbf{\Lambda} = \mathrm{diag}\,[1, 1]$, $\mathbf{K}_D = \mathrm{diag}\,[2, 2]$, $\alpha = \mathrm{diag}\,[10, \ldots, 10]$, $\tau_C = 10$, $\beta = 0.1$, $\lambda_C = 0.85$, $\delta = 16$, $\Phi = 0.8$, $\eta = [0.01, 0.01]^T$, $\mathbf{F} = [2, 2]^T$. The neural network, shown in Fig. 8.10, applied in the simulation consists of six bipolar sigmoid neurons and is a network with functional extensions, whose weights of first layer (D-R) were generated from the interval $[-0.1, 0.1]$ by a random number generator. Relations (8.53) and (8.57) with network weights shown in Fig. 8.10b were applied respectively in order to approximate the critic (element ACE) and actor (element ASE). Calculations were carried out in Matlab/Simulink within Euler's integration method with discretization step of time 0.01 [s].

The results, in the form of selected motion and control parameters, are shown in Figs. 8.11, 8.12, 8.13 and 8.14. The values of two-wheeled mobile robot overall control signals are presented in Fig. 8.11a for wheels 1 and 2, respectively. The overall control, in accordance with the applied control law, consists of nonlinearity compensation in the form of a signal generated by the ACE-ASE structure, Fig. 8.11b, and a supervisory control, Fig. 8.11c. The consequence of applying the neural network with zero weights values in the initial phase of adaptation process in order to

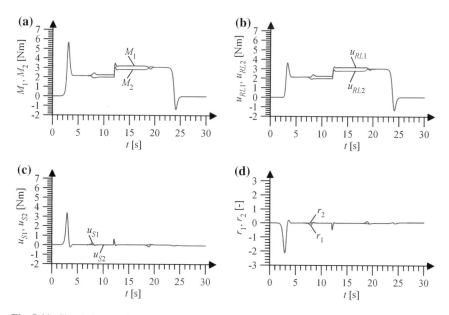

**Fig. 8.11**  Simulation results

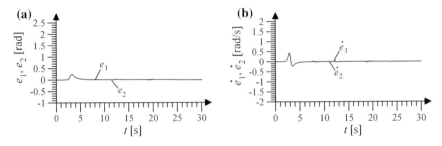

**Fig. 8.12**  Tracking errors

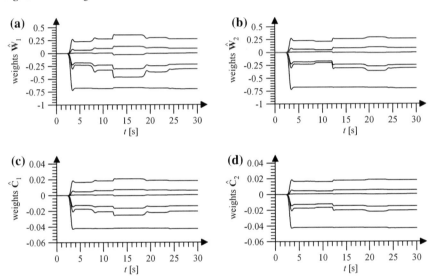

**Fig. 8.13**  Weights of ASE, ACE elements

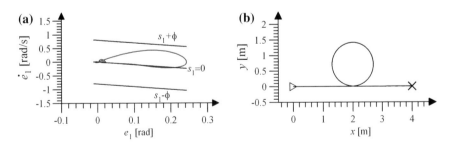

**Fig. 8.14**  Evaluation of motion execution of robot's point $A$, **a** phase plane of errors, **b** desired and executed path

compensate for the nonlinearities of the controlled object is a variable presence of individual component signals in the overall control. During the initial stage of the network weights adaptation a major role in control plays the supervisory regulator, whose presence in the overall control decreases along with the progress of the actor's - ASE NN weights adaptation process. In the analyzed algorithm of actor and critic's structure, the two-layer NNs with sigmoid neurons activation functions and random selection of input layer weights in the network initialization phase, were applied. Figure 8.11d presents reinforcement signal $r_1$ and $r_2$.

The most important issue in the tracking systems is the precise execution of the desired trajectory, and that is why the primary objective in designing the algorithms of tracking control is to minimize the tracking error. Figure 8.12a presents tracking errors $e_1$ and $e_2$, Fig. 8.12b presents errors $\dot{e}_1$ and $\dot{e}_2$ of angular velocities, which are limited.

In the initialization process of NN individual weights of the network neurons assume zero initial values, then they are optimized based on the progress of network's adaptation process, in order to reach constant values. Changes of actor's (ASE) weights values are shown, respectively, for wheel 1 in Fig. 8.13a and for wheel 2 in Fig. 8.13b. Values of critic's (ACE) NN weights are presented in Fig. 8.13c, d. In both cases, weights of the NN, in accordance with the theory of stability, are limited. Errors $e_1$ and $\dot{e}_1$, $\forall t \geq 0$, $|s_1(t)| \leq \Phi \Rightarrow \forall t \geq 0$, $|e_1(t)| \leq 0.8$ and $|\dot{e}_1| \leq 1.6$, were calculated by applying relation (8.71), and their values are shown in Fig. 8.14a on the phase plane. By comparing those values with sequences presented in Fig. 8.12 we can conclude that these are conservative calculation. Figure 8.14b presents the desired and executed path of point $A$ of the WMR; with no evident difference. The maximum error of the desired path execution which amounts to 0.02 [m] is provided in Table 8.8.

For quantitative assessment of results obtained after conducted numerical test the values of individual performance indexes for execution of tracking motion with control generated by the actor-critic system are shown in Tables 8.8 and 8.9.

**Table 8.8** Values of selected performance indexes for the control system

| Index | $e_{maxj}$ [rad] | $\varepsilon_j$ [rad] | $\dot{e}_{maxj}$ [rad/s] | $\dot{\varepsilon}_j$ [rad/s] | $\sigma_j$ [rad/s] |
|---|---|---|---|---|---|
| wheel 1, $j = 1$ | 0.2428 | 0.04424 | 0.4306 | 0.05893 | 0.0724 |
| wheel 2, $j = 2$ | 0.2428 | 0.04231 | 0.4306 | 0.05859 | 0.0720 |

**Table 8.9** Values of selected performance indexes for the assessment of the desired path execution

| Index | $d_{max}$ [m] | $\rho$ [m] |
|---|---|---|
| Value | 0.02 | 0.0038 |

### 8.3.3   Conclusions

In contrast to the traditional approach in learning with a critic the ASE-ACE structure, based on two artificial NNs, was applied in order to approximate the nonlinearity and changeable working conditions of the WMR. The proposed algorithms are operating on-line without any knowledge of the model of the controlled object. Also, they do not require a preliminary learning phase. Stability of the closed system was achieved by applying an additional supervisory control in the control law. Simulation tests have confirmed the correctness of adopted assumptions and efficacy of proposed algorithm in the WMR control.

## 8.4   Reinforcement Learning of Actor-Critic Type in the Optimal Adaptive Control

In this section, the presented approach focuses on the implementation of the actor-critic (ACE-ASE) [3, 4, 7, 16, 18] parametric structure with the use of learning with a critic. The architecture of systems built in accordance with the adaptive dynamic programming concept, in comparison to other control algorithms, uses two separate parametric structures: actor to generate control law and critic to generate approximation of the value function. This approach minimizes the computational cost incurred in connection with the control of the object. Taking into account the previous achievements in the field of control with a critic, the proposed tracking control algorithm of the two-wheeled mobile robot is based on the optimal adaptive control [8, 10, 15, 17] architecture which is built with the use of two artificial NNs. The proposed algorithm does not require an initial phase of learning. It operates on-line, without any knowledge of mathematical model of the controlled object.

### 8.4.1   Control Synthesis, Stability of a System, Reinforcement Learning Algorithm

The dynamic equations of a WMR motion can be written in the following form

$$\mathbf{M}\ddot{\mathbf{q}} + \mathbf{C}(\dot{\mathbf{q}})\dot{\mathbf{q}} + \mathbf{F}(\dot{\mathbf{q}}) + \boldsymbol{\tau}_d(t) = \mathbf{u}. \tag{8.72}$$

The task of a control system is to generate a vector of control $\mathbf{u} = [M_1, M_2]^T$ which guarantees the tracking motion of point $A$ of the WMR along the desired trajectory $\mathbf{q}_d = [\alpha_{d1}, \alpha_{d2}]^T$. Let's define the tracking error $\mathbf{e}(t)$ and the quality measure of tracking control $\mathbf{s}(t)$

$$\mathbf{s} = \dot{\mathbf{e}} + \boldsymbol{\Lambda}\mathbf{e}, \tag{8.73}$$

$$\mathbf{e} = \mathbf{q}_d - \mathbf{q} , \tag{8.74}$$

On the basis of the relation (8.73) and (8.74), the dynamic equations of the mobile robot's motion, expressed in the filtered tracking error space, will be written in the form of

$$\mathbf{M}\dot{\mathbf{s}} = -\mathbf{Cs} + \mathbf{f}(\mathbf{x}) - \mathbf{u} + \boldsymbol{\tau}_d(t) , \tag{8.75}$$

and we will define the unknown nonlinear function $\mathbf{f}(\mathbf{x})$ as

$$\mathbf{f}(\mathbf{x}) = \mathbf{M}\left(\ddot{\mathbf{q}}_d + \boldsymbol{\Lambda}\dot{\mathbf{e}}\right) + \mathbf{C}(\dot{\mathbf{q}})\left(\dot{\mathbf{q}}_d + \boldsymbol{\Lambda}\mathbf{e}\right) + \mathbf{F}(\dot{\mathbf{q}}) , \tag{8.76}$$

and the vector $\mathbf{x}$ is defined as $\mathbf{x} = \left[\dot{\mathbf{q}}_d^T, \ddot{\mathbf{q}}_d^T, \mathbf{e}^T, \dot{\mathbf{e}}^T\right]^T$. Let's assume the control law formulated as

$$\mathbf{u} = \hat{\mathbf{f}} + \mathbf{Ks} - \mathbf{v}(t) , \tag{8.77}$$

where $\hat{\mathbf{f}}$ is a estimate of function $\mathbf{f}(\mathbf{x})$, $\mathbf{K} = \mathbf{K}^T$ is a design positive-definite matrix, $\mathbf{Ks} = \mathbf{K}\dot{\mathbf{e}} + \mathbf{K}\boldsymbol{\Lambda}\mathbf{e}$ is a equation of the PD controller which is placed in the external loop control system feedback, while $\mathbf{v}(t)$ is an auxiliary signal introduced into the control law in order to provide robustness of the designed algorithm to disturbances and parametric inaccuracies. In order to approximate the nonlinear function (8.76) the NN, described by the following equation, was applied

$$\mathbf{f}(\mathbf{x}) = \mathbf{W}^T\boldsymbol{\varphi}(\mathbf{x}) + \varepsilon , \tag{8.78}$$

where $\mathbf{W}$ is a vector of network's ideal weights, $\boldsymbol{\varphi}(\mathbf{x})$ is a function of neurons activation, $\varepsilon$ is an approximation error limited to the interval $\|\varepsilon\| \leq z_\varepsilon$.

For such a case, the estimate of function $\mathbf{f}(\mathbf{x})$ is in the following form

$$\hat{\mathbf{f}}(\mathbf{x}) = \hat{\mathbf{W}}^T\boldsymbol{\varphi}(\mathbf{x}) , \tag{8.79}$$

while the control law (8.77) will be in the form of

$$\mathbf{u} = \hat{\mathbf{W}}^T\boldsymbol{\varphi}(\mathbf{x}) + \mathbf{Ks} - \mathbf{v}(t) . \tag{8.80}$$

Description of the closed control system in the filtered tracking error space is expressed in the following form

$$\mathbf{M}\dot{\mathbf{s}} + (\mathbf{K} + \mathbf{C})\mathbf{s} - \tilde{\mathbf{f}} - \mathbf{v}(t) = \mathbf{0} . \tag{8.81}$$

One of the application method of control with a critic in approximation of nonlinear mappings is implementation of a structure consisting of two NNs which operate as actor and critic. The tracking control system of the two-wheeled mobile robot with the implementation of an actor-critic structure, in order to compensate for the nonlinearities of the controlled object, is schematically presented in Fig. 8.15. In

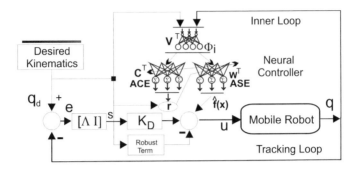

**Fig. 8.15**  Tracking control system of the wheeled mobile robot with actor-critic structure

the presented control system the actor (ASE) is responsible for generating control law. Its weights are adapted on the basis of the signal generated by the second parametric structure - the critic (ACE). The outputs from ASE and ACE systems can be formulated as $\hat{\mathbf{W}}^T \boldsymbol{\varphi}$ and $\hat{\mathbf{C}}^T \boldsymbol{\varphi}$, where the output from the critic is called a reinforcement signal **r**, which is implemented in order to optimize the values of weights of the actor's NN output layer $\hat{\mathbf{W}}$. In order to perform the tasks of individual ACE and ASE structures in the presented implementation of the actor-critic system in tracking control of the two-wheeled mobile robot, the two-layer artificial NNs with a sigmoid functions of neurons activation of the hidden layer, whose weights of the input layer **V** are randomly generated in the phase of network initialization and remain constant in the learning process, were used. The proposed algorithm is based on the reinforcement learning method, operates on-line, without any knowledge of the controlled object model. Weights of the NNs $\hat{\mathbf{W}}$ and $\hat{\mathbf{C}}$ are determined on-line without initial learning phase.

The task of the tracking control system of the two-wheeled mobile robot is to generate control signals required to execute the desired, at least two times differentiated, trajectory $\mathbf{q}_d$, for which all motion parameters are limited, and the trajectory is executable for the controlled object. Norms for ideal vectors of $\|\mathbf{W}\|$ and $\|\mathbf{C}\|$ weights are limited by the constant values $W_m$ and $C_m$. For the synthesis of weights adaptation law of the actor-critic structure NNs, having decisive influence on the stability of the closed control system, and in order to determine relations which calculate the reinforcement signal **r**, the proposition of Lyapunov function in the following form was applied

$$V = \frac{1}{2}\mathbf{s}^T \mathbf{M}\mathbf{s} + \frac{1}{2}\mathrm{tr}\tilde{\mathbf{W}}^T \mathbf{F}_W^{-1} \tilde{\mathbf{W}} + \frac{1}{2}\mathrm{tr}\tilde{\mathbf{C}}^T \mathbf{F}_C^{-1} \tilde{\mathbf{C}} . \tag{8.82}$$

By differentiating Eq. (8.82) and applying the learning law of weights of the ASE and ACE NNs in the form of

$$\dot{\hat{\mathbf{W}}} = \mathbf{F}_W \boldsymbol{\varphi}\mathbf{r}^T - \gamma \mathbf{F}_W \|\mathbf{s}\|\hat{\mathbf{W}} , \tag{8.83}$$

$$\dot{\hat{\mathbf{C}}} = -\mathbf{F}_C \|\mathbf{s}\| \boldsymbol{\varphi} \left( \hat{\mathbf{W}}^T \boldsymbol{\varphi} \right)^T - \gamma \mathbf{F}_C \|\mathbf{s}\| \hat{\mathbf{C}} , \tag{8.84}$$

where $\mathbf{F}_W$, $\mathbf{F}_C$ are positive-definite diagonal matrices, and by applying the reinforcement signal

$$\mathbf{r} = \mathbf{s} + \|\mathbf{s}\| \hat{\mathbf{C}}^T \boldsymbol{\varphi} , \tag{8.85}$$

and signal that provides the robustness of the designed system

$$\mathbf{v}(t) = z_0 \mathbf{s} / \|\mathbf{s}\| , \tag{8.86}$$

we receive

$$\dot{V} \leq -\mathbf{s}^T \mathbf{K} \mathbf{s} + \mathbf{s}^T \boldsymbol{\varepsilon} + \|\mathbf{s}\| \mathrm{tr} \left[ -\tilde{\mathbf{W}}^T \boldsymbol{\varphi} \left( \hat{\mathbf{C}}^T \boldsymbol{\varphi} \right)^T + \gamma \tilde{\mathbf{W}}^T \hat{\mathbf{W}} + \tilde{\mathbf{C}}^T \boldsymbol{\varphi} \left( \hat{\mathbf{W}}^T \boldsymbol{\varphi} \right)^T + \gamma \tilde{\mathbf{C}}^T \hat{\mathbf{C}} \right] . \tag{8.87}$$

If there occurs a relation $\tilde{\mathbf{W}} \hat{\mathbf{W}} \leq \|\tilde{\mathbf{W}}\| W_m - \|\tilde{\mathbf{W}}\|^2$ and $\tilde{\mathbf{C}} \hat{\mathbf{C}} \leq \|\tilde{\mathbf{C}}\| C_m - \|\tilde{\mathbf{C}}\|^2$, with $\|\boldsymbol{\varphi}\|^2 \leq m$, relation (8.87) will have the following form

$$\dot{V} \leq -\lambda_{min} \|\mathbf{s}\|^2 - \gamma \|\mathbf{s}\| \left\{ \left( \|\tilde{\mathbf{W}}\| - (mC_m + \gamma W_m) / 2\gamma \right)^2 + \left( \|\tilde{\mathbf{C}}\| - (mW_m + \gamma C_m) / 2\gamma \right)^2 - \left[ ((mC_m + \gamma W_m) / 2\gamma)^2 + ((mW_m + \gamma C_m) / 2\gamma)^2 + z_\varepsilon / \gamma \right] \right\} . \tag{8.88}$$

The derivative of Lyapunov function is negatively definite when the relations are satisfied

$$\|\mathbf{s}\| > \gamma \left[ ((mC_m + \gamma W_m) / 2\gamma)^2 + ((mW_m + \gamma C_m) / 2\gamma)^2 + z_\varepsilon / \gamma \right] / \lambda_{min} = b_S , \tag{8.89}$$

or

$$\|\tilde{\mathbf{W}}\| > (mC_m + \gamma W_m) / 2\gamma + \sqrt{\left[ ((mC_m + \gamma W_m) / 2\gamma)^2 + ((mW_m + \gamma C_m) / 2\gamma)^2 + z_\varepsilon / \gamma \right]} = b_W , \tag{8.90}$$

or

$$\|\tilde{\mathbf{C}}\| > (mW_m + \gamma C_m) / 2\gamma + \sqrt{\left[ ((mC_m + \gamma W_m) / 2\gamma)^2 + ((mW_m + \gamma C_m) / 2\gamma)^2 + z_\varepsilon / \gamma \right]} = b_C , \tag{8.91}$$

where $\lambda_{min}$ is a minimum eigenvalue of matrix $\mathbf{K}$. In this particular case, the derivative of Lyapunov function $\dot{V}$ is negative definite outside the set

$$\boldsymbol{\Omega} : \boldsymbol{\Omega} = \left\{ \left( \|\mathbf{s}\|, \|\tilde{\mathbf{W}}\|, \|\tilde{\mathbf{C}}\| \right), 0 \leq \|\mathbf{s}\| \leq b_s, 0 \leq \|\tilde{\mathbf{W}}\| \leq b_W, 0 \leq \|\tilde{\mathbf{C}}\| \leq b_C \right\} , \tag{8.92}$$

which guarantees the convergence of estimation errors of the actor's $\tilde{\mathbf{W}}$ and critic's $\tilde{\mathbf{C}}$ NN weights to the set $\boldsymbol{\Omega}$ which is limited by $b_s, b_W, b_C$ constants.

## 8.4.2  Simulation Tests

Simulation of the designed control system was carried out for the motion of selected point $A$ of the WMR along the desired path in the shape of a loop consisting of five characteristic motion stages described in Sect. 8.1.2. Zero initial conditions of the controlled object, $\boldsymbol{\alpha}(0) = \mathbf{0}$, $\dot{\boldsymbol{\alpha}}(0) = \mathbf{0}$, and other data, $\boldsymbol{\Lambda} = jag[1, 1]$, $\mathbf{K}_D = diag[2, 2]$, $\gamma = 0.001$, $z_o = 0.01$, $\mathbf{F}_C = diag[1, \ldots, 1]^T$, $\mathbf{F}_W = diag[25, \ldots, 25]^T$, were applied in the simulation. The NN, shown in Fig. 8.10, applied in the simulation consists of six bipolar sigmoid neurons and is a network with functional extensions, whose first layer weights were generated by the random number generator from the interval $[-0.1, 0.1]$. Relations (8.83) and (8.84) with weight of network, shown in Fig. 8.10b, were applied in order to approximate the critic (ACE element) and actor (ASE element). The calculations were conducted in Matlab/Simulink by adopting the Euler's integration method with a discretization step of time 0.01 [s]. Obtained results, in the form of selected parameters and control, were shown in Figs. 8.16 and 8.17. Figure 8.16a presents values of overall control of the mobile robot for wheels 1 and 2, respectively. The overall control, according to the relation (8.80), consists of a PD controller (Fig. 8.16c), compensation control generated by the ASE-ACE structure (Fig. 8.16b) and a robust control (Fig. 8.16d). The consequence of application of NNs with weights of zero values in the initial phase of adaptation process, in the compensation of nonlinearities of the controlled

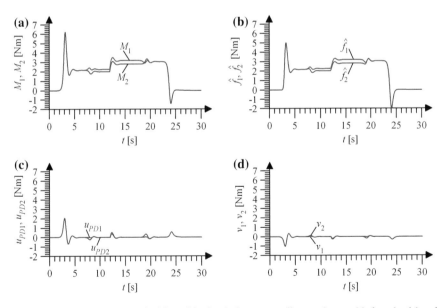

**Fig. 8.16**  Results of the control with a critic simulation: **a** overall control $u_1 = M_1$ for wheel 1 and overall control $u_2 = M_2$ for wheel 2, **b** compensation control $\hat{f}_1$ and $\hat{f}_2$, **c** PD control for wheel 1 and wheel 2, **d** robust control $v_1$ and $v_2$

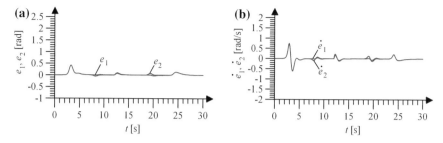

**Fig. 8.17** Tracking errors

**Table 8.10** Values of selected performance indexes for a control system

| Index | $e_{maxj}$ [rad] | $\varepsilon_j$ [rad] | $\dot{e}_{maxj}$ [rad/s] | $\dot{\varepsilon}_j$ [rad/s] | $\sigma_j$ [rad/s] |
|---|---|---|---|---|---|
| wheel 1, $j = 1$ | 0.423 | 0.07194 | 0.7559 | 0.1213 | 0.1405 |
| wheel 2, $j = 2$ | 0.423 | 0.07219 | 0.7559 | 0.1221 | 0.1413 |

**Table 8.11** Values of selected performance indexes for the assessment of the desired path execution

| Index | $d_{max}$ [m] | $\rho$ [m] |
|---|---|---|
| Value | 0.0349 | 0.008 |

object, is a changeable presence of component signals in the overall control. During the initial stage of network weights adaptation the most important role in control plays the PD controller whose importance in the overall control decreases along with the progress of actor's - ASE NN weights adaptation process.

Figures 8.17a, b presents the tracking errors $e_1$ and $e_2$ for wheel 1 and wheel 2, and the corresponding errors of tracking velocities $\dot{e}_1$ and $\dot{e}_2$. It can be noted, that all values of errors are bounded.

The performance indexes, specified in Sect. 8.1.2 were applied in order to perform the quantitative assessment of the numerical test results. The values of individual indexes are presented in Tables 8.10 and 8.11.

Changes in values of actor's (ASE) weights, which are equal to zero before the learning process starts and remain limited when the learning process progresses are shown in Figs. 8.18a, b. Critic's (ACE) NN weights are shown in Fig. 8.18c, d. They are limited according to the results of the stability analysis.

Figure 8.19 presents the desired and executed path of point $A$ of the mobile robot. Because the error is small the difference is unnoticeable. The maximum error of the desired path execution, which amounts to 0.0349 [m], is provided in Table 8.11.

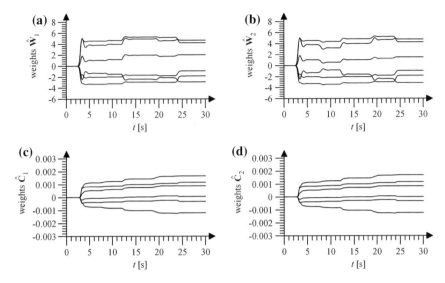

**Fig. 8.18** Neural network of actor and critic's weights: **a** ASE weights for wheel 1, **b** ASE weights for wheel 2, **c** ACE weights for wheel 1, **d** ACE weights for wheel 2

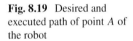

**Fig. 8.19** Desired and executed path of point $A$ of the robot

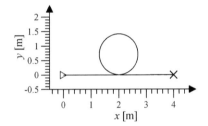

## 8.4.3   Conclusions

In this example, simulation of the proposed control algorithm of tracking motion of the two-wheeled mobile robot with reinforcement learning in the form of ASE-ACE structure implemented with the use of an artificial NNs was carried out. The applied algorithm does not require the initial learning phase, it operates on-line without any knowledge of the controlled object model. Results of the simulation confirm the correctness of the adopted assumptions and efficacy of the designed control system in the execution of tracking motion. The obtained signals are limited and the stability of the system is ensured.

## 8.5 Implementation of Critic's Adaptive Structure in Optimal Control

The chapter focuses on the implementation of an adaptive critic structure in a tracking control of the WMR. The applied method does not require knowledge of nonlinearity of the robot and the proposed solution determines the suboptimal control in the signal generated by a critic function, whose task is to compensate for nonlinearities of the robot and the variable working conditions. The proposed control system includes a critic, implemented by the NN, whose task is to approximate the value function which is dependent from the quality measure of control and a control signal. Stability of the control system was determined based on the Lyapunov stability theory.

Implementation of reinforcement learning in the form of adaptive critic structure in the control synthesis of nonlinear continuous systems is much more difficult than its implementation in the discrete systems. These issues were discussed by many authors [1, 5, 6, 14, 16, 22]. In order to discuss them we will apply theoretical considerations provided in [16]. Let the nonlinear continuous object be described by the mathematical model in the following form $\dot{\mathbf{x}}(t) = \mathbf{f}(\mathbf{x}(t)) + \mathbf{b}(\mathbf{x}(t))\mathbf{u}(\mathbf{x}(t))$, $\mathbf{x}(0) = \mathbf{x}_0$ where $\mathbf{x}(t) \in \Re^n$ is a state vector, $\mathbf{f}(\mathbf{x}(t)) \in \Re^n$, $\mathbf{b}(\mathbf{x}(t)) \in \Re^{n \times m}$ is a vector function and control matrix, respectively, $\mathbf{u}(\mathbf{x}(t)) \in \Re^m$ is a control vector. The integral form of the objective function $V(\mathbf{x}(t)) = \int_t^\infty r(\mathbf{x}(\tau), \mathbf{u}(\tau)) d\tau$ was given for the particular stabilizing control. After the differentiation of this relation we will obtain Hamilton–Jacobie–Bellman equation (HJB) $r(\mathbf{x}(t), \mathbf{u}(\mathbf{x}(t))) + (\nabla V_x)^T (\mathbf{f}(\mathbf{x}(t)) + \mathbf{b}(\mathbf{x}(t))\mathbf{u}(\mathbf{x}(t))) = 0$, $V(0) = 0$, where $\nabla V_x$ is the gradient of the objective function with respect to $\mathbf{x}$. It is easy to notice that this equation includes vector functions $\mathbf{f}(\mathbf{x})$, $\mathbf{u}(\mathbf{x})$ which are not included in the discrete version of Bellman's equation which complicates, for example, application of ADHDP procedure. In addition, Bellman's equation for discrete systems requires the determination of value of the objective function in two periods $k$ and $k + 1$. This enables the application of iterative procedures in order to determine the value of objective function. However, the value of the objective function occurs only once in the HJB equation and that is why the implementation of the iterative procedures in determining its value is difficult. A practical approach to solve the problem of continuous dynamic systems control results in the application of approximation of derivative of the value function or transformations of partial derivatives of the value function. In this subchapter, the synthesis of adaptive critic in the tracking control of the WMR was conducted on the basis of theoretical considerations provided in [6, 21].

### 8.5.1 Control Synthesis, Critic's Learning Algorithm, Stability of a System

Similarly to the previous sections of this chapter, a WMR is the controlled object. Its dynamic equations of motion, taking into account Eq. (8.1), will be written in the state space as

$$\dot{\mathbf{x}}_1 = \mathbf{x}_2 \ , \tag{8.93}$$
$$\dot{\mathbf{x}}_2 = \mathbf{g}_n \left( \mathbf{x} \right) + \mathbf{u} + \mathbf{z} \ ,$$

where $\mathbf{x}_1 = [\alpha_1, \alpha_2]^T$, $\mathbf{x}_2 = [\dot{\alpha}_1, \dot{\alpha}_2]^T$, $\mathbf{u} = \mathbf{M}^{-1}\boldsymbol{\tau} \in \mathfrak{R}^2$, $\mathbf{z} = -\mathbf{M}^{-1}\boldsymbol{\tau}_d \in \mathfrak{R}^2$, and the nonlinear function is in the following form

$$\mathbf{g}_n \left( \mathbf{x} \right) = -\mathbf{M}^{-1} \left( \mathbf{C} \left( \mathbf{x}_2 \right) \mathbf{x}_2 + \mathbf{F} \left( \mathbf{x}_2 \right) \right) \in \mathfrak{R}^2 \ , \tag{8.94}$$

where all symbols are marked as presented in Sect. 8.1. It was assumed, that the matrix of inertia of the WMR is known and the moments of force of drive wheels of the WMR will be determined from relation $\boldsymbol{\tau} = \mathbf{M}\mathbf{u}$. Let's define the tracking error $\mathbf{e}$, the filtered tracking error $\mathbf{s}$ interpreted as a measure of control quality in the form of

$$\mathbf{e} \left( t \right) = \mathbf{x}_d \left( t \right) - \mathbf{x} \left( t \right) \ , \tag{8.95}$$

$$\mathbf{s} \left( t \right) = \dot{\mathbf{e}} \left( t \right) + \boldsymbol{\Lambda}\mathbf{e} \left( t \right) \ , \tag{8.96}$$

where $\boldsymbol{\Lambda}$ is a positive-definite diagonal matrix. Then equation (8.93) can be transformed into the following form

$$\dot{\mathbf{s}} = \mathbf{f} \left( \mathbf{x}, \mathbf{x}_d \right) + \mathbf{u} + \mathbf{z} \ , \tag{8.97}$$

where $\mathbf{f} \left( \mathbf{x}, \mathbf{x}_d \right)$ includes the nonlinearities of control object (8.94), $\mathbf{u} = \mathbf{u}_C - \mathbf{u}_S$ is a control signal comprising of a control signal with critic $\mathbf{u}_C$ and supervisory control $\mathbf{u}_S$, which is determined in the further part of this section and whose goal is to compensate for the disturbances and ensuring stability of the closed system. Let's define the optimal value of value function in the form of [16]

$$V^o \left( \mathbf{s} \left( t \right) \right) = \min_{u(\tau)} \int_t^{\infty} r \left( \mathbf{s} \left( \varsigma \right), \mathbf{u}_C \left( \mathbf{s} \left( \tau \right) \right) \right) d\varsigma \ , \tag{8.98}$$

where

$$r \left( \mathbf{s}, \mathbf{u}_C \right) = \mathbf{s}^T \mathbf{Q}\mathbf{s} + \mathbf{u}_C^T \mathbf{R}\mathbf{u}_C \ , \tag{8.99}$$

is a local cost with positive-definite symmetrical matrices $\mathbf{Q} \in \mathfrak{R}^{2 \times 2}$, $\mathbf{R} \in \mathfrak{R}^{2 \times 2}$. The optimal control is defined in the form of

$$\mathbf{u}_C^o \left( \mathbf{s} \right) = -\frac{1}{2}\mathbf{R}^{-1}\frac{\partial V^o \left( \mathbf{s} \right)^T}{\partial \mathbf{s}} \ , \tag{8.100}$$

assuming a continuous differentiability of value function $V^o \left( \mathbf{s} \right)$ with condition $V^o \left( \mathbf{0} \right) = 0$. For an object described by an equation (8.97) the Hamiltonian is formulated as

$$H \left( \mathbf{s}, \mathbf{u}_C, V_s \right) = V_s \left( \mathbf{f} \left( \mathbf{x}, \mathbf{x}_d \right) + \mathbf{u}_C \right) + r \left( \mathbf{s}, \mathbf{u}_C \right) \ , \tag{8.101}$$

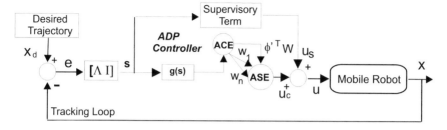

**Fig. 8.20** Tracking control system of the wheeled mobile robot with the use of adaptive critic's structure

where $V_s = \frac{\partial V}{\partial s} \in \Re^{1\times 2}$ is a gradient of the value function. By introducing optimal value function (8.98) and optimal control (8.100) into the Hamiltonian equation (8.101), we receive.

$$H\left(\mathbf{s}, \mathbf{u}_C^o, V_s^o\right) = V_s^o\left(\mathbf{f}\left(\mathbf{x}, \mathbf{x}_d\right) + \mathbf{u}_C^o\right) + r\left(\mathbf{s}, \mathbf{u}_C^o\right) . \qquad (8.102)$$

If we will replace the optimal solutions of Eq. (8.101) with their approximation, then Eq. (8.101) will be formulated as

$$\hat{H}\left(\mathbf{s}, \hat{\mathbf{u}}_C, \hat{V}_s\right) = \hat{V}_s\left(\mathbf{f}\left(\mathbf{x}, \mathbf{x}_d\right) + \hat{\mathbf{u}}_C\right) + r\left(\mathbf{s}, \hat{\mathbf{u}}_C\right) . \qquad (8.103)$$

Hamiltonian, for the optimal solutions of control and the value function, is equal to zero, so when defining Bellman's error as Eqs. (8.102) and (8.103) difference, we will receive

$$E_{HJB} = \hat{V}_s\left(\mathbf{f}\left(\mathbf{x}, \mathbf{x}_d\right) + \hat{\mathbf{u}}_C\right) + r\left(\mathbf{s}, \hat{\mathbf{u}}_C\right) . \qquad (8.104)$$

The presented approach to tracking control of the WMR is presented in Fig. 8.20.

By adopting the control structure with adaptive critic and on the basis of Eq. (8.104), we will determine the approximation of the optimal control and value function. The optimal value of value function occurring in Eq. (8.98) and the optimal control signal may be determined by the NN in the form of

$$V^o\left(\mathbf{s}\right) = \mathbf{W}^T \boldsymbol{\varphi}\left(\mathbf{s}\right) + \varepsilon\left(\mathbf{s}\right) , \qquad (8.105)$$

$$\mathbf{u}_C^o\left(\mathbf{s}\right) = -\frac{1}{2}\mathbf{R}^{-1}\left(\dot{\boldsymbol{\varphi}}\left(\mathbf{s}\right)^T \mathbf{W} + \varepsilon\left(\mathbf{s}\right)\right) , \qquad (8.106)$$

where $\mathbf{W} \in \Re^N$ are unknown ideal weights, $N$ is a number of network's neurons, functions describing neurons $\boldsymbol{\varphi}\left(\mathbf{s}\right) = \left[\varphi_1\left(\mathbf{s}\right), \varphi_2\left(\mathbf{s}\right), \ldots, \varphi_N\left(\mathbf{s}\right)\right]^T \in \Re^N$ and $\dot{\boldsymbol{\varphi}}\left(\mathbf{s}\right) = \frac{\partial \boldsymbol{\varphi}}{\partial s} \in \Re^{N\times 2}$ should satisfy the following relations $\varphi_i\left(0\right) = 0$, $\dot{\varphi}_i\left(0\right) = 0$, $\forall i = 1, 2, \ldots, N$ and $\varepsilon\left(\mathbf{s}\right) \in \Re$ is a mapping error. By applying the Weierstrass approximation theorem it was assumed that the value function, its derivative and signal of an optimal control can be uniformly approximated by the NN (8.105), which means that when $N \to \infty$ then the approximation error $\varepsilon$ (s) $\to 0$. Taking into account the above mentioned the estimate of value function will be formulated as follows

$$\hat{V} (s) = \hat{W}^T \varphi (s) , \qquad (8.107)$$

where $\hat{W} \in \Re^N$ weights are an estimates of network's ideal weights. Since the optimal control signal (8.106) is determined in the gradient function of value function, the second parametric structure, which approximates the control signal, is not necessary. Therefore, the approximation of control signal was determined from the following equation

$$\hat{u}_C (s) = -\frac{1}{2} R^{-1} \left( \dot{\varphi} (s)^T \hat{W} \right) , \qquad (8.108)$$

Such a problem is called an optimal control with a critic [14]. The iterative algorithm of the least squares method was applied in order to adapt the NN weights approximating the value function. Implementation of this approach is justified by the fact that the equation describing Bellman error (8.104) is a linear function of the critic's weights.

The following relation was applied in order to determine the learning law of the critic's network weights

$$E_C = \int_0^t (E_{HJB} (\varsigma))^2 \, d\varsigma . \qquad (8.109)$$

By determining the error gradient (8.109) and comparing it to zero, we received

$$\frac{\partial E_C}{\partial \hat{W}} = 2 \int_0^t E_{HJB} (\varsigma) \frac{\partial E_{HJB} (\varsigma)}{\partial \hat{W}} d\varsigma = 0 . \qquad (8.110)$$

The iterative version of the least squares normalized method [19], in which the assessment of network weights were determined from the following relation, was applied for the implementation of the adopted solution in real time (on-line)

$$\dot{\hat{W}} = \eta G \frac{\delta}{1 + v \delta^T G \delta} E_{HJB} , \qquad (8.111)$$

where $\eta, v \in \Re$ are a positive coefficients, $\delta = \dot{\varphi} \left( f (x, x_d) + \hat{u}_c \right)$, and $G = \left( \int_0^t \delta (\varsigma) \delta (\varsigma)^T d\varsigma \right)^{-1} \in \Re^{N \times N}$ is a symmetric matrix determined from the equation

$$\dot{G} = \eta G \frac{\delta \delta^T}{1 + v \delta^T G \delta} G , \qquad (8.112)$$

where $G (0) = \alpha_c I$, $\alpha_c \gg 0$ and which satisfies the condition of good determination $\dot{G} \geq 0$.

The designed control algorithm with a critic must comply with the condition of closed system stability. This means that the control signal $\mathbf{u}$ must be designed in such a way to ensure stability of the closed control system, so the filtered tracking error $\mathbf{s}$ must be constrained i.e.

$$|s_i| \leq \Phi_i , \quad \forall t > 0 , \tag{8.113}$$

where $\Phi_i$, $i = 1, 2$, assumes constant values. For this purpose, the determined control signal $\hat{\mathbf{u}}_C$ was enriched with the supervisory control $\mathbf{u}_S$ [23] generated by the additional supervisory term.

For this particular problem let's introduce the following control law

$$\mathbf{u} = -\hat{\mathbf{u}}_C - \mathbf{u}_S . \tag{8.114}$$

By substituting (8.114) into (8.97) we receive a description of the closed system in the form of

$$\dot{\mathbf{s}} = \mathbf{f}(\mathbf{x}, \mathbf{x}_d) - \hat{\mathbf{u}}_C - \mathbf{u}_S + \mathbf{z} . \tag{8.115}$$

Applying additional signal

$$\mathbf{u}^* = \mathbf{f}(\mathbf{x}, \mathbf{x}_d) + \mathbf{Ks} , \tag{8.116}$$

relation (8.115) will be formulated as

$$\dot{\mathbf{s}} = \mathbf{u}^* - \mathbf{Ks} - \hat{\mathbf{u}}_C - \mathbf{u}_S + \mathbf{z} . \tag{8.117}$$

Let's define the positive-definite function

$$L = \frac{1}{2}\mathbf{s}^T \mathbf{s} . \tag{8.118}$$

By determining the derivative of function (8.118) and applying relation (8.117), we receive

$$\dot{L} = \mathbf{s}^T \dot{\mathbf{s}} = -\mathbf{s}^T \mathbf{Ks} + \mathbf{s}^T \left[ \mathbf{u}^* - \hat{\mathbf{u}}_C + \mathbf{z} \right] - \mathbf{s}^T \mathbf{u}_S . \tag{8.119}$$

Derivative (8.119) is in the following form

$$\dot{L} \leq -\mathbf{s}^T \mathbf{Ks} + \sum_{i=1}^{2} |s_i| \left[ |f_i(\mathbf{x}, \mathbf{x}_d) - \hat{u}_{Ci} + k_i s_i + z_i| \right] - \sum_{i=1}^{2} s_i u_{Si} . \tag{8.120}$$

If we will select a supervisory control signal in the following form

$$u_{Si} = [F_{0i} + |k_i s_i| + Z_i + \eta_i] \, \mathrm{sgn} s_i , \tag{8.121}$$

with the following assumption

$$|f_i(\mathbf{x}, \mathbf{x}_d) - \hat{f}_i(\mathbf{x}, \mathbf{x}_d)| \leq F_{0i}, \quad |z_i| \leq Z_i, \tag{8.122}$$

when $i = 1, 2$, $F_{0i} > 0$, $\eta_i > 0$, we will finally receive

$$\dot{L} \leq -\mathbf{s}^T \mathbf{K} \mathbf{s} - \sum_{i=1}^{2} \eta_i |s_i|. \tag{8.123}$$

The derivative of Lyapunov function is negative semidefinite, hence the closed system (8.117) is stable in relation to the state vector $\mathbf{s}$. The control signal (8.121) includes a $\mathrm{sgn}s_i$ function, which activates the high frequency of switching of the actuator. In order to eliminate the high frequency of switching of the actuator, we introduce the approximation of $\mathrm{sgn}(s_i)$ function in its neighbourhood $\Phi_i$, by the function [20]

$$\mathrm{sat}(s_i) = \begin{cases} \mathrm{sgn}(s_i) & for \ |s_i| > \Phi_i, \\ \frac{s_i}{\Phi_i} & for \ |s_i| \leq \Phi_i. \end{cases} \tag{8.124}$$

If $\mathbf{e}(0) = \mathbf{x}_d(0) - \mathbf{x}(0) = \mathbf{0}$, then the accuracy of control is ensured in a finite time

$$\forall t \geq 0, \ |s_i(t)| \leq \Phi_i \ \Rightarrow \ \forall t \geq 0, \ |e_i^{(i)}(t)| \leq (2\lambda_i)^i \varepsilon_i, \tag{8.125}$$

where $i = 0, 1, \ldots, n_i - 1$, $\varepsilon_i = \Phi_i / \lambda_i^{n_i-1}$, $n_i = 2$ is the rank ot the $i$-th system and the superscript $(i)$ is an error derivative.

### 8.5.2  Simulation Tests

Similarly to the previous sections of this chapter, the simulation of the designed control system was carried out for the motion of selected point $A$ of the WMR along the desired path in the shape of a loop consisting of five characteristic stages of motion as described in Sect. 8.1.2. For time $t \geq 12$ [s] the following parametric disturbance $\mathbf{a} + \Delta\mathbf{a}$ occurs where $\Delta\mathbf{a} = [0, 0, 0, 0, 1, 1]^T$, associated with a change of robot's resistances to motion. Zero initial conditions of the controlled object $\boldsymbol{\alpha}(0) = \mathbf{0}$, $\dot{\boldsymbol{\alpha}}(0) = \mathbf{0}$ and other data $\mathbf{Q} = \mathrm{diag}[2000, 2000]$, $\mathbf{R} = \mathrm{diag}[0.002, 0.002]$, $\boldsymbol{\Lambda} = \mathrm{diag}[1, 1]$, $\mathbf{K}_D = \mathrm{diag}[2, 2]$, $\eta = 0.001$, $\nu = 0.1$, $\alpha_c = 10$, $\mathbf{Z} = [0.01, 0.01]^T$ were applied in the simulation. Since the value function, in the analyzed case, is a quadratic function, therefore the basic functions of neurons $\varphi_i(\mathbf{s})$ were assumed as a quadratic functions of state vector $\mathbf{s} \otimes \mathbf{s}$ elements, where $\otimes$ is a Kronecker product reduced to dimension 3, i.e. $\boldsymbol{\varphi}(\mathbf{s}) = [s_1^2, s_1 s_2, s_2^2]^T$. The obtained results, in the form of values of control signals and obtained errors of kinematic parameters, are shown, respectively, in Figs. 8.21 and 8.22. Figure 8.21a presents values of overall control of the

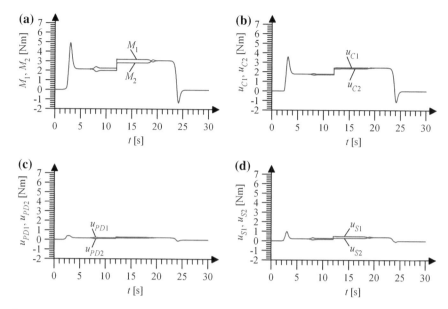

**Fig. 8.21** Results of the control with a critic simulation: **a** overall control $u_1$ for wheel 1 and overall control $u_2$ for wheel 2, **b** compensation control $u_{C1}$ and $u_{C2}$, **c** PD control for wheel 1 and wheel 2, **d** supervisory control $u_{S1}$ and $u_{S2}$

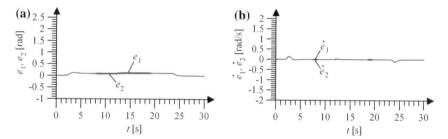

**Fig. 8.22** Tracking errors

WMR, respectively, for wheels 1 and 2. The overall control consists of the PD control Fig. (8.21c), compensation control $\mathbf{u}_C$ (Fig. 8.21b) and the supervisory control (Fig. 8.21). In the initial phase of the control process, due to the zero initial conditions of network weights (Fig. 8.23a), the main role in control plays the supervisory control term (Fig. 8.21d), which consists of the PD controller (Fig. 8.21c). Presence of the PD control signals in the overall control decreases within the control progress which minimizes the value function and whose task is to compensate for the nonlinearities and changeable working conditions of the robot.

Figure 8.22a, b presents tracking errors, $e_1$ for wheel 1 and $e_2$ for wheel 2, and relevant errors of tracking velocity $\dot{e}_1$, $\dot{e}_2$. It can be noticed, that the values of errors

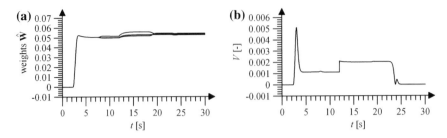

**Fig. 8.23**  **a** Critic's weights, **b** values of the value function

**Table 8.12**  Values of selected performance indexes for control system

| Index | $e_{maxj}$ [rad] | $\varepsilon_j$ [rad] | $\dot{e}_{maxj}$ [rad/s] | $\dot{\varepsilon}_j$ [rad/s] | $\sigma_j$ [rad/s] |
|---|---|---|---|---|---|
| wheel 1, $j = 1$ | 0.1319 | 0.0953 | 0.1435 | 0.027 | 0.099 |
| wheel 2, $j = 2$ | 0.122 | 0.0778 | 0.1435 | 0.0268 | 0.0824 |

**Table 8.13**  Values of selected performance indexes for the assessment of the desired path execution

| Index | $d_{max}$ [m] | $\rho$ [m] |
|---|---|---|
| Value | 0.01 | 0.0061 |

are constrained and according to the assumed indexes their values are provided in Tables 8.12 and 8.13.

The performance indexes, listed in Sect. 8.1.2, were applied in order to perform the quantitative assessment of the numerical test results.

The changes in values of the critic's weights, which before the learning process begins are equal to zero and during the progress of the learning process remain limited, are shown in Fig. 8.23a. Figure 8.23b, on the other hand, presents values of the value function.

Figure 8.24b presents values of desired and executed path of point $A$ of the WMR. The maximum error of the desired path execution amounts to 0.01.[m] and is provided in Table 8.13. Errors $e_1$ and $\dot{e}_1$, amount to $|e_1(t)| \leq 0.8$ and $|\dot{e}_1(t)| \leq 1.6$, were calculated by applying relation (8.125), and their values are shown in Fig. 8.24a on the phase plane. By comparing those values with sequences presented in Fig. 8.22 we can conclude that these are conservative calculation.

Figure 8.24b presents the values of desired and executed path of point $A$ of the mobile robot. Because the error is small, the difference in the path values is unnoticeable. The maximum error of the desired path execution, which amounts to 0.01.[m], is provided in Table 8.13.

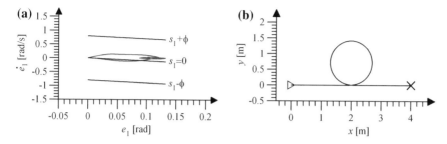

**Fig. 8.24** **a** Phase plane of errors of wheel 1, **b** desired and executed path of point $A$ of the robot

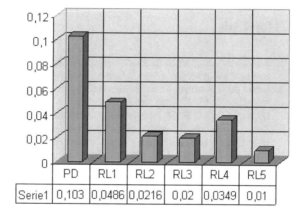

**Fig. 8.25** Values of maximum errors $d_{max}$ $(i)$ [m] during execution of desired path

| | PD | RL1 | RL2 | RL3 | RL4 | RL5 |
|---|---|---|---|---|---|---|
| Serie1 | 0,103 | 0,0486 | 0,0216 | 0,02 | 0,0349 | 0,01 |

## 8.5.3 Conclusions

For the synthesis of tracking control of the WMR the optimal control with an adaptive critic was applied in this subchapter. The linear NN due to the output weights was applied in order to approximate the value function. In the applied approach procedures for determining the values of network's initial weights and network's pre-learning are not required. Approximation of value function is performed in real time and on the basis of NN's weights adaptive learning. Since the optimal control is determined on the basis of the optimal value function gradient the additional NN, which approximates the actor, is not required. The applied solution enables obtaining a small error of the desired path execution which maximum value amounts to 0.01 [m]. The asymptotic stability of control system was demonstrated by applying Lyapunov's stability theory due to the filtered tracking error and preservation of the control signals and NN's weights limitation.

In summary, Fig. 8.25 presents values of maximum errors, $d_{max}$, of execution of the desired path of a selected point $A$ of the WMR for methods implemented in Sects. 8.1–8.5.

Results obtained in the algorithms of the reinforcement learning RL1, where 1 marks the point of the chapter in which the individual method was analyzed, were presented on the horizontal axis. The biggest error was obtained during the implementation of a PD control, which was analyzed in subchapter 8.1. The best solution was obtained for the structure of an adaptive critic in optimal control (RL5). The maximum errors obtained in methods provided in Sects. 8.2 and 8.3 (RL2, RL3) are very similar and relatively small. However, bigger errors were obtained in classical reinforcement learning (RL1) and actor-critic reinforcement learning (RL4) procedures. It should be noted, that the assessment of the individual methods should be verified during the implementation of discussed solutions on a real object, in verification tests.

# References

1. Abu-Khalaf, M., Lewis, F.: Nearly optimal control laws for nonlinear systems with saturating actuators using a neural network HJB approach. Automatica **41**, 779–791 (2005)
2. Barto, A.G., Anandan, P.: Pattern-recognizing stochastic learning automata. IEEE Trans. Syst. Man Cybern. **15**, 360–375 (1985)
3. Barto, A., Sutton, R.: Reinforcement learning: an introduction. MIT Press, Cambridge (1998)
4. Barto, A., Sutton, R., Anderson, C.: Neuronlike adaptive elements that can solve difficult learning problems. IEEE Trans. Syst. Man Cybern. Syst. **13**, 834–846 (1983)
5. Beard, R.W., Saridis, G.N., Wen, J.T.: Galerkin approximations of the generalized Hamilton–Jacobi–Bellman equation. Automatica **33**, 2159–2178 (1997)
6. Bhasin, S., Kamalapurkar, R., Johnson, M., Vamvoudakis, K.G., Lewis, F.L., Dixon, W.E.: A novel actorcriticidentifier architecture for approximate optimal control of uncertain nonlinear systems. Automatica **49**, 82–92 (2013)
7. Doya, K.: Reinforcement learning in continuous time and space. Neural Comput. **12**, 219–245 (2000)
8. Hendzel, Z.: Fuzzy reactive control of wheeled mobile robot. J. Theor. Appl. Mech. **42**, 503–517 (2004)
9. Hendzel, Z.: Adaptive critic neural networks for identification of wheeled mobile robot. Lect. Notes Artif. Int. **4029**, 778–786 (2006)
10. Hendzel, Z.: An adaptive critic neural network for motion control of a wheeled mobile robot. Nonlinear Dyn. **50**, 849–855 (2007)
11. Hendzel, Z.: Control of wheeled mobile robot using approximate dynamic programming. In: Proceedings of Conference on Dynamical Systems Theory and Applications, pp. 735–742 (2007)
12. Hendzel, Z., Szuster, M.: Approximate dynamic programming in robust tracking control of wheeled mobile robot. Arch. Mech. Eng. **56**, 223–236 (2009)
13. Kim, Y.H., Lewis, F.L.: High-Level Feedback Control with Neural Networks. World Scientific Publishing, River Edge (1998)
14. Lewis, F.L., Campos, J., Selmic, R.: Neuro-Fuzzy Control of Industrial Systems with Actuator Nonlinearities. Society for Industrial and Applied Mathematics, Philadelphia (2002)
15. Lewis, F.L., Yesildirek, A., Jagannathan, S.: Neural Network Control of Robot Manipulators and Nonlinear Systems. Taylor and Francis, Bristol (1999)
16. Lewis, F.L., Vrabie, D., Syroms, V.L.: Optimal Control, 3rd edn. Wiley and Sons, New Jersey (2012)
17. Lin, C.-K.: A reinforcement learning adaptive fuzzy controller for robots. Fuzzy Set. Syst. **137**, 339–352 (2003)
18. Prokhorov, D., Wunch, D.: Adaptive critic designs. IEEE Trans. Neural Netw. **8**, 997–1007 (1997)

19. Sastry, S., Bodson, M.: Adaptive control: stability, convergence, and robustness. Prentice-Hall, Englewood Cliffs (1989)
20. Slotine, J.J., Li, W.: Applied Nonlinear Control. Prentice Hall, New Jersey (1991)
21. Vamvoudakis, K.G., Lewis, F.L.: Online actor-critic algorithm to solve the continuous-time infinite horizon optimal control problem. Automatica **46**, 878–888 (2010)
22. Vrabie, D., Lewis, F.: Neural network approach to continuous-time direct adaptive optimal control for partially unknown nonlinear systems. Neural Netw. **22**, 237–246 (2009)
23. Wang, L.: A Course in Fuzzy Systems and Control. Prentice-Hall, New York (1997)

# Chapter 9
# Two-Person Zero-Sum Differential Games and H∞ Control

As an extension of the topics discussed in Chap. 8, here we are going to apply the differential game theory as an element of decentralised control, widely used in complex mechatronic systems. We will discuss neuro-dynamic programming in the context of the solution of a 2-person zero-sum game, using the Nash saddle point theory [8]. The two-person zero-sum differential game theory and H-infinity control results from the solution of the Hamilton–Jacobi–Isaac equation (HJI) which is a generalisation of the Hamilton–Jacobi–Bellman equation, and these topics are present in the optimal dynamic object control theory [1–4, 6, 7, 12]. The topic of $H_\infty$ control was presented at first in [14] and it concerns designing a control system in the complex space which would be little sensitive to disturbance. In the domain of real variables, the problem of $H_\infty$ control is of the minimax type, i.e. it can be solved by providing a solution of a two-person zero-sum differential game, in which a control system minimising the cost function in the presence of the most unfavourable disturbance is designed [1, 5, 11]. In order to solve the two-person zero-sum differential game occurring in optimal $H_\infty$ control, it is necessary to determine control in a feedback system resulting from the HJI equation. The HJI equation is difficult to solve or its solutions do not exist in the case of nonlinear control systems. In such a case, approximate solutions of the HJI equation are looked for. We will apply the issues described in Sects. 9.1 and 9.2 of this Chapter in Sect. 9.3 to control the drive unit of a mobile robot, and in Sect. 9.4 in a WMR tracking control.

## 9.1 H∞ control

Let us consider a nonlinear, stationary, and linear - in terms of control and disturbance - dynamic object in the following form

© Springer International Publishing AG 2018
M. Szuster and Z. Hendzel, *Intelligent Optimal Adaptive Control for Mechatronic Systems*, Studies in Systems, Decision and Control 120, https://doi.org/10.1007/978-3-319-68826-8_9

$$\dot{\mathbf{x}} = \mathbf{f}(\mathbf{x}) + \mathbf{g}(\mathbf{x})\mathbf{u}(\mathbf{x}) + \mathbf{k}(\mathbf{x})\mathbf{z}(\mathbf{x}) \ ,$$
$$\mathbf{y} = \mathbf{h}(\mathbf{x}) \ , \tag{9.1}$$

where $\mathbf{x}(t) \in \mathfrak{R}^n$ is the state of the object, $\mathbf{y}(t) \in \mathfrak{R}^s$ is the object output, $\mathbf{u}(\mathbf{x}) \in \mathfrak{R}^m$ is the control signal, $\mathbf{z}(\mathbf{x}) \in \mathfrak{R}^p$ is the disturbance signal, and the functions $\mathbf{f}(\mathbf{x}) \in \mathfrak{R}^n$, $\mathbf{g}(\mathbf{x}) \in \mathfrak{R}^{n \times m}$, $\mathbf{k}(\mathbf{x}) \in \mathfrak{R}^{n \times p}$, $\mathbf{h}(\mathbf{x}) \in \mathfrak{R}^{s \times n}$ are smooth. Let us also assume that the function $\mathbf{f}(\mathbf{x})$ satisfies locally the Lipschitz condition and that $\mathbf{f}(0) = \mathbf{0}$, which means that $\mathbf{x} = \mathbf{0}$ is the object's equilibrium.

On the basis of [11] we can say that the controlled object (9.1) has a gain of $L_2$ type less than or equal to $\gamma$, if the following inequality

$$\int_0^\infty \left( \mathbf{h}^T\mathbf{h} + \mathbf{u}^T\mathbf{R}\mathbf{u} \right) dt \leq \gamma^2 \int_0^\infty \|\mathbf{z}\|^2 dt \ , \tag{9.2}$$

is satisfied for every $\mathbf{z} \in L_2[0, \infty)$ and a positive $\gamma$, where $\mathbf{R} = \mathbf{R}^T > \mathbf{0}$. In H∞ control, it is necessary to determine the lowest $\gamma^* > 0$ such that for any $\gamma$, the inequality $\gamma > \gamma^*$ is true, and the $L_2$ type gain has a solution. In H∞ control, the performance index shall be defined as

$$J(\mathbf{x}(0), \mathbf{u}, \mathbf{z}) = \int_0^\infty \left( \mathbf{x}^T\mathbf{Q}\mathbf{x} + \mathbf{u}^T\mathbf{R}\mathbf{u} - \gamma^2\|\mathbf{z}\|^2 \right) dt \equiv \int_0^\infty r(\mathbf{x}, \mathbf{u}, \mathbf{z}) dt \ , \tag{9.3}$$

where $\mathbf{Q} \in \mathfrak{R}^{n \times n} > \mathbf{0}$ and $r$ is the local control cost. Let us define, for fixed control and disturbance signals, the value function

$$V(\mathbf{x}(0), \mathbf{u}, \mathbf{z}) = \int_0^\infty \left( \mathbf{x}^T\mathbf{Q}\mathbf{x} + \mathbf{u}^T\mathbf{R}\mathbf{u} - \gamma^2\|\mathbf{z}\|^2 \right) dt \ . \tag{9.4}$$

Based on the Leibniz rule, the relation (9.4) has an equivalent form describing a two-person zero-sum differential game as a Bellman equation [7]

$$0 = r(\mathbf{x}, \mathbf{u}, \mathbf{z}) + (\nabla V)^T \left( \mathbf{f}(\mathbf{x}) + \mathbf{g}(\mathbf{x})\mathbf{u}(\mathbf{x}) + \mathbf{k}(\mathbf{x})\mathbf{z}(\mathbf{x}) \right) \ , \tag{9.5}$$

where $V(0) = 0$, $\nabla V = \partial V / \partial \mathbf{x} \in \mathfrak{R}^n$ is a gradient. For a problem formulated in such a way, the Hamiltonian will be as follows

$$H(\mathbf{x}, \nabla V, \mathbf{u}, \mathbf{z}) = r(\mathbf{x}, \mathbf{u}, \mathbf{z}) + (\nabla V)^T \left( \mathbf{f}(\mathbf{x}) + \mathbf{g}(\mathbf{x})\mathbf{u}(\mathbf{x}) + \mathbf{k}(\mathbf{x})\mathbf{z}(\mathbf{x}) \right) \ . \tag{9.6}$$

In optimal control, it is necessary so that $\mathbf{u}(\mathbf{x})$ control stabilises the controlled object (9.1) and ensures a finite value of the function (9.4), which means that it has to be an admissible control [1, 6].

## 9.2 A Two-Person Zero-Sum Differential Game

According to the differential games theory, to determine control from the state for which $L_2$ gain is less than or equal to $\gamma$, it is sufficient to solve a two-person zero-sum differential game. In this case, the task of $\mathbf{u}(\mathbf{x})$ control is to minimise the value function (9.4), while the other player (disturbance) aims to generate the most unfavourable disturbance scenario in the process of value function minimisation. As it was said in Sect. 9.1, determination of the $L_2$ type gain is equivalent to solving a zero-sum differential game [1, 2, 12]

$$V^*(\mathbf{x}(0)) = \min_{\mathbf{u}} \max_{\mathbf{z}} \int_0^\infty \left( \mathbf{x}^T \mathbf{Q}\mathbf{x} + \mathbf{u}^T \mathbf{R}\mathbf{u} - \gamma^2 \|\mathbf{z}\|^2 \right) dt \ . \tag{9.7}$$

An unambiguous solution to this problem exists if there exists a saddle point $(\mathbf{u}^*, \mathbf{z}^*)$ of the differential game, that is if the Nash condition is fulfilled [4, 8, 11]

$$J\left(\mathbf{x}(0), \mathbf{u}^*, \mathbf{z}\right) \leq J\left(\mathbf{x}(0), \mathbf{u}^*, \mathbf{z}^*\right) \leq J\left(\mathbf{x}(0), \mathbf{u}, \mathbf{z}^*\right) \ . \tag{9.8}$$

On the basis of the Bellman optimality equation [6, 7]

$$\min_{\mathbf{u}} \max_{\mathbf{z}} \left[ H\left(\mathbf{x}, \nabla V^*, \mathbf{u}, \mathbf{z}\right) \right] = 0 \ , \tag{9.9}$$

the optimal control and the most unfavourable disturbance are determined as follows

$$\mathbf{u}^* = -\frac{1}{2}\mathbf{R}^{-1}\mathbf{g}(\mathbf{x})^T \nabla V^* \ , \tag{9.10}$$

$$\mathbf{z}^* = \frac{1}{2\gamma^2}\mathbf{k}(\mathbf{x})^T \nabla V^* \ . \tag{9.11}$$

By introducing relations (9.10), (9.11) to the Eq. (9.5), we obtain a HJI equation in the following form

$$H(\mathbf{x}, \nabla V^*, \mathbf{u}, \mathbf{z}) = \mathbf{x}^T \mathbf{Q}\mathbf{x} + \nabla V^{*T}\mathbf{f}(\mathbf{x}) - \tfrac{1}{4}\nabla V^{*T}\mathbf{g}(\mathbf{x})\,\mathbf{R}^{-1}\mathbf{g}(\mathbf{x})^T \nabla V^* + \\ + \tfrac{1}{4\gamma^2}\nabla V^{*T}\mathbf{k}(\mathbf{x})\,\mathbf{k}(\mathbf{x})^T \nabla V^* = 0 \ . \tag{9.12}$$

In order to solve a two-person differential game, it is necessary to determine the value function $V^*(\mathbf{x})$ on the basis of the Eq. (9.12), and then introduce the function to Eqs. (9.10) and (9.11). As it can be seen, determination of an optimal control on the basis of the HJI equation is difficult. Therefore, optimal control has to be approximated. In order to illustrate the discussed theory of $H_\infty$ control and two-person zero-sum differential game, in the next section of this Chapter we will solve a linear problem on the basis of the control of the drive unit of a WMR on the basis of an approximate solution of the HJI equation.

## 9.3  Application of a Two-Person Zero-Sum Differential Game in Control of the Drive Unit of a WMR

In this section, to easily verify the generalised Riccati equation, we will apply the considerations contained in Sects. 9.1 and 9.2 to control of a linear object - the drive unit of a WMR. In a general case, dynamic equations of motion of a linear object we write down as follows

$$\dot{\mathbf{x}} = \mathbf{A}\mathbf{x} + \mathbf{B}\mathbf{u} + \mathbf{E}\mathbf{z} ,$$
$$\mathbf{y} = \mathbf{C}\mathbf{x} ,$$

(9.13)

where the matrices and vectors are of relevant dimension. As the value function, we assume the relation (9.4)

$$V\left(\mathbf{x}\left(t\right), \mathbf{u}, \mathbf{z}\right) = \frac{1}{2} \int_{t}^{\infty} \left(\mathbf{x}^{T}\mathbf{Q}\mathbf{x} + \mathbf{u}^{T}\mathbf{R}\mathbf{u} - \gamma^{2}\|\mathbf{z}\|^{2}\right) dt \equiv \int_{t}^{\infty} r\left(\mathbf{x}, \mathbf{u}, \mathbf{z}\right) dt .$$

(9.14)

As we know, the value function in the case of a linear problem is a quadratic form of the state of the analysed object [6, 7, 12], written as follows

$$V\left(\mathbf{x}\right) = \frac{1}{2}\mathbf{x}^{T}\mathbf{P}\mathbf{x} ,$$

(9.15)

where $\mathbf{P} > \mathbf{0}$ is a positive-definite matrix. We select the control and disturbance signals in a linear form from the state

$$\mathbf{u}\left(\mathbf{x}\right) = -\mathbf{L}_{s}\mathbf{x} ,$$

(9.16)

$$\mathbf{z}\left(\mathbf{x}\right) = \mathbf{L}_{z}\mathbf{x} .$$

(9.17)

By introducing the above discussed relations to the Eq. (9.5) we arrive at the Lyapunov equation for the matrix $\mathbf{P}$ in the function of control and disturbance

$$\mathbf{P}\left(\mathbf{A} - \mathbf{B}\mathbf{L}_{s} + \mathbf{E}\mathbf{L}_{z}\right) + \left(\mathbf{A} - \mathbf{B}\mathbf{L}_{s} + \mathbf{E}\mathbf{L}_{z}\right)^{T}\mathbf{P} + \mathbf{Q} + \mathbf{L}_{s}\mathbf{R}\mathbf{L}_{s} - \gamma^{2}\mathbf{L}_{z}^{T}\mathbf{L}_{z} = 0 .$$

(9.18)

By determining the stationary point of the Eq. (9.18) we will obtain

$$\mathbf{u}\left(\mathbf{x}\right) = -\mathbf{R}^{-1}\mathbf{B}^{T}\mathbf{P}\mathbf{x} ,$$

(9.19)

$$\mathbf{z}\left(\mathbf{x}\right) = \frac{1}{\gamma^{2}}\mathbf{E}^{T}\mathbf{P}\mathbf{x} .$$

(9.20)

If the matrix $\mathbf{P}\left(\mathbf{A} - \mathbf{B}\mathbf{L}_{s} + \mathbf{E}\mathbf{L}_{z}\right)$ is stable, the pair $(\mathbf{A}, \mathbf{C})$ is observable and $\gamma > \gamma^{*} > 0$, then there exists a positive-definite matrix $\mathbf{P}$, and (9.15) is a value of the function (9.14) [12].

By introducing (9.19), (9.20) to the Eq. (9.18) we get a generalised algebraic Riccati equation

$$\mathbf{A}^T\mathbf{P} + \mathbf{P}\mathbf{A} + \mathbf{Q} - \mathbf{P}\mathbf{B}\mathbf{R}^{-1}\mathbf{B}^T\mathbf{P} + \frac{1}{\gamma^2}\mathbf{P}\mathbf{E}\mathbf{E}^T\mathbf{P} = 0 \ . \tag{9.21}$$

In order to solve the problem of a two-person zero-sum differential game, first of all it is necessary to determine the matrix $\mathbf{P} > \mathbf{0}$ on the basis of the Eq. (9.21). Then we determine the optimal control based on the relation (9.19) at the most unfavourable disturbance (9.20). A solution of a differential game exists when matrix $(\mathbf{A}, \mathbf{B})$ is stable, pair $(\mathbf{A}, \mathbf{C})$ is observable, and $\gamma > \gamma^* > 0$ is a gain of $H_\infty$ type.

### 9.3.1 Simulation Tests

To model dynamic properties of the drive unit, the block scheme shown in Fig. 9.1 was assumed.

We write down the kinetic energy of the system as

$$E = \frac{1}{2}I_s\omega_s^2 + \frac{1}{2}I_w\omega_w^2 \ , \tag{9.22}$$

where $I_s$, $I_w$ is an adequately reduced mass moment of inertia of the shaft, respectively from the side of the motor and from the side of the wheel. The gear ratio is defined as

$$i_{sw} = \frac{\omega_s}{\omega_w} \ , \tag{9.23}$$

where $\omega_s$, $\omega_w$ are, respectively, angular velocity of the motor shaft and the shaft connected to the wheel.

Assuming Lagrange's mathematical formalism, we will obtain dynamic equations of motion of the drive unit

$$\frac{d}{dt}\left(\frac{\partial E}{\partial\dot\varphi_s}\right) - \frac{\partial E}{\partial\varphi_s} = Q \ , \tag{9.24}$$

**Fig. 9.1** Scheme of the drive unit, where S - motor, P - transmission, K - road wheel of the WMR

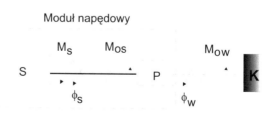

where $Q$ stands for the generalised force, determined on the basis of the relation

$$Q\delta\varphi_s = -a\dot{\varphi}_s\delta\varphi_s - b\dot{\varphi}_w\delta\varphi_w + M_s\delta\varphi_s , \tag{9.25}$$

where $a$ – medium resistance on the motor shaft, $b$ – medium resistance on the drive shaft, $M_s$ – torque on the motor shaft. By assuming, on the basis of (9.23), that

$$\delta\varphi_w = \frac{\delta\varphi_s}{i_{sw}} , \tag{9.26}$$

we will obtain an expression of generalised force

$$Q = -\left(a + b\frac{1}{i_{sw}^2}\right)\dot{\varphi}_s + M_s . \tag{9.27}$$

By carrying out operations on the Eq. (9.24), we will arrive at

$$\left(I_s + \frac{1}{i_{sw}^2}I_w\right)\ddot{\varphi}_s + \left(a + b\frac{1}{i_{sw}^2}\right)\dot{\varphi}_s = M_s . \tag{9.28}$$

The dynamics of a direct current motor with permanent magnets can be described by a differential equation of the rotor current

$$L\frac{di_a}{dt} + Ri_a + K_b\dot{\varphi}_s = V(t) , \tag{9.29}$$

where $L$ – rotor inductance, $i_a$ – rotor current, $R$ – rotor resistance, $K_b$ – counter-electromotive force constant, $V(t)$ – rotor voltage, and by an equation describing the torque of the motor $M_s$

$$M_s = K_m i_a , \tag{9.30}$$

where $K_m$ – the torque constant [Nm/A]. By dividing the Eq. (9.29) by $R$ and assuming a small value of the electrical time constant $L/R \ll$, we obtain

$$i_a = \frac{1}{R}V(t) - \frac{K_b}{R}\dot{\varphi}_s . \tag{9.31}$$

By introducing (9.30), (9.31) to (9.28), we obtain

$$\left(I_s + \frac{1}{i_{sw}^2}I_w\right)\ddot{\varphi}_s + \left(a + b\frac{1}{i_{sw}^2} + \frac{K_m K_b}{R}\right)\dot{\varphi}_s = \frac{K_m}{R}V(t) . \tag{9.32}$$

By dividing the Eq. (9.32) by the parameter expression at the angular velocity of the motor shaft, we obtain a mathematical model of the drive unit of a mobile robot

$$T\ddot{\varphi}_s + \dot{\varphi}_s = KV(t) , \tag{9.33}$$

where $T$ is the time constant, $K$ is the velocity gain constant of the unit. Using the notations $\varphi_s = x_1$, $\dot{\varphi}_s = x_2$ we will obtain a description of dynamic properties of the unit in the state space

$$\dot{x} = Ax + Bu + Ez ,$$
$$y = x ,$$
$$(9.34)$$

where $u$ is the control signal, and $z$ is the disturbance signal. Following parametric identification of the constants $T$ and $K$ on a real object, for the purposes of further investigations it was assumed as follows

$$A = \begin{bmatrix} 0 & 1 \\ 0 & -2 \end{bmatrix}, \quad B = \begin{bmatrix} 0 \\ 300 \end{bmatrix}, \quad E = \begin{bmatrix} 0 \\ 2 \end{bmatrix}. \qquad (9.35)$$

In simulation tests, the matrix $Q$ and the constant $R$ which occur in the value function (9.14), were assumed as follows: $Q = \mathrm{diag}\,[1, 1]$, $R = 70$, a $\gamma = 0.5$.

In the problem of a quadratic linear game, the solution of the HJI equation results from the solution of the Riccati equation

$$A^T P + PA + Q - PBR^{-1}B^T P + \frac{1}{\gamma^2}PEE^T P = 0 . \qquad (9.36)$$

The solution of the Riccati equation was found using the Matlab *care(.)* procedure, with the resulting matrix $P$ in the following form

$$P = \begin{bmatrix} 1.0033 & 0.0033 \\ 0.0033 & 0.0033 \end{bmatrix} . \qquad (9.37)$$

Assuming the control signal (9.19) and the disturbance signal (9.20) in the Eq. (9.33), the eigenvalues of the closed system $s_1 = -1$, $s_2 = -300.0033$ were determined. In accordance with theoretical investigations contained in Sect. 9.3, the optimal value of the value function (9.14) was determined on the basis of the relation (9.15). In Fig. 9.2, results of a differential game simulation are presented. The aim of the game was to determine optimal control reducing a nonzero initial condition of the module $x(0) = [10\Pi, 0]^T$ to zero in the conditions of the most unfavourable disturbance.

In Fig. 9.3, the sequences of the optimal value function for the case under analysis is presented.

The determination of the optimal control and disturbance signal in a two-person zero-sum game confirms that the Nash saddle point exists, which is shown in Fig. 9.4a. In Fig. 9.4b, the projection of the value function on the subspace $x_1$, $x_2$ and the trajectory of the optimum solution $u^*$, $z^*$ are shown. In order to modify the difference of scale, the control signal was multiplied by 40, and the disturbance signal by 15.

The obtained solution of the two-person zero-sum differential game of control and disturbance made it possible to solve the problem of $H_\infty$ control. Within this control, the determined optimal control (9.19) was applied, with the assumption of a zero initial condition $x(0) = [0, 0]^T$ and the disturbance signal $z(t) = 1500\exp(-0.06t) \in$

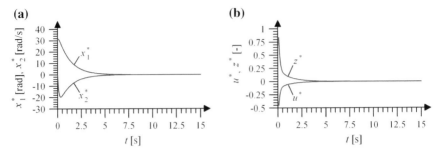

**Fig. 9.2** Optimal solutions of the differential game: **a** optimal sequences of the state vector, **b** optimal sequences of control and disturbance

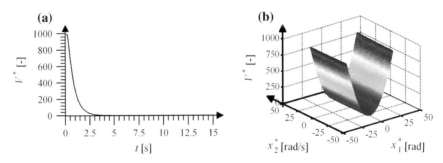

**Fig. 9.3** Diagram of the optimal value function: **a** sequences in time, **b** sequences on a phase surface

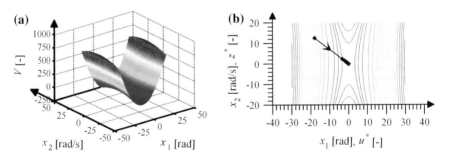

**Fig. 9.4** **a** Interpretation of the Nash condition in three dimensions, **b** projection onto the subspace with the sequences of the trajectory of the optimal control and disturbance

$L_2 [0, \infty)$ for time $t > 3$ [s]. The results obtained are shown in Figs. 9.5, 9.6 and 9.7. In Fig. 9.5a, the sequences of the state vector in the conditions of disturbance is presented. Despite the disturbance, a convergent solution was obtained.

As we know, there exists a solution of $H_\infty$ type control when it is possible to determine such a gain so that the condition $\gamma > \gamma^*$ is satisfied. In Fig. 9.6a, a result is shown in which gain of $L_2$ type meets this condition. While in Fig. 9.6b, a control signal ensuring convergence of the solution in the conditions of disturbance is shown.

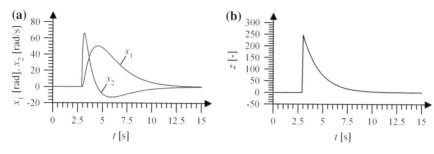

Fig. 9.5  **a** State vector, **b** assumed disturbance

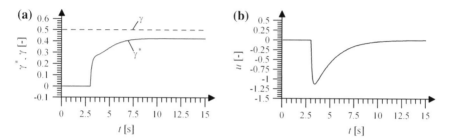

Fig. 9.6  **a** Sequences of gain $\gamma > \gamma^*$, **b** control signal

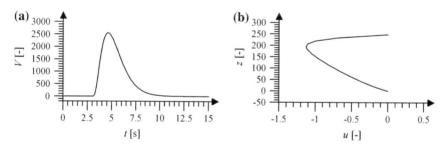

Fig. 9.7  **a** Value function, **b** control and disturbance trajectory

In this solution, the value function assumes values in time, as show in Fig. 9.7a. While in Fig. 9.7b, the sequences of the trajectory of control and the assumed disturbance in the $H_\infty$ problem are presented.

In Fig. 9.8, a solution case is shown where $\mathbf{x}(0) \neq \mathbf{0}$, $z(t) \neq 0$. The values obtained confirm that there exist a limited gain of $L_2$ type (Fig. 9.8c) given the concurrent asymptotic stability of the closed system, which is confirmed by simulation results shown in Fig. 9.8a, obtained for the optimal control presented in Fig. 9.8d, with disturbance occurring for time $t > 3$ [s], presented in Fig. 9.8b.

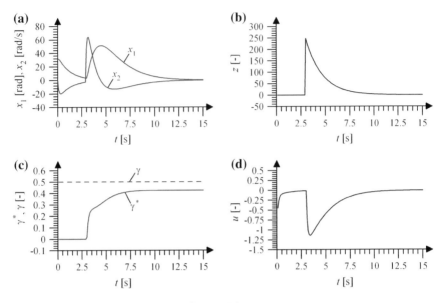

**Fig. 9.8** Results of H$_\infty$ control, $\mathbf{x}\,(0) \neq \mathbf{0}$, $z\,(t) \neq 0$

## 9.3.2  Conclusions

In this section, simulation tests were carried out for optimal control of H$_\infty$ type. The drive unit of a WMR was the controlled object. To solve the tasks, the theory of the two-person zero-sum differential game was applied. A minimax problem was solved in which a control minimising the cost function in the presence of the most unfavourable disturbance was designed. In the procedure subject to analysis, the lowest value of the gain $\gamma$ was determined by trial and error, by choosing the relevant value of matrix $\mathbf{Q}$ and constant $R$ so that the solution of Riccati equation would have positive solution. As it was shown in the discussed example, designing a control system minimising the gain of $L_2$ type is equivalent to finding a solution of a two-person zero-sum differential game for the lowest possible value $\gamma$. This problem is connected with optimal control of H$_\infty$ type.

## 9.4  Application of a Neural Network in the Two-Person Zero-Sum Differential Game in WMR Control

In this section we will apply theoretical considerations contained in Sects. 9.1 and 9.2 in WMR tracking control. To solve the HJI equation

$$H\left(\mathbf{x}, \nabla V^*, \mathbf{u}, \mathbf{z}\right) = \mathbf{x}^T \mathbf{Q} \mathbf{x} + \nabla V^{*T} \mathbf{f}\left(\mathbf{x}\right) - \frac{1}{4} \nabla V^{*T} \mathbf{g}\left(\mathbf{x}\right) \mathbf{R}^{-1} \mathbf{g}\left(\mathbf{x}\right)^T \nabla V^* +$$
$$+ \frac{1}{4\gamma^2} \nabla V^{*T} \mathbf{k}\left(\mathbf{x}\right) \mathbf{k}\left(\mathbf{x}\right)^T \nabla V^* = 0 , \tag{9.38}$$

we will use a NN whose aim will be to approximate the value function $V^*\left(\mathbf{x}\right)$. Let the WMR be described by dynamic equations of motion in the state space

$$\begin{aligned} \dot{\mathbf{x}}_1 &= \mathbf{x}_2 , \\ \dot{\mathbf{x}}_2 &= \mathbf{g}_n\left(\mathbf{x}\right) + \mathbf{u} + \mathbf{u}_z , \end{aligned} \tag{9.39}$$

where $\mathbf{x}_1 = [\alpha_1, \alpha_2]^T$, $\mathbf{x}_2 = [\dot{\alpha}_1, \dot{\alpha}_2]^T$, $\mathbf{u} = \mathbf{M}^{-1}\tau \in \mathfrak{R}^2$, $\mathbf{u}_z = -\mathbf{M}^{-1}\tau_d \in \mathfrak{R}^2$, and the nonlinear function has the following form

$$\mathbf{g}_n\left(\mathbf{x}\right) = -\mathbf{M}^{-1}\left(\mathbf{C}\left(\mathbf{x}_2\right)\mathbf{x}_2 + \mathbf{F}\left(\mathbf{x}_2\right)\right) \in \mathfrak{R}^2 , \tag{9.40}$$

where all notations have the meaning specified in Sect. 8.1. We assume that the inertia matrix of a mobile robot $\mathbf{M}$ is known, and the torques propelling the driving wheels of the WMR will be determined on the basis of the relation $\tau = \mathbf{M}\mathbf{u}$. Let us define tracking error $\mathbf{e}$, filtered tracking error $\mathbf{s}$, understood as a control quality measure in the following form

$$\mathbf{e}\left(t\right) = \mathbf{x}_d\left(t\right) - \mathbf{x}\left(t\right) , \tag{9.41}$$

$$\mathbf{s}\left(t\right) = \dot{\mathbf{e}}\left(t\right) + \mathbf{\Lambda}\mathbf{e}\left(t\right) , \tag{9.42}$$

where $\mathbf{\Lambda}$ is a positive-definite diagonal matrix. Then the Eq. (9.39) can be transformed into

$$\dot{\mathbf{s}} = \mathbf{f}\left(\mathbf{x}, \mathbf{x}_d\right) + \mathbf{u} + \mathbf{u}_z , \tag{9.43}$$

where $\mathbf{f}\left(\mathbf{x}, \mathbf{x}_d\right)$ contains nonlinearities (9.40) of the controlled object. The object contains two input signals, $\mathbf{u}$ is the control signal determined on the basis of the Bellman optimality principle (9.9), while $\mathbf{u}_z$ is the most unfavourable disturbance signal. Let us define, for established control and disturbance signals, the value function

$$V\left(\mathbf{s}\left(0\right), \mathbf{u}, \mathbf{u}_z\right) = \int_0^\infty \left(\mathbf{s}^T \mathbf{Q} \mathbf{s} + \mathbf{u}^T \mathbf{R} \mathbf{u} - \gamma^2 \|\mathbf{u}_z\|^2\right) dt . \tag{9.44}$$

As it was said in Sect. 9.2, determination of optimal control on the basis of the equation HJI (9.12) is difficult and needs approximation. By assuming that the value function, its derivative, and optimal control (9.10) and disturbance signal (9.11) can be uniformly approximated by a NN (8.105), the values in consideration can be expressed as

$$\hat{V}\left(\mathbf{s}\right) = \hat{\mathbf{W}}^T \varphi\left(\mathbf{s}\right) , \tag{9.45}$$

$$\hat{\mathbf{u}}\left(\mathbf{s}\right) = \frac{1}{2}\mathbf{R}^{-1}\left(\dot{\varphi}\left(\mathbf{s}\right)^T \hat{\mathbf{W}}\right) , \tag{9.46}$$

$$\hat{\mathbf{u}}_z = -\frac{1}{2\gamma^2} \left( \dot{\varphi}(\mathbf{s})^T \hat{\mathbf{W}} \right) , \qquad (9.47)$$

where the notations, value dimension, are interpreted in Sect. 8.5.1. By respectively replacing the values occurring in the HJI equation (9.6) with their estimations, the Hamiltonian will have the following form

$$H\left(\mathbf{s}, \hat{V}_s, \hat{\mathbf{u}}, \hat{\mathbf{u}}_z\right) = \left(\mathbf{s}^T \mathbf{Q} \mathbf{s} + \hat{\mathbf{u}}^T \mathbf{R} \hat{\mathbf{u}} - \gamma^2 \|\hat{\mathbf{u}}_z\|^2\right) + \left(\dot{\varphi}(\mathbf{s})^T \hat{\mathbf{W}}\right)\left(\mathbf{f}(\mathbf{x}, \mathbf{x}_d) + \hat{\mathbf{u}} + \hat{\mathbf{u}}_z\right) . \qquad (9.48)$$

Since for the optimal solutions resulting from the theory of differential games the Hamiltonian is zero, then - when we define the Bellman error as the difference of the Eqs. (9.6) and (9.48), we obtain

$$E_{HJI} = \left(\mathbf{s}^T \mathbf{Q} \mathbf{s} + \hat{\mathbf{u}}^T \mathbf{R} \hat{\mathbf{u}} - \gamma^2 \|\hat{\mathbf{u}}_z\|^2\right) + \left(\dot{\varphi}(\mathbf{s})^T \hat{\mathbf{W}}\right)\left(\mathbf{f}(\mathbf{x}, \mathbf{x}_d) + \hat{\mathbf{u}} + \hat{\mathbf{u}}_z\right) . \quad (9.49)$$

For the purposes of adaptive learning of weights of the NN approximating the value function as a critic and two structures (9.46) and (9.47) approximating an optimal solution of the Nash saddle point, an iterative algorithm of the method of least squares was applied. Such an approach is justified by the fact that the equation describing the Bellman error (9.49) is a linear function of critic's weights.

To determine the rule of learning of the critic network's weights, the following relation was assumed

$$E_c = \int_0^t (E_{HJI}(\rho))^2 \, d\rho . \qquad (9.50)$$

By determining the error gradient (9.50) and equalling it to zero, the following was obtained

$$\frac{\partial E_c}{\partial \hat{\mathbf{W}}} = 2 \int_0^t E_{HJI}(\rho) \frac{\partial E_{HJI}(\rho)}{\partial \hat{\mathbf{W}}} d\rho = 0 . \qquad (9.51)$$

To implement the adopted solution in real time (online), an iterative version of the normalised least squares method [11] was applied, pursuant to which the estimations of the network's weights were determined on the basis of the following relation

$$\dot{\hat{\mathbf{W}}} = \eta \mathbf{G} \frac{\delta}{1 + \nu \delta^T \mathbf{G} \delta} E_{HJI} , \qquad (9.52)$$

where $\eta, \nu \in \Re$ are positive coefficients, $\delta = \dot{\varphi}\left(\mathbf{f}(\mathbf{x}, \mathbf{x_d}) + \hat{\mathbf{u}} + \hat{\mathbf{u}}_z\right)$, while $\mathbf{G} = \left(\int_0^t \delta(\rho)\delta(\rho)^T \, d\rho\right)^{-1} \in \Re^{N \times N}$ is a symmetrical matrix determined on the basis of the equation

$$\dot{\mathbf{G}} = \eta \mathbf{G} \frac{\delta \delta^T}{1 + \nu \delta^T \mathbf{G} \delta} \mathbf{G} , \qquad (9.53)$$

where $\mathbf{G}(0) = \alpha_c \mathbf{I}, \alpha_c \gg 0$, and fulfilling the condition of well-conditioning $\dot{\mathbf{G}} \geq \mathbf{0}$. Such an approach to the problem allows to solve the two-person zero-sum differential game in real time, on the basis of data obtained from the measurements of the state vector of the controlled object. As we know [10], the method of least squares is efficient in terms of convergence if the condition of constant system excitation is fulfilled, which means convergence of network weights $\hat{\mathbf{W}}$ to their unknown current values $\mathbf{W}$. The determined estimates of the weights of the network $\hat{\mathbf{W}}$ make it possible in turn to arrive at an approximate solution of the Bellman equation (9.49) given an adequately determined control $\hat{\mathbf{u}}$ and disturbance $\hat{\mathbf{u}}_z$. The proof of convergence of the solution is presented in [12]. However, in this section, to demonstrate the stability of the closed system and to limit the control errors $\mathbf{e}, \dot{\mathbf{e}}$ within the assumed maximal value, the control signal $\hat{\mathbf{u}}$ was enriched by supervisory control $\mathbf{u}_S$ [13], generated by an additional supervisory term in order to limit the filtered tracking error $\mathbf{s}$, i.e.

$$|s_i| \leq \Phi_i , \quad \forall t > 0 , \tag{9.54}$$

where $\Phi_i, i = 1, 2$, assumes constant values. The control $\mathbf{u}_S$ will activate only if the threshold limit (9.54) is exceeded. Let us introduce, for the tasks formulated in such a way, a control law

$$\mathbf{u} = -\hat{\mathbf{u}}_{GR} - \mathbf{u}_S , \tag{9.55}$$

where control signal $\hat{\mathbf{u}}_{GR} = \hat{\mathbf{u}}$ is the signal (9.46) resulting from the control synthesis with the application of a two-person zero-sum differential game. By introducing control (9.55) to (9.43), we obtain a description of a closed system in the following form

$$\dot{\mathbf{s}} = \mathbf{f}(\mathbf{x}, \mathbf{x}_d) + \mathbf{u}_z - \hat{\mathbf{u}}_{GR} - \mathbf{u}_S . \tag{9.56}$$

By assuming an additional signal

$$\mathbf{u}^* = \mathbf{f}(\mathbf{x}, \mathbf{x}_d) + \mathbf{u}_z , \tag{9.57}$$

we will write down the relation (9.56) as

$$\dot{\mathbf{s}} = \mathbf{u}^* - \hat{\mathbf{u}}_{GR} - \mathbf{u}_S . \tag{9.58}$$

Let us define a positive-definite function

$$L = \frac{1}{2}\mathbf{s}^T \mathbf{s} . \tag{9.59}$$

By determining the derivative of the function (9.59) and using the relation (9.58), we arrive at

$$\dot{L} = \mathbf{s}^T \dot{\mathbf{s}} = \mathbf{s}^T \left[\mathbf{u}^* - \hat{\mathbf{u}}_{GR}\right] - \mathbf{s}^T \mathbf{u}_S , \tag{9.60}$$

and after normalisation, the derivative (9.60) will have the following form

$$\dot{L} \leq \sum_{i=1}^{2} |s_i| \left[ |f_i(\mathbf{x}, \mathbf{x}_d) + u_{zi} - \hat{u}_{GRi}| \right] - \sum_{i=1}^{2} s_i u_{Si} . \qquad (9.61)$$

If we assume that the control signal $\hat{\mathbf{u}}_{GR}$ was designed for the most unfavourable disturbance, i.e. nonlinearities and changing operating conditions of the WMR, then when selecting the supervisory control signal in the form of

$$u_{Si} = [F_{0i} + \eta_i] \, \mathrm{sgn} s_i , \qquad (9.62)$$

assuming that $|f_i(\mathbf{x}, \mathbf{x}_d) + u_{zi} - \hat{u}_{GRi}| \leq F_{0i}, \ i = 1, 2, \ F_{0i} > 0, \ \eta_i > 0$, we eventually arrive at

$$\dot{L} \leq -\sum_{i=1}^{2} \eta_i |s_i| . \qquad (9.63)$$

The derivative of the Lyapunov function is negative semidefinite, hence the closed system (9.58) is stable in terms of the state vector $\mathbf{s}$. In the control signal (9.62) there is function $\mathrm{sgn} s_i$ which activates high frequencies of the actuating system. In order to eliminate high frequencies of switching of the actuating system, an approximation of the function $\mathrm{sgn} s_i$ is introduced in its neighbourhood $\Phi_i$ with the function [9]

$$\mathrm{sat}(s) = \begin{cases} \mathrm{sgn}(s) \ dla \ |s| \geq \Phi , \\ 0 \qquad dla \ |s| < \Phi . \end{cases} \qquad (9.64)$$

If $\mathbf{s}(0) = \mathbf{0}$, then in a finite time the accuracy of control is ensured

$$\forall t \geq 0 , \quad |s_i(t)| \leq \Phi . \qquad (9.65)$$

If the condition (9.54) is satisfied, the basis control signal is the control (9.46).

### 9.4.1  Simulation Tests

In this section, to solve the task of tracking control of a mobile robot - or the motion of a selected point of a mobile robot, $A$, along a desired path in the form of a loop described in Sect. 8.1.2, we will apply the theory of two-person zero-sum differential game described in Sect. 9.4. We assume the form of dynamic equation of WMR's motion as in (9.39), with notations described in Sect. 8.1. Its form in the state space can be presented according to the relation (9.39), and after introducing relations (9.41) and (9.42) with the matrix $\mathbf{\Lambda} = [1, 1]$, it will assume the form (9.43), similarly as in Sect. 8.5. Due to the complexity of the nonlinear function $\mathbf{f}(\mathbf{x}, \mathbf{x}_d)$, the function is not presented here. As the value function, we assume the relation (9.44) with matrix values $\mathbf{Q} = \text{diag}\,[2000, 2000]$, $\mathbf{R} = \text{diag}\,[0.002, 0.002]$ and coefficient $\gamma = 5.3$. The approximation of the value function (9.45) and its gradient, occurring in control (9.46) and disturbance (9.47), is implemented by a NN comprised of 3 neurons with basic functions of the following form, respectively $\varphi\,(\mathbf{s}) = \left[s_1^2, s_1 s_2, s_2^2\right]^T$.

Taking into account the task of control, a zero initial condition was assumed, $\mathbf{s} = [0, 0]^T$. The learning rule of the NN was implemented in line with the relation (9.52) with parameters $\eta = 0.001$, $v = 0.1$, $\alpha_c = 10$. In Fig. 9.9, results obtained in the process of implementation of the two-person zero-sum differential game are presented. In Fig. 9.9a, b state vector errors, denoted, correspondingly, by $e_1$, $e_2$ and $\dot{e}_1$, $\dot{e}_2$ are presented, obtained in the process of execution of the desired trajectory. While in Fig. 9.9c, d, a suboptimal implementation of the Nash saddle point is shown. In Fig. 9.9e, f, critic weights and approximation of the value function are presented. Values of the critic weights, which are equal zero before the learning process is started, remain limited when the process of learning progresses.

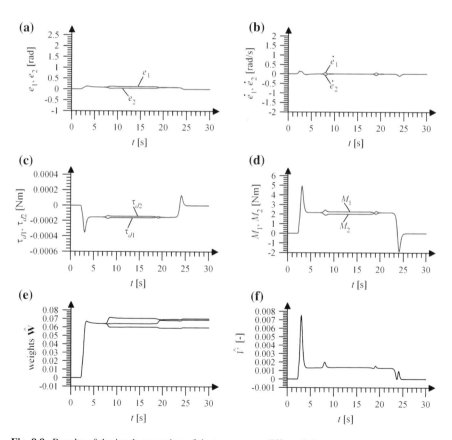

**Fig. 9.9** Results of the implementation of the two-person differential game

From the solutions found, it results that the control obtained, as shown in Fig. 9.9d. is a stabilising control and ensures a finite value of the value function, which is confirmed by Fig. 9.9f. Therefore, an admissible control was obtained. In order to verify the $H_\infty$ control, and therefore, the approximate solution of the Hamilton–Jacobi–Isaac equation, the obtained control signal resulting from the solution of the minimax task was applied in the control system, when for the time $t \geq 12$ [s] there is a parametric disturbance $\mathbf{a} + \Delta\mathbf{a}$ where $\Delta\mathbf{a} = [0\,0\,0\,0\,1\,1]^T$, connected with the change of the robot's resistances to motion. The results of the obtained kinematic parameter errors are appropriately presented in Fig. 9.10a, b. Despite the disturbance, a convergent solution was found. While in Fig. 9.10c, d, the introduced disturbance $z_1 = (a_5 + \Delta a_5)\,\mathrm{sgn}x_{21}$, $z_2 = (a_6 + \Delta a_6)\,\mathrm{sgn}x_{22}$ and the control signal determined on the basis of the game are presented, respectively. If there exists a solution of $H_\infty$ control, it is possible to determine such a gain so that the condition $\gamma > \gamma^*$ is satisfied. In Fig. 9.10e a result is shown in which gain of $L_2$ type meeting the

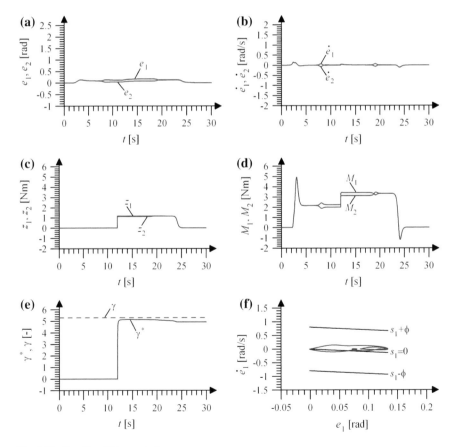

**Fig. 9.10**  Results of $H_\infty$ control

**Table 9.1**  Values of the selected quality ratings of the control system

| Index | $e_{maxj}$ [rad] | $\varepsilon_j$ [rad] | $\dot{e}_{maxj}$ [rad/s] | $\dot{\varepsilon}_j$ [rad/s] | $\sigma_j$ [rad/s] |
|---|---|---|---|---|---|
| Wheel 1, $j = 1$ | 0.1665 | 0.1108 | 0.1502 | 0.03109 | 0.115 |
| Wheel 2, $j = 2$ | 0.1336 | 0.07757 | 0.1502 | 0.03159 | 0.0838 |

**Table 9.2**  Values of quality ratings for the assessment of the execution of the desired path

| Index | $d_{max}$ [m] | $\rho$ [m] |
|---|---|---|
| Value | 0.02446 | 0.01141 |

above condition was obtained. Figure 9.10f confirms that the only control is optimal control, and that the supervisory control is inactive.

To carry out the quantitative evaluation of the simulation example, performance indexes specified in Sect. 8.1 were applied. The values of individual indexes are presented in Tables 9.1 and 9.2.

The quantitative assessments reveal small task execution errors in the light of the theory of two-person zero-sum differential game. The results presented confirm the stability of the proposed control method.

## 9.4.2  Conclusions

In this sub-section, to synthesise the WMR tracking control, the theory of differential games were applied with the use of neuro-dynamic programming in solving a two-person zero-sum game, using the Nash saddle point theory. By determining an approximate solution of the HJI equation, also a $H_\infty$ control was determined. To illustrate the discussed theory of $H_\infty$ control and the two-person zero-sum differential game, numerical tests were carried out. In the first test, a linear problem was solved on the basis of control of the drive unit of a WMR with the use of an approximate solution of the HJI equation. Another numerical test concerned the implementation of the approximate HJI solution in control of a nonholonomic, nonlinear object (a 2-wheeled mobile robot). Taking account of the obtained level of accuracy in the execution of the desired trajectory, in order to compare the methods applied in Chaps. 8 and 9, Fig. 8.25 was supplemented with results obtained in this chapter, as shown in Fig. 9.11.

On the horizontal axis, results obtained in solving the tracking motion problem with the use of control (9.46) were determined by DG. As it stems from the analysis of Fig. 9.11, the best solution was obtained for an adaptive critic's structure in optimal control of WMR RL5. It should also be noted that the structures of control algorithms of RL2, RL3, and DG type lead to similar results relating to the maximal error of execution of the desired trajectory. Nevertheless, the assessment of the methods used should be verified in experimental research.

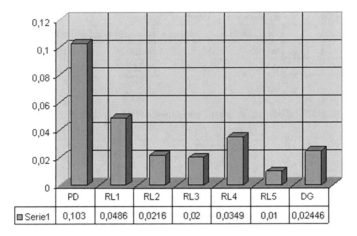

**Fig. 9.11** Maximal error values $d_{max}$ $(i)$ [m] in the execution of the desired trajectory

# References

1. Basar, T., Bernard, P.: H∞ - Optimal Control and Related Minimax Design Problems. Birkhäuser, Boston (1995)
2. Basar, T., Olsder, G.J.: Dynamic Noncooperative Game Theory, 2nd edn. Society for Industrial and Applied Mathematics, Philadelphia (1999)
3. Beard, R.W., Saridis, G.N., Wen, J.T.: Approximate solutions to the time-invariant Hamilton–Jacobi–Bellman equation. J. Optim. Theory Appl. **96**, 589–626 (1998)
4. Isaacs, R.: Differential Games. Wiley, New York (1965)
5. Isidori, A., Kwang, W.: H∞ control via measurement feedback for general nonlinear systems. IEEE Trans. Autom. Control **40**, 466–472 (1995)
6. Lewis, F.L., Liu, D. (eds.): Reinforcement Learning and Approximate Dynamic Programming for Feedback Control. Wiley, Hoboken (2013)
7. Lewis, F.L., Vrabie, D., Syroms, V.L.: Optimal Control, 3rd edn. Wiley, New Jersey (2012)
8. Nash, J.: Non-cooperative games. Ann. Math. **2**, 286–295 (1951)
9. Slotine, J.J., Li, W.: Applied Nonlinear Control. Prentice Hall, New Jersey (1991)
10. Soderstrom, T., Stoica, P.: System Identification. Prentice Hall, Upper Saddle River (1988)
11. Van Der Schaft, A.: $L_2$ - Gain and Passivity Techniques in Nonlinear Control. Springer, Berlin (1996)
12. Vrabie, D., Vamvoudakis, K.G., Lewis, F.L.: Optimal Adaptive Control and Differential Games by Reinforcement Learning Principles. Control Engineering Series. IET Press, London (2013)
13. Wang, L.: A Course in Fuzzy Systems and Control. Prentice-Hall, New York (1997)
14. Zames, G.: Feedback and optimal sensitivity: model reference transformations, multiplicative seminorms, and approximate inverses. IEEE Trans. Autom. Control **26**, 301–320 (1981)

# Chapter 10
# Experimental Verification of Control Algorithms

The following chapter includes the outcomes of conducted research and the description of laboratory stands, which were used to conduct verification research of the designed control algorithms. The scientific research included: the analysis of tracking control algorithms of WMR's point $A$, the research of behavioral control algorithms for WMR motion in an unknown environment, and the analysis of tracking control algorithms of the operating arm of the RM.

## 10.1 Description of Laboratory Stands

Verification tests of the proposed control algorithms were carried out on two laboratory stands. The first one allows to control the motion of WMR, and the second one allows to control the motion of the RM. Each stand includes a PC with dSpace digital signal processing (DSP) board, a power supply module, and a control object. DSP boards allow the device to realize programmed control algorithms in real-time, and make the acquisition of selected sequences.

### 10.1.1 WMR Motion Control Stand

Verification tests of WMR's motion control algorithms have been carried out on a laboratory stand, which is schematically illustrated in Fig. 10.1. It includes a PC with dSpace DS1102 DSP board, a power supply, and WMR Pioneer 2-DX. Software provided by board's manufacturer, Control Desk 3.2.1 [3], has been used to handle the board's interface and to conduct research experiments in real-time with the use of resources from the DSP board. The control algorithms were programmed

© Springer International Publishing AG 2018
M. Szuster and Z. Hendzel, *Intelligent Optimal Adaptive Control
for Mechatronic Systems*, Studies in Systems, Decision and Control 120,
https://doi.org/10.1007/978-3-319-68826-8_10

**Fig. 10.1** Scheme for WMR
motion control stand

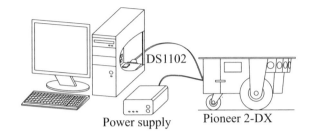

Power supply        Pioneer 2-DX

using Matlab 5.2.1/Simulink 2.2 software environment [9], with the use of dSPACE
RTI1102 Board Library 3.3.1 toolbox [4]. Then it was compiled with the use of RTI
3.3.1 libraries [5], supported by dSpace DS1102 board's software.

dSpace DS1102 DSP board [1, 2], is the enhancement board for PCs. It communicates with the mother board using ISA (Industrial Standard Architecture), and it
only requires half of the width of a standard 16-bit port. TMS320C31 digital signal
processor from Texas Instruments is the main calculation unit and operates at 60
MFlops (Mega Floating Point Operations Per Second - millions of floating point
operations per second). TMS320C31 processor used in the board is characterized by
the following parameters (the most important ones):

- frequency: 60 [MHz],
- the duration of processing a single instruction: 33.3 [ns],
- two blocks of built-in 32-bit memory with 2KB capacity each,
- 32-bit instruction word,
- 24-bit addresses,
- serial port, four external interrupts,
- two 32-bit timers.

DS1102 DSP board includes:

- Processor:

  – TMS320C31.

- Memory:

  – 128 KB 32-bit memory.

- Analog inputs:

  – two 16-bit A/D converters with the conversion time of 4 [μs],
  – two 12-bit A/D converters with the conversion time of 1.25 [μs],
  – input signal voltage range: ± 10 [V],
  – signal-to-noise ratio >80 [dB] (16-bit)/65 [dB] (12-bit).

- Analog outputs:

  – four 12-bit D/A converters,
  – conversion time 4 [μs],
  – output signal voltage range: ± 10 [V].

- Digital inputs/outputs:

  – sub-system of programmable inputs/outputs built based on a 32-bit Texas Instruments TMS320P14 25 [MHz] signal processor,
  – 16 digital inputs/outputs,
  – 6 PWM channels,
    1. 8-bit resolution for 100 [kHz] frequency,
    2. 10-bit resolution for 25 [kHz] frequency,
    3. 12-bit resolution for 6.26 [kHz] frequency,
    4. 14-bit resolution for 1.506 [kHz] frequency,
  – external user interrupts.

- Incremental rotary encoders interface:

  – ability to connect two incremental rotary encoders,
  – maximum frequency for impulse counting: 8.3 [MHz],
  – measurement noise filter,
  – 24-bit counter.

- Communication interfaces:

  – ISA (half of the 16-bit connector),
  – connecting signals with 62-pin D-SUB connector,
  – RS232 serial port,
  – JTAG interface.

The described laboratory stand allows to quickly prototype WMR motion control systems, and to verify programmed control algorithms in real-time, with the use of hardware resources of DSP board.

## 10.1.2 RM Motion Control Stand

Verification analyses of the RM's motion control algorithms were carried out on a laboratory stand shown in Fig. 10.2 schematically. The stand includes a PC with dSpace DS1104 DSP card, power supply and Scorbot-ER 4pc RM. In order to use the card's interfaces and to conduct research in real-time with the use of the resources of the DSP board the software was used, provided by the manufacturer - Control Desk 4.0 [6]. The control algorithms were programmed in Matlab 6.5.1/Simulink 5.1 [9] computing environment, with the use of dSPACE RTI1102 Board Library 4.4.7 toolbox, and compiled with the use of RTI 4.5 [8] libraries into the format that is supported by dSpace DS1104 software.

The dSpace DS1104 [7] DSP board is the enhancement card for PCs, which communicates with motherboard using 32-bit PCI bus. The main computational unit is the MPC8240 integrated microprocessor, equipped with the PowerPC 603e core. The MPC8240 processor is characterized by the following parameters (the most important ones):

**Fig. 10.2** Scheme for RM
motion analysis' stand

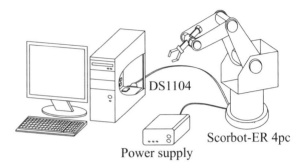

DS1104

Scorbot-ER 4pc

Power supply

- frequency: 250 [MHz],
- most of the instructions are executed in one clock cycle,
- two blocks of cache memory with 16KB capacity each (for data and instructions),
- 32/64-bit word length,
- 32-bit memory addressing,
- PCI controller, DMA, I2C, expanded interrupt system,
- six counters, including one 64-bit, others are 32-bit.

   DS1104 controller card includes:

- Processor:

  – MPC8240,

- Memory:

  – 32 MB of 32-bit memory,

- Analog inputs:

  – one 16-bit A/D converter, 4 channels (multiplexed),
  – four 12-bit A/D converters,
  – input voltage range: ± 10 [V],
  – SNR >80 [dB] (16-bit)/70 [dB] (12-bit).

- Analog outputs:

  – eight 16-bit channels of D/A converter,
  – output voltage range: ± 10 [V],
  – SNR >80 [dB].

- Digital inputs/outputs:

  – 20 digital input-output lines, controlled individually by the master processor,
  – subsystem of programmable inputs/outputs, built based on an additional 32-bit
    signal slave processor - Texas Instruments TMS320F240 20 [MHz],
  – 14 digital input-output lines,
  – six PWM channels,

    – user external interrupt.

- Incremental rotary encoders interface:

    – ability to connect two incremental rotary encoders,
    – maximum frequency for impulse counting: 1.65 [MHz],
    – it is possible to power the encoders,
    – 24-bit counter.

- Communication interfaces:

    – PCI bus (32-bit),
    – connecting signals through two 50-pin D-SUB sockets,
    – RS232/RS422/RS485 serial port,
    – SPI interface.

The described research stand allows to quickly prototype the motion control systems of a RM and to verify programmed algorithms in real-time, with the use of the DSP board's hardware resources.

## 10.2  Analysis of the PD Control

Verification tests of the PD controller in the task of a tracking control of point $A$ of the WMR Pioneer 2-DX were carried out on a laboratory stand described in Sect. 10.1.1. Among the results of carried out experimental tests two sequaences with the trajectory with path in the shape of digit 8, with parametric disturbance in the form of additional mass carried by the WMR Pioneer 2-DX, and without disturbance were presented. Verification tests of the PD controller in the task of tracking control of point $C$ of the RM Scrobot-ER 4pc end-effector were carried out on a laboratory stand described in Sect. 10.1.2. The results of experiment carried out with the use of trajectory in the shape of a semicircle with parametric disturbance, which occurs during the motion of the robot's arm executed in the form of loading the gripper with a mass, were presented.

The same parameters applied in the numerical tests described in Sect. 7.1.2 were applied in all experimental tests on tracking control system consisting of a PD controller.

The performance indexes presented in Sect. 7.1.2 were applied in order to evaluate the quality of execution of tracking motion.

### 10.2.1  Analysis of the WMR Motion Control

The analyses of the tracking control of point $A$ of the WMR Pioneer 2-DX with the use of a PD controller were carried out in the configuration without disturbance Sect. 10.2.1.1, and with disturbance implemented in the form of loading the WMR's frame during its motion Sect. 10.2.1.2. Presentation of results of two experiments

**Fig. 10.3** Desired and
executed path of point *A* of
the WMR

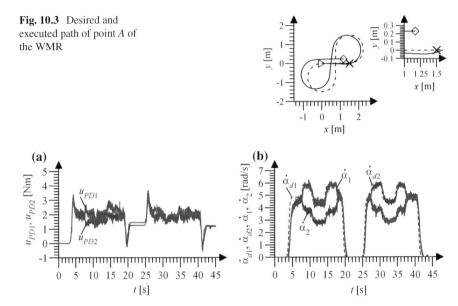

**Fig. 10.4  a** Values of control signal $u_{PD1}$ and $u_{PD2}$, **b** values of desired ($\dot{\alpha}_{d1}$, $\dot{\alpha}_{d2}$) and executed ($\dot{\alpha}_1$, $\dot{\alpha}_2$) angular velocities of drive wheels

enables a comparison of sequences of individual motion parameters and illustration of the influence of the loading of robot's frame on the execution of tracking motion.

### 10.2.1.1  Analysis of the PD Control, Experiment Without Disturbance

Verification tests were carried out on the tracking control system of the WMR, consisting of a PD controller, with implementation of a trajectory with path in the shape of digit 8 as discussed in Sect. 2.1.1.2. The values of motion parameters obtained after solving the inverse kinematics problem were marked by the index *d* and were implemented as desired motion parameters. The parametric disturbances in the form of loading of the WMR with mass were not implemented in the presented experiment.

Figure 10.3a presents the desired (dashed line) and executed (continuous line) path of point *A* of the WMR. The initial location of point *A* was marked by a triangle, the desired end-position by "**X**", and the reached end-position of point *A* of the WMR by a rhombus. Enlargement of the selected area of the chart (on the left side) is presented on the right side.

The executed trajectory of point *A* of the WMR result from introducing, into the robot's actuator systems, control signals generated by the PD controller. The values of control signals $u_{PD1}$ and $u_{PD2}$ of wheels 1 and 2 are shown in Fig. 10.4a, values of desired ($\dot{\alpha}_{d1}$, $\dot{\alpha}_{d2}$) and executed ($\dot{\alpha}_1$, $\dot{\alpha}_2$) angular velocities of the drive wheels are shown in Fig. 10.4b.

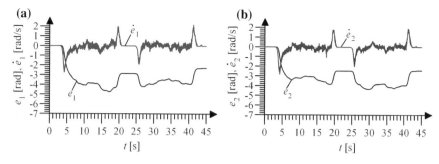

**Fig. 10.5** **a** Values of errors of desired motion parameters $e_1$ and $\dot{e}_1$ execution, **b** values of errors of desired motion parameters $e_2$ and $\dot{e}_2$ execution

**Table 10.1** Values of selected performance indexes

| Index | $e_{maxj}$ [rad] | $\varepsilon_j$ [rad] | $\dot{e}_{maxj}$ [rad/s] | $\dot{\varepsilon}_j$ [rad/s] | $s_{maxj}$ [rad/s] | $\sigma_j$ [rad/s] |
|---|---|---|---|---|---|---|
| Wheel 1, $j = 1$ | 4.73 | 3.44 | 2.75 | 0.53 | 3.68 | 1.82 |
| Wheel 2, $j = 2$ | 4.34 | 3.23 | 2.83 | 0.5 | 3.52 | 1.71 |

**Table 10.2** Values of selected performance indexes of path execution

| Index | $d_{max}$ [m] | $d_n$ [m] | $\rho_d$ [m] |
|---|---|---|---|
| Value | 0.525 | 0.409 | 0.314 |

The difference between the desired and executed path of point $A$ of the WMR results from the execution of the desired trajectory with errors. Values of errors of the desired angles of rotation $(e_1, e_2)$, and errors of the desired angular velocities $(\dot{e}_1, \dot{e}_2)$ of the WMR drive wheels 1 and 2 are shown in Fig. 10.5a, b.

It should be noted that the task of tracking control algorithm in presented approach is to minimize the tracking errors. The values of individual performance indexes of tracking motion execution with control generated by the PD controller are presented in Tables 10.1 and 10.2.

The calculations were carried out on the value of selected performance indexes by dividing the executed trajectory with path in the shape of digit 8 into two stages, according to the methodology described in Sect. 7.1.2.2. The values of selected performance indexes of individual stages of motion are presented in Table 10.3.

In the case of an absence of the element compensating for the nonlinearity of the controlled object, e.g., the adaptive algorithm or NN, in the structure of the control system of the motion execution of point $A$ of the WMR, the values of selected performance indexes reach similar values in both stages of the experiment.

**Table 10.3** Values of selected performance indexes in I and II stage of motion

| Index | $e_{maxj}$ [rad] | $\varepsilon_j$ [rad] | $\dot{e}_{maxj}$ [rad/s] | $\dot{\varepsilon}_j$ [rad/s] | $s_{maxj}$ [rad/s] | $\sigma_j$ [rad/s] |
|---|---|---|---|---|---|---|
| Stage I, $k = 1, \ldots, 2250$ | | | | | | |
| Wheel 1, $j = 1$ | 4.73 | 3.42 | 2.75 | 0.6 | 3.22 | 1.86 |
| Wheel 2, $j = 2$ | 3.83 | 2.81 | 2.83 | 0.56 | 3.42 | 1.56 |
| Stage II, $k = 2251, \ldots, 4500$ | | | | | | |
| Wheel 1, $j = 1$ | 4.2 | 3.35 | 2.18 | 0.44 | 3.68 | 1.71 |
| Wheel 2, $j = 2$ | 4.34 | 3.52 | 2.0 | 0.42 | 3.52 | 1.81 |

#### 10.2.1.2  Analysis of the PD Control, Experiment with Disturbance

The next analyses were carried out on tracking control of the WMR with the use of a
PD controller with parametric disturbance. A desired trajectory with path in the shape
of digit 8, discussed in Sect. 2.1.1.2, was implemented in the experimental tests. The
parametric disturbance, occurring twice, was introduced in the form of a changed
mass carried by the WMR. The WMR was loaded with a $m_{RL} = 4.0$ [kg] mass at time
$t_{d1} = 12.5$ [s], and at time $t_{d2} = 32.5$ [s] the mass was removed. The approximate
time of disturbance occurrence results from conditions of the experiment execution,
where the human factor had an influence on time of loading of the robot's frame with
additional mass.

Figure 10.6a presents the desired (dashed line) and executed (continuous line)
trajectory of point $A$ of the WMR.

Values of control signals $u_{PD1}$ and $u_{PD2}$ of wheel 1 and 2 are shown in Fig. 10.7a,
values of desired $(\dot{\alpha}_{d1}, \dot{\alpha}_{d2})$ and executed $(\dot{\alpha}_1, \dot{\alpha}_2)$ angular velocities of drive wheels
are presented in Fig. 10.7b. The influence of the loading of the WMR frame with
additional mass on the obtained sequences are marked by ellipses. After occurrence
of the first disturbance at time $t_{d1}$ an increase in the value of the control signals
caused by the change in the dynamics of the controlled object, and the temporary
reduction of values of execution angular velocities can be noticed. Reduction of the
carried mass by the WMR at time $t_{d2}$ results in reduction of control signals values,
and a temporary increase in the value of executed angular velocities of drive wheels
rotation.

**Fig. 10.6** Desired and
executed path of point $A$ of
the WMR

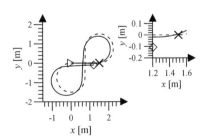

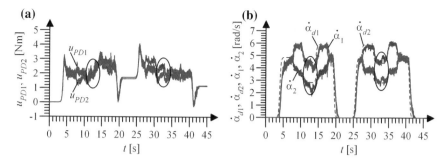

**Fig. 10.7** **a** Values of control signal $u_{PD1}$ and $u_{PD2}$, **b** values of desired ($\dot{\alpha}_{d1}$, $\dot{\alpha}_{d2}$) and executed ($\dot{\alpha}_1$, $\dot{\alpha}_2$) angular velocities of drive wheels

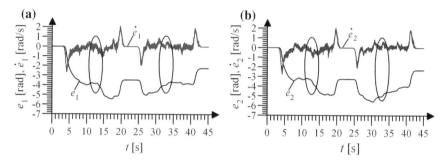

**Fig. 10.8** **a** Values of errors of desired motion parameters $e_1$ and $\dot{e}_1$ execution, **b** values of errors of desired motion parameters $e_2$ and $\dot{e}_2$ execution

Values of errors of desired angles of rotation ($e_1$, $e_2$) execution and errors of desired angular velocities ($\dot{e}_1$, $\dot{e}_2$) of drive wheels 1 and 2 of the WMR are shown in Fig. 10.8a, b. By analyzing the values of errors of the trajectory execution an increase of values of errors after time $t_{d1}$ and their reduction after time $t_{d2}$, in relation to the sequences during the experiment without disturbance.

The values of individual performance indexes of tracking motion execution with control generated by the PD controller in the experiment with parametric disturbance are presented in Tables 10.4 and 10.5.

Values of performance indexes obtained during a trial with disturbance are larger than those obtained during the experiment without disturbance.

**Table 10.4** Values of selected performance indexes

| Index | $e_{maxj}$ [rad] | $\varepsilon_j$ [rad] | $\dot{e}_{maxj}$ [rad/s] | $\dot{\varepsilon}_j$ [rad/s] | $s_{maxj}$ [rad/s] | $\sigma_j$ [rad/s] |
|---|---|---|---|---|---|---|
| Wheel 1, $j = 1$ | 5.44 | 3.76 | 2.67 | 0.55 | 4.07 | 1.97 |
| Wheel 2, $j = 2$ | 5.57 | 3.82 | 2.76 | 0.54 | 4.08 | 2.0 |

**Table 10.5** Values of selected performance indexes of path execution

| Index | $d_{max}$ [m] | $d_n$ [m] | $\rho_d$ [m] |
|-------|---------------|-----------|--------------|
| Value | 0.499         | 0.319     | 0.268        |

**Table 10.6** Values of selected performance indexes in I and II stage of motion

| Index | $e_{maxj}$ [rad] | $\varepsilon_j$ [rad] | $\dot{e}_{maxj}$ [rad/s] | $\dot{\varepsilon}_j$ [rad/s] | $s_{maxj}$ [rad/s] | $\sigma_j$ [rad/s] |
|-------|------------------|-----------------------|--------------------------|-------------------------------|--------------------|--------------------|
| Stage I, $k = 1, \ldots, 2250$ | | | | | | |
| Wheel 1, $j = 1$ | 5.44 | 3.69 | 2.67 | 0.62 | 3.51 | 2.01 |
| Wheel 2, $j = 2$ | 4.9  | 3.37 | 2.76 | 0.59 | 3.42 | 1.85 |
| Stage II, $k = 2251, \ldots, 4500$ | | | | | | |
| Wheel 1, $j = 1$ | 5.03 | 3.69 | 1.97 | 0.46 | 4.07 | 1.86 |
| Wheel 2, $j = 2$ | 5.57 | 4.12 | 2.09 | 0.47 | 4.08 | 2.09 |

Similarly as in experiment Sect. 10.2.1, the calculations of values of selected performance indexes were carried out by dividing the executed trajectory with path in the shape of digit 8 into two stages. The values of selected performance indexes of individual stages of motion are presented in Table 10.6.

The values of performance indexes in the first and second stages of motion are slightly different. The values of obtained performance indexes will be applied in the further part of this book as a base to compare the accuracy of the tracking motion execution with the use of other control algorithms.

## 10.2.2  Analysis of the RM Motion Control

The experimental tests were carried out on the tracking control system of point $C$ of the RM's end-effector. The control system consisted of the PD controller only.

### 10.2.2.1  Analysis of the PD Control, Experiment with Disturbance

In the next experiment, the analyses were carried out on the RM motion with the use of a PD controller, assuming a desired trajectory with path in the shape of a semi-circle discussed in Sect. 2.2.1.2. The introduced parametric disturbance consist in loading of the robot's gripper with $m_{RL} = 1.0$ [kg] mass at time $t_{d1} \in < 21; 27 >$ [s] and $t_{d2} \in < 33; 40 >$ [s]. The mass was attached to the gripper, by the operator, when the end-effector was stopped in turning point of executed path which corresponds to the point $G$ of the desired path, and disassembled when the end-effector was in point that corresponds to point $S$ of the desired path.

In Fig. 10.9a the green line represents the desired path of point $C$ of the RM's end-effector and the red line the executed path of point $C$. The initial position (point $S$) is market by a triangle, the desired revers position of the path (point $G$) by "**X**", and rhombus represents the end-position reached by point $C$ of the RM's end-effector. Figure 10.9b presents the error values of the desired path execution of point $C$ of the RM's end-effector, $e_x = x_{Cd} - x_C$, $e_y = y_{Cd} - y_C$, $e_z = z_{Cd} - z_C$.

The values of control signals of drive units of individual links of the RM $u_{PD1}$, $u_{PD2}$ and $u_{PD3}$ are shown in Fig. 10.10a, the absolute values of desired ($|v_{Cd}|$) and executed ($|v_C|$) velocity of point $C$ of the end-effector are shown in Fig. 10.10b. The area of charts illustrating the influence of parametric disturbances on the obtained values are marked by rounded rectangles. During the occurrence of parametric disturbance $t_{d1}$ and $t_{d2}$ the changes in values of individual signals in comparison to obtained sequences for the not-loaded manipulator can be noticed. The most evident are changes in value of the control signal $u_{PD3}$.

The values of errors of the desired angles of rotation ($e_1$, $e_2$, $e_3$) execution and errors of the desired angular velocities ($\dot{e}_1$, $\dot{e}_2$, $\dot{e}_3$) execution of links 1, 2 and 3 of the RM are presented in Fig. 10.11a and b, respectively. By analyzing the values of errors of the trajectory execution the increase in values of errors during $t_{d1}$ and $t_{d2}$, in relation to the experiment process without a disturbance, can be noticed.

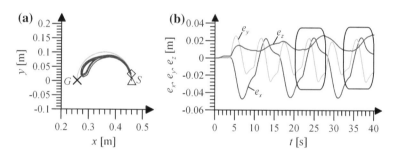

**Fig. 10.9** **a** Desired and executed path of point $C$ of the robotic manipulator's end-effector, **b** error values of trajectory $e_x$, $e_y$, $e_z$ execution

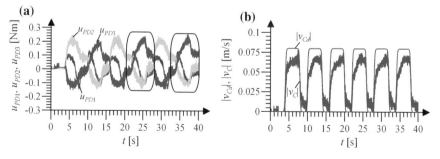

**Fig. 10.10** **a** Values of control signals $u_{PD1}$, $u_{PD2}$ and $u_{PD3}$, **b** absolute value of desired ($|v_{Cd}|$) and executed ($|v_C|$) velocity of point $C$ of the end-effector

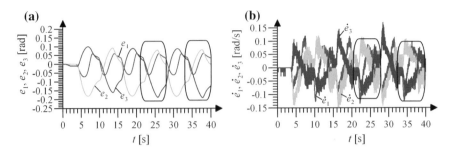

**Fig. 10.11** **a** Error values of desired angles of rotation $e_1$, $e_2$ end $e_3$ execution **b** error values of desired angular velocities $\dot{e}_1$, $\dot{e}_2$ and $\dot{e}_3$ execution

**Table 10.7** Values of selected performance indexes

| Index | $e_{maxj}$ [rad] | $\varepsilon_j$ [rad] | $\dot{e}_{maxj}$ [rad/s] | $\dot{\varepsilon}_j$ [rad/s] | $s_{maxj}$ [rad/s] | $\sigma_j$ [rad/s] |
|---|---|---|---|---|---|---|
| Link 1, $j = 1$ | 0.072 | 0.042 | 0.135 | 0.048 | 0.148 | 0.064 |
| Link 2, $j = 2$ | 0.183 | 0.098 | 0.148 | 0.053 | 0.248 | 0.111 |
| Link 3, $j = 3$ | 0.191 | 0.094 | 0.169 | 0.049 | 0.255 | 0.107 |

**Table 10.8** Values of selected performance indexes of path execution

| Index | $d_{max}$ [m] | $d_n$ [m] | $\rho_d$ [m] |
|---|---|---|---|
| Value | 0.0489 | 0.0319 | 0.0318 |

The values of individual performance indexes of tracking motion execution of the RM with control generated by the PD controller during the experiment with parametric disturbance are shown in Tables 10.7 and 10.8.

The obtained values of performance indexes are slightly different from the values obtained in the numerical test Sect. 7.1.2.5.

## 10.2.3  Conclusions

The calculated values of performance indexes of tracking motion execution with PD controller are similar to those obtained in numerical tests. The quality of control is low, which is evidenced by the large value of assumed performance indexes. Errors of execution of desired angles of rotation assume large values in the initial stage of motion - acceleration. During the parametric disturbance it is possible to notice an increase in the values of errors of desired trajectory execution. Increased values of errors occur throughout the phase of motion of the controlled object with additional mass.

## 10.3 Analysis of the Adaptive Control

The verification tests of adaptive tracking control algorithm of point $A$ of the WMR, which synthesis is presented in Sect. 7.2.1, were carried out. Experiments were carried out on laboratory stand described in Sect. 10.1.1. The implemented trajectory with path in the shape of digit 8 is presented in Sect. 2.1.1.2. One experiment was selected, from results of the conducted tests, in which the trajectory with path in the shape of digit 8 in version with parametric disturbance in the form of additional mass carried by the WMR Pioneer 2-DX was implemented.

In all experimental tests of adaptive tracking control system the same parametric values, as in the numerical test provided in Sect. 7.2.2, were applied.

The performance indexes presented in Sect. 7.1.2 were applied in order to evaluate the quality of tracking motion execution.

Selected results of conducted verification test, in version with trajectory with path in the shape of digit 8 execution with parametric disturbance, are listed below.

### 10.3.1 Analysis of the WMR Motion Control

Experimental test was carried out on adaptive tracking control algorithm of point $A$ of the WMR Pioneer 2-DX in the task of trajectory execution with path in the shape of digit 8 by introducing the double parametric disturbance in the form of changes in the parameters of the WMR corresponding to the loading of robot's frame with $m_{RL} = 4.0$ [kg] mass at time $t_{d1} = 12.5$ [s], and the removing of additional mass at time $t_{d2} = 33$. The estimated time of disturbance occurrence results from the conditions of the carried out experiment, where the human factor had an influence on time of loading of the robot's frame with additional mass. The markings were the same as in the previous experiments.

The values of overall control signals of the adaptive control system $u_1$ and $u_2$ are shown in Fig. 10.12a. The overall control signals consist of signals compensating for the nonlinearities of the WMR ($\hat{f}_1$ and $\hat{f}_2$) and control signals generated by the PD controller ($u_{PD1}$ and $u_{PD2}$) presented in Fig. 10.12b and c, respectively. The inverse values of the PD controller control signals ($-u_{PD1}$ and $-u_{PD2}$) were presented, in order to compare them with sequences of control signals of the PD regulators of other algorithms and more suitable evaluation of their influence on the overall control signals of the adaptive control system. The values of the WMR parameters estimates are shown in Fig. 10.12d. Zero initial values of the parameter estimates were assumed. The highest values are obtained by the parameters of the adaptive system $\hat{a}_5$ and $\hat{a}_6$, which are estimates of the parameters modeling the resistance to motion of the WMR. During the disturbances occurrence the values of parameters are changed in order to compensate for the changes in the controlled object dynamics.

Because the compensation for nonlinearities of the WMR Pioneer 2-DX in the initial phase of the first stage of motion is not sufficient , the influence of control

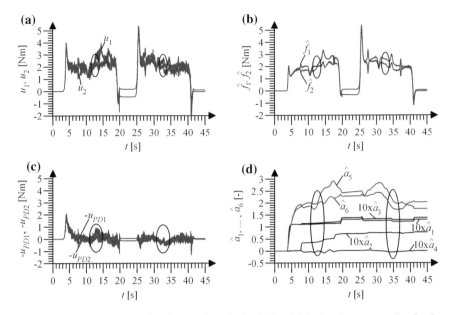

**Fig. 10.12** **a** Overall control signals $u_1$ and $u_2$ of wheels 1 and 2, **b** signals compensating for the nonlinearities of the WMR $\hat{f}_1$ and $\hat{f}_2$, **c** control signals $-u_{PD1}$ and $-u_{PD2}$, **d** values of parameters estimates $\hat{a}_1, \ldots, \hat{a}_6$

signals generated by the PD controller in overall control signals is large, due to the assumption of zero initial values of the adaptive algorithm parameters. Progress of the parameters adaptation process increases the presence of the signals compensating for the nonlinearities in the generated overall control signals, at the expense of control signals of the PD controller. During the occurrence of parametric disturbances the values of control signals of the adaptive algorithm compensate for changes in the controlled object dynamics.

In the first phase of the second stage of motion, acceleration, control signals of the adaptive algorithm ensure good compensation for the nonlinearities of the controlled object. Control signals $u_{PD1}$ and $u_{PD2}$ of the PD controller assume values close to zero. The changes in the dynamics of the controlled object, resulting from the parametric disturbances at time $t_{d1}$ and $t_{d2}$, are compensated by the adaptive algorithm. This modifies the generated values of the control signals $\hat{f}_1$ and $\hat{f}_2$ after introducing the disturbances. During the disturbance $t_{d1}$, an increase in the value of the control signals connected to the compensation for the changes in the dynamics of the WMR can be noticed. During the occurrence of the second disturbance $t_{d2}$, the reduction in the value of control signals is noticeable. This results from the removal of the additional mass carried by the WMR.

The values of errors of desired angles of rotation $(e_1, e_2)$ execution and desired angular velocities $(\dot{e}_1, \dot{e}_2)$ execution of the WMR drive wheels 1 and 2 are presented in Fig. 10.13a and b, respectively. The highest values of errors of trajectory execution occur in the initial phase of the first stage of motion, acceleration, due to the applica-

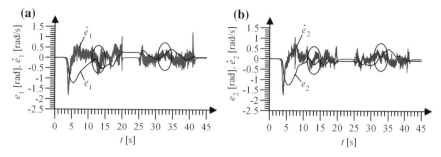

**Fig. 10.13** **a** Error values of the desired motion parameters $e_1$ and $\dot{e}_1$, **b** error values of the desired motion parameters $e_2$ and $\dot{e}_2$

tion of zero initial values of the adaptive system parameters. In this stage of motion, the overall control signals are generated by the PD controller and the adaptation of the WMR nonlinearity does not take place. Because of that, the tracking errors are the largest. During the parameters adaptation process of system which generates the control signals compensating for the nonlinearities of the WMR, the values of errors of the trajectory execution are reduced. In the initial phase of the second stage of motion, acceleration, the values of errors of the trajectory execution are smaller in comparison to the acceleration phase of the first stage of motion, which results from the selection of parameters values of the algorithm compensating for the nonlinearities of the WMR in the adaptation process. By analyzing the sequences of errors of the trajectory execution it is possible to notice an increase in their value after time $t_{d1}$ and $t_{d2}$, due to the parametric disturbance occurrence. The influence of the parametric disturbances on the process of tracking motion execution is compensated by the adaptive system and generation of control signals including a change in the WMR dynamics.

The sequences of signals values, obtained during experimental tests, are disturbed. This is a result of many factors, among others, operation of optical encoders or non-parametric disturbances in the form of irregularity of the laboratory track structure. All sequences in the discrete steps of the process, with discretization step $h = 0.01$ [s], were recorded during the experimental tests. Figure 10.13 presents a characteristic "measurement noise"of recorded sequences of error values of the angular velocities of the WMR drive wheels rotation execution. The filters smoothing the $\dot{\alpha}_1$, $\dot{\alpha}_2$ values were not used in the measurement system, because they introduce a time delay in the processed signal and can cause an additional error of the desired trajectory execution. The values of control signals of the adaptive algorithm are clearly "smoother"than control signals of the PD controller, which are generated on the basis of sequences of errors of the desired motion parameters execution.

Figure 10.14 presents the desired and executed path of point $A$ of the WMR with the same markings as in the previous experiments. Enlargement of the selected area of the chart (on the left side) is presented on the right side.

The values of individual performance indexes of tracking motion execution with adaptive control system are provided in Tables 10.9 and 10.10.

**Fig. 10.14** Desired and
executed trajectory of point
A of the WMR

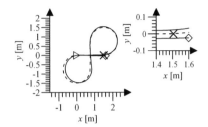

**Table 10.9** Values of selected performance indexes

| Index | $e_{maxj}$ [rad] | $\varepsilon_j$ [rad] | $\dot{e}_{maxj}$ [rad/s] | $\dot{\varepsilon}_j$ [rad/s] | $s_{maxj}$ [rad/s] | $\sigma_j$ [rad/s] |
|---|---|---|---|---|---|---|
| Wheel 1, $j = 1$ | 1.22 | 0.38 | 1.81 | 0.26 | 2.06 | 0.32 |
| Wheel 2, $j = 2$ | 1.26 | 0.36 | 1.75 | 0.25 | 2.07 | 0.31 |

**Table 10.10** Values of selected performance indexes of path execution

| Index | $d_{max}$ [m] | $d_n$ [m] | $\rho_d$ [m] |
|---|---|---|---|
| Value | 0.166 | 0.097 | 0.087 |

The obtained values of all performance indexes, in comparison to the execution
of trajectory with control system built only from a PD controller, were lower.

The values of selected performance indexes of the first and second stage of motion
along the path in the shape of digit 8, calculated according to the methodology
described in the numerical test Sect. 7.1.2.3, are provided in Table 10.11.

The values of the performance indexes, in the second stage of motion execution,
are lower than those obtained in the first stage. This is due to the application of
knowledge of the controlled object, stored in the adaptive system parameters, in
order to generate a control signal at the beginning of the second stage of motion. The
algorithm, compensating for the nonlinearities of the controlled object, generates a
control signals that ensures a good quality of tracking motion execution in the second
stage of motion, while the control signals of the PD controller assume values close
to zero.

**Table 10.11** Values of selected performance indexes in I and II stage of motion

| Index | $e_{maxj}$ [rad] | $\varepsilon_j$ [rad] | $\dot{e}_{maxj}$ [rad/s] | $\dot{\varepsilon}_j$ [rad/s] | $s_{maxj}$ [rad/s] | $\sigma_j$ [rad/s] |
|---|---|---|---|---|---|---|
| Stage I, $k = 1, \ldots, 2250$ | | | | | | |
| Wheel 1, $j = 1$ | 1.22 | 0.49 | 1.81 | 0.32 | 2.06 | 0.41 |
| Wheel 2, $j = 2$ | 1.26 | 0.46 | 1.75 | 0.32 | 2.07 | 0.39 |
| Stage II, $k = 2251, \ldots, 4500$ | | | | | | |
| Wheel 1, $j = 1$ | 0.43 | 0.18 | 1.27 | 0.17 | 1.3 | 0.18 |
| Wheel 2, $j = 2$ | 0.5 | 0.21 | 0.62 | 0.15 | 0.65 | 0.19 |

## 10.3.2  Conclusions

Adaptive algorithm of tracking control of the WMR ensures better quality of control, which is expressed by lower values of performance indexes when compared to the PD controller. When the parametric disturbance occurs during the motion, the adaptive algorithm is compensating for the influence of changes in the dynamics of the controlled object caused by loading the robot with additional mass. The biggest errors of trajectory execution occur in the initial stage of motion. This results from assuming the worst case scenario of zero initial values of estimated parameters. Application of non-zero initial values of estimated parameters, e.g., the values of end parameters from the carried out experiment, provides better quality of control and reduction of errors of trajectory execution in the initial stage of motion. Rich system stimulation through execution of more diversified trajectory has a positive influence on the parameters adaptation process.

## 10.4  Analysis of the Neural Control

The verification tests of the neural control algorithm, which synthesis is presented in Sect. 7.3.1, were carried out. Experiments were conducted on laboratory stand described in Sect. 10.1.1. Trajectory with path in the shape of digit 8, described in Sect. 2.1.1.2, was implemented. From results of conducted experimental tests selected were sequences obtained by application of trajectory with path in the shape of digit 8 in version with parametric disturbance in the form of an additional mass carried by the WMR Pioneer 2-DX.

The same values of parameters, as in the numerical tests described in Sect. 7.1.2, were applied in all experimental tests of neural tracking control system.

The performance indexes, presented in Sect. 7.1.2, were implemented in order to evaluate the quality of the tracking motion execution.

The following are selected results of the neural control verification tests conducted with parametric disturbance and trajectory with path in the shaped of digit 8.

### 10.4.1  Analysis of the WMR Motion Control

The experimental tests were carried out on neural tracking control algorithm of point $A$ of the WMR Pioneer 2-DX in the task of execution of the trajectory with path in the shape of digit 8, with double parametric disturbance in the form of change in the parameters of the WMR corresponding to loading of the robot's frame with $m_{RL} = 4.0$ [kg] mass at time $t_{d1} = 12$ [s], and removal of the additional mass at time $t_{d2} = 33$ [s]. The same markings, as in the previous numerical tests, were applied.

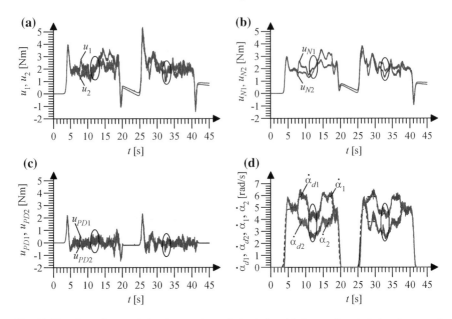

**Fig. 10.15** **a** Overall control signals $u_1$ and $u_2$ of wheels 1 and 2, **b** neural control signals $u_{N1}$ and $u_{N2}$, **c** control signal of PD controller $u_{PD1}$ and $u_{PD2}$, **d** values of desired ($\dot{\alpha}_{d1}, \dot{\alpha}_{d2}$) and executed ($\dot{\alpha}_1, \dot{\alpha}_2$) angular velocities of the WMR's drive wheels rotation

The values of overall control signals of the neural control system are shown in Fig. 10.15a. They consist of compensating control signals generated by the NN (Fig. 10.15b) and control signals of the PD controller (Fig. 10.15c). Values of desired ($\dot{\alpha}_{d1}, \dot{\alpha}_{d2}$) and executed ($\dot{\alpha}_1, \dot{\alpha}_2$) angular velocities of the WMR's drive wheels rotation are presented in Fig. 10.15d.

In the initial phase of the first stage of motion the control signals generated by the PD controller have a large percentage in the generated overall control signals due to the lack of insufficient compensation for the WMR Pioneer 2-DX nonlinearities implemented by the NN. The progress of the NN output layer weights adaptation process increases, at the PD controller signals expense, the percentage of control signals $u_{N1}$ and $u_{N2}$ in the generated overall control signals. The values of NN control signals, during the parametric disturbances, are adopted to changes in the dynamics of the controlled object. In the first phase of the second stage of motion, acceleration, the control signals of the PD controller assume large values, which may indicate that the knowledge stored in the NN weights does not provide a full compensation for the nonlinearities of the controlled object whose dynamics was changed by the carried additional mass. In contrast, the values of control signals in the breaking phase of the second stage of motion are significantly lower than in the breaking phase of the first stage of motion.

Figure 10.16 presents tracking errors of the WMR Pioneer 2-DX trajectory execution.

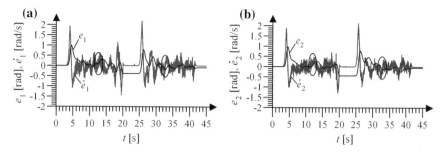

**Fig. 10.16** **a** Error values of the desired motion parameters $e_1$ and $\dot{e}_1$, **b** error values of the desired motion parameters $e_2$ and $\dot{e}_2$

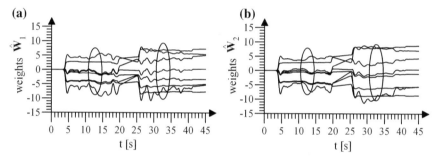

**Fig. 10.17** **a** Values of the NN 1 $\hat{\mathbf{W}}_1$ output layer weights, **b** values of the NN 2 $\hat{\mathbf{W}}_2$ output layer weights

The values of errors of the trajectory execution reach the largest values in acceleration phases of the first and second stage of motion. The occurrence of the parametric disturbances during $t_{d1}$ and $t_{d2}$ causes a temporary increase in the errors values of the trajectory execution, which is minimized by the NN weights adaptation adjusting the control signals to changing working conditions of the controlled object. The difference in the values of errors between the breaking phase of the first stage of motion and the breaking phase of the second stage of motion are noticeable. In the first case the values of errors are higher.

The values of NN output layer weights are presented in Fig. 10.17. Zero initial values of output layer weights were adjusted in the NN initialization process. NN output layer weights remain bounded and are stabilizing around certain values. Changes in the values of the NN output layer weights occur during the parametric disturbances. The changes result from the course of the adaptation process and are implemented in order to compensate for the changes in the dynamics of the controlled object.

The desired (dashed line) and executed (continuous line) path of point $A$ of the WMR is located on the left side of the Fig. 10.18. The same markings, as in the previous experiments, were applied. Enlargement of the selected area of the chart (on the left side) is presented on the right side.

**Fig. 10.18** Desired and
executed path of point $A$ of
the WMR

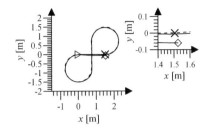

**Table 10.12** Values of selected performance indexes

| Index | $e_{maxj}$ [rad] | $\varepsilon_j$ [rad] | $\dot{e}_{maxj}$ [rad/s] | $\dot{\varepsilon}_j$ [rad/s] | $s_{maxj}$ [rad/s] | $\sigma_j$ [rad/s] |
|---|---|---|---|---|---|---|
| Wheel 1, $j = 1$ | 1.0 | 0.25 | 2.19 | 0.38 | 2.33 | 0.41 |
| Wheel 2, $j = 2$ | 0.97 | 0.27 | 2.14 | 0.34 | 2.26 | 0.37 |

**Table 10.13** Values of selected performance indexes of path execution

| Index | $d_{max}$ [m] | $d_n$ [m] | $\rho_d$ [m] |
|---|---|---|---|
| Value | 0.085 | 0.062 | 0.039 |

**Table 10.14** Values of selected performance indexes in I and II stage of motion

| Index | $e_{maxj}$ [rad] | $\varepsilon_j$ [rad] | $\dot{e}_{maxj}$ [rad/s] | $\dot{\varepsilon}_j$ [rad/s] | $s_{maxj}$ [rad/s] | $\sigma_j$ [rad/s] |
|---|---|---|---|---|---|---|
| Stage I, $k = 1, \ldots, 2250$ | | | | | | |
| Wheel 1, $j = 1$ | 1.0 | 0.28 | 1.94 | 0.39 | 2.2 | 0.42 |
| Wheel 2, $j = 2$ | 0.97 | 0.27 | 1.89 | 0.35 | 2.15 | 0.38 |
| Stage II, $k = 2251, \ldots, 4500$ | | | | | | |
| Wheel 1, $j = 1$ | 0.79 | 0.2 | 2.19 | 0.38 | 2.33 | 0.39 |
| Wheel 2, $j = 2$ | 0.78 | 0.23 | 2.14 | 0.34 | 2.26 | 0.35 |

The values of individual performance indexes of the tracking motion execution
with the use of neural control system are presented in Tables 10.12 and 10.13.

The values of selected performance indexes were lower in comparison to those
obtained within the execution of trajectory with the use of tracking control system
built only from a PD controller. The values of performance indexes are comparable
to those obtained in the process of verification of the adaptive control system.

Table 10.14 presents the values of selected performance indexes of the first and
second stage of motion along the path in the shape of digit 8, which were calculated
according to the methodology described in Sect. 7.1.2.3.

The values of selected performance indexes obtained in the second stage of motion
are slightly lower than those calculated for the first stage of motion.

## 10.4.2 Conclusions

The designed neural tracking control system of point $A$ of the WMR Pioneer 2-DX, implemented with the use of RVFL NN, in order to compensate for the nonlinearity of the controlled object, guarantees high accuracy of tracking motion execution. The influence of parametric disturbances on the WMR's point $A$ motion is compensated during the NN output layer weights adaptation process. NN output layer weights remain bounded and are stabilizing around certain values.

## 10.5 Analysis of the HDP Control

Experimental tests were carried out on tracking control system with the use of HDP structures, which synthesis is shown in Sect. 7.4.1. The tests were conducted on a laboratory stand described in Sect. 10.1.1. The same trajectory with the path in the shape of digit 8, as described in Sect. 2.1.1.2, was implemented. Two experiments were selected among the carried out tests. The first one with implemented trajectory with path in the shape of digit 8 in version with parametric disturbance in the form of additional mass carried by the WMR Pioneer 2-DX and zero initial values of actor and critic's NN output layer weights, and the second one with non-zero initial values of critic's NN output layer weights only.

In all experimental tests of the tracking control system with HDP structures, the same values of parameters, as in the numerical tests described in Sect. 7.4.2, were applied. In order to evaluate the quality of generated control and tracking motion execution the performance indexes, presented in Sect. 7.4.2, were applied.

The description of variables, in the description of verification tests of algorithms operating in the discrete time domain, was simplified by resigning from the index of the $k$-th step of the process. In order to facilitate the analysis of results and enable comparison against the continuous sequences of previously presented tracking control systems of point $A$ of the WMR Pioneer 2-DX, the abscissa axis of charts of discrete sequences was calibrated in time $t$.

## 10.5.1 Analysis of the WMR Motion Control

This subchapter presents selected results of experimental tests carried out on WMR Pioneer 2-DX motion with the use of a trajectory with the path in the shape of digit 8, with parametric disturbance. Experiment was conducted with zero initial values of actor and critic's NN output layer weights Sect. 10.5.1.1 and with non-zero initial values of critic's NN output layer weights only Sect. 10.5.1.2. A significant influence of implementing the non-zero values of critic's NN output layer weights on the quality of the WMR's tracking motion execution was noticed.

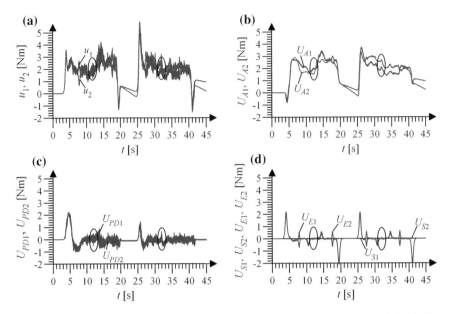

**Fig. 10.19** **a** Overall control signals $u_1$ and $u_2$ for wheels 1 and 2, **b** actor's NNs control signals $U_{A1}$ and $U_{A2}$, $\mathbf{U}_A = -\frac{1}{h}\mathbf{Mu}_A$, **c** PD control signals $U_{PD1}$ and $U_{PD2}$, $\mathbf{U}_{PD} = -\frac{1}{h}\mathbf{Mu}_{PD}$, **d** supervisory term's control signals $U_{S1}$ and $U_{S2}$, $\mathbf{U}_S = -\frac{1}{h}\mathbf{Mu}_S$, and additional control signals $U_{E1}$ and $U_{E2}$, $\mathbf{U}_E = -\frac{1}{h}\mathbf{Mu}_E$

### 10.5.1.1 Analysis of the HDP Control, Zero Initial Values of Critic's NN Output Layer Weights

Verification tests were carried out on tracking control system of point $A$ of the WMR with the use of NDP structures in HDP configuration assuming zero initial values of NN output layer weights of actor and critic's structures. The task of the control system was to generate control signals providing execution of the desired trajectory of point $A$ of the WMR with path in the shape of digit 8, with double parametric disturbance resulting from loading of the WMR Pioneer 2-DX frame with $m_{RL} = 4.0$ [kg] mass during $t_{d1} = 12$ [s], and removal of additional mass during $t_{d2} = 32$ [s].

The values of signals of individual structures included in the control system are presented in Fig. 10.19. The overall control signals $u_1$ and $u_2$ of wheels 1 and 2 are shown in Fig. 10.19a and according to the relation (7.66) consist of control signals generated by HDP structures $u_{A1}$, $u_{A2}$ (Fig. 10.19b), control signals of the PD controller $u_{PD1}$, $u_{PD2}$ (Fig. 10.19c), supervisory control signals $u_{S1}$, $u_{S2}$, and additional control signals $u_{E1}$, $u_{E2}$ (Fig. 10.19d).

In the initial phase of the first stage of motion, acceleration, a large role in the overall control signals $\mathbf{u}$ play signals generated by the PD controller $\mathbf{u}_{PD}$ and signals $\mathbf{u}_E$. Since the worst case scenario of zero initial values of NNs output layer weights of NDP structures was applied, actor's NNs control signals after a while start to play a dominant role in the overall control signals. In the second stage of motion, in NNs output layer weights of NDP structures, a certain knowledge of the controlled object dynamics is stored and the control signals of actor's NNs play a dominant role in overall signals controlling the tracking motion of point $A$ of the WMR. Occurrence of the first parametric disturbance at time $t_{d1}$ increases the value of control signals generated by the structures of adaptive critic, which compensates for the changes in the dynamics of the controlled object. Similarly, the occurrence of the second disturbance during $t_{d2}$ reduces the values of control signals of the HDP structures. The values of control signals $\mathbf{u}_E$ depend mainly on $\mathbf{z}_{d3}$. Changes in the values of tracking errors have a small influence on the sequences of these signals. The values of supervisory control signals $\mathbf{u}_S$ remain equal to zero, because the generalized tracking errors $\mathbf{s}$ does not exceed the assumed acceptable level.

The values of desired and executed angular velocities of the WMR drive wheels 1 and 2 are shown in Fig. 10.20a and b, respectively. The biggest differences in values between the desired and executed angular velocities of the WMR drive wheels rotation occure in the initial phase of the first stage of motion. They result from the insufficient compensation for the WMR nonlinearity by neural critic system in the initial adaptation phase, due to the application of zero initial values of the NNs output layer weights. During the occurrence of parametric disturbance at $t_{d1}$ and $t_{d2}$, a temporary increase in the values of errors of the desired angular velocities of drive wheels can be noticed. However, these differences are minimized due to the NNs output layer weights adaptation process of HDP structures.

The values of errors of the desired trajectory execution of wheels 1 and 2 of the WMR are shown in Fig. 10.21a, b. Figure 10.21c presents the sequence of value of filtered tracking error $s_1$, and Fig. 10.21d a sliding manifold of wheel 1. The largest values of errors of trajectory execution occur in the initial phase of motion. They are minimized due to the NNs output layer weights adaptation process of HDP structures. During the occurrence of parametric disturbances the absolute values of tracking errors are increasing, however, due to the NNs output layer weights adaptation process the control signals of NDP structures are compensating for the influence of changes in the dynamics of the controlled object. Errors of trajectory

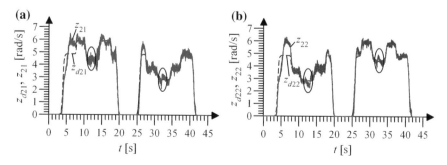

**Fig. 10.20  a** Values of desired ($z_{d21}$) and executed ($z_{21}$) angular velocity of drive wheel 1 of the WMR, **b** values of desired ($z_{d22}$) and executed ($z_{22}$) angular velocity of drive wheel 2 of the WMR

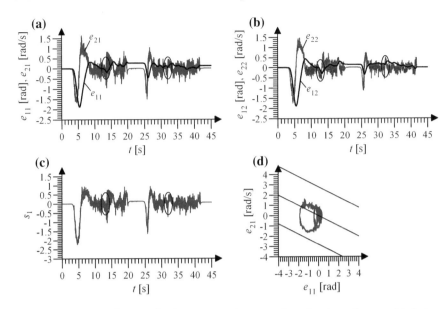

**Fig. 10.21** **a** Values of errors of trajectory execution $e_{11}$ and $e_{21}$, **b** values of errors of trajectory execution $e_{12}$ and $e_{22}$, **c** values of filtered tracking error $s_1$, **d** sliding manifold of wheel 1

execution, in the initial phase of the second stage of motion along the trajectory in the shape of curved path in the rightward direction have smaller values than at the beginning of the first stage of motion, due to the information regarding the dynamics of the controlled object contained in the structures of actor and critic's NNs output layer weights.

The values of actor and critic's NNs output layer weights of the first HDP structure are shown in Fig. 10.22a, b. The values of NNs output layer weights of the second HDP structure, generating the control signal of wheel 2, are similar. Zero initial values of the NNs output layer weights were assumed. The values of output layer weights stabilize around certain values and remain bounded. During the occurrence of parametric disturbances the changes in actor's NNs output layer weights, resulting from the adjustments of generated signals compensating for the nonlinearity of the WMR to the changed parameters of the controlled object, are noticeable. The RVFL NNs were implemented in actor's structures. In contrast, the critic's structures were implemented with the use of NNs with activation functions of Gaussian curves type.

The desired (dashed line) and executed (continuous line) path of point $A$ of the WMR is shown in Fig. 10.23. The same markings were implemented as in the previous experiments. Enlargement of the selected area of the chart (on the left side) is presented on the right side.

The values of individual performance indexes of tracking motion execution with the use of control system with adaptive critic structures in HDP configuration are presented in Tables 10.15 and 10.16.

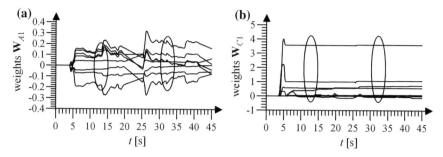

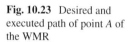

Fig. 10.22 **a** Values of actor 1 $\mathbf{W}_{A1}$ NN output layer weights, **b** values of critic 1 $\mathbf{W}_{C1}$ NN output layer weights

Fig. 10.23 Desired and executed path of point $A$ of the WMR

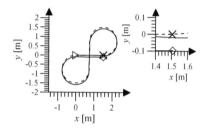

Table 10.15 Values of selected performance indexes

| Index | $e_{max1j}$ [rad] | $\varepsilon_{1j}$ [rad] | $e_{max2j}$ [rad/s] | $\varepsilon_{2j}$ [rad/s] | $s_{maxj}$ [rad/s] | $\sigma_j$ [rad/s] |
|---|---|---|---|---|---|---|
| Wheel 1, $j = 1$ | 1.88 | 0.38 | 1.67 | 0.39 | 2.22 | 0.43 |
| Wheel 2, $j = 2$ | 1.88 | 0.37 | 1.75 | 0.37 | 2.26 | 0.42 |

Table 10.16 Values of selected performance indexes of path execution

| Index | $d_{max}$ [m] | $d_n$ [m] | $\rho_d$ [m] |
|---|---|---|---|
| Value | 0.159 | 0.096 | 0.085 |

Obtained values of selected performance indexes were lower in comparison to the execution of the trajectory with the use of tracking control system built from a PD controller only. However, the values of performance indexes are larger than those obtained during the verification test of adaptive and neural control system.

Table 10.17 presents the values of selected performance indexes of first and second stage of motion execution along the path in the shape of digit 8, calculated according to the methodology described in the numerical test Sect. 7.1.2.2.

The values of implemented performance indexes, obtained in the second stage of motion, were lower than those obtained in the first stage. A significant difference occurs in the case of performance index $\varepsilon_{1j}$. This results from the application of non-zero values of NNs output layer weights in the initial phase of motion of the second

**Table 10.17** Values of selected performance indexes in I and II stage of motion

| Index | $e_{max1j}$ [rad] | $\varepsilon_{1j}$ [rad] | $e_{max2j}$ [rad/s] | $\varepsilon_{2j}$ [rad/s] | $s_{maxj}$ [rad/s] | $\sigma_j$ [rad/s] |
|---|---|---|---|---|---|---|
| Stage I, $k = 1, \ldots, 2250$ | | | | | | |
| Wheel 1, $j = 1$ | 1.88 | 0.51 | 1.67 | 0.47 | 2.22 | 0.54 |
| Wheel 2, $j = 2$ | 1.88 | 0.5 | 1.75 | 0.48 | 2.26 | 0.55 |
| Stage II, $k = 2251, \ldots, 4500$ | | | | | | |
| Wheel 1, $j = 1$ | 0.45 | 0.14 | 1.5 | 0.27 | 1.64 | 0.28 |
| Wheel 2, $j = 2$ | 0.34 | 0.14 | 1.12 | 0.2 | 1.24 | 0.22 |

stage of the experiment. Knowledge of the controlled object dynamics, included in the NNs output layer weights of HDP structures, enables generation of signal compensating for the WMR nonlinearities from the beginning of the second stage of motion, which improves the quality of tracking motion control.

### 10.5.1.2   Analysis of the HDP Control, Non-zero Initial Values of Critic's NN Output Layer Weights

Experimental tests were carried out on the algorithm controlling the tracking motion of point $A$ of the WMR with the use of NDP structures in HDP configuration assuming non-zero initial values of the critic's ($\mathbf{W}_{Cj\{k=0\}} = \mathbf{1}$) NNs output layer weights. Initial values of actor's NNs output layer weights were assumed zero. The task of the control system was to generate control signals providing the execution of the desired trajectory of point $A$ of the WMR with path in the shape of digit 8, with double parametric disturbance implemented in the form of loading of the robot's (Pioneer 2-DX) frame with $m_{RL} = 4.0$ [kg] mass at time $t_{d1} = 12$ [s], and removal of the additional mass during $t_{d2} = 32$ [s].

Values of the overall signals controlling the tracking motion and control signals of the individual structures included in the system of motion execution of the WMR Pioneer 2-DX with the use of NDP structures in HDP configuration are presented in Fig. 10.24, with the same markings as in the previous experimental test.

Application of non-zero initial values of critic's NNs output layer weights resulted in acceleration of the HDP structures adaptation process. Values of the PD controller's control signals in the initial phase of the first stage of motion are smaller than those obtained in experimental tests with zero initial values of the critic's NNs output layer weights in the experiment Sect. 10.5.1.1. In subsequent stages of motion they are reduced to values close to zero. A slight changes in the PD controller's control signals, during the disturbance occurrence, are noticeable. The influence of changes of the WMR's parameters on the object's dynamics is compensated by adjustment of values of the actor's NNs control signals of HDP structures.

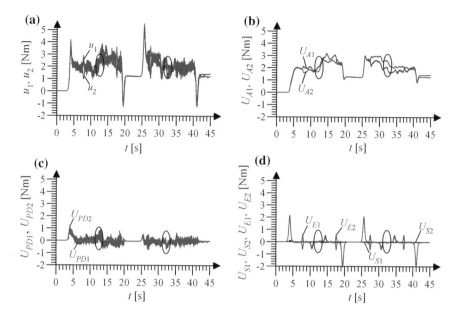

**Fig. 10.24** **a** Overall control signals $u_1$ and $u_2$ for wheels 1 and 2, **b** actor's NNs control signals $U_{A1}$ and $U_{A2}$, $\mathbf{U}_A = -\frac{1}{h}\mathbf{Mu}_A$, **c** PD control signals $U_{PD1}$ and $U_{PD2}$, $\mathbf{U}_{PD} = -\frac{1}{h}\mathbf{Mu}_{PD}$, **d** supervisory term's control signals $U_{S1}$ and $U_{S2}$, $\mathbf{U}_S = -\frac{1}{h}\mathbf{Mu}_S$, and additional control signals $U_{E1}$ and $U_{E2}$, $\mathbf{U}_E = -\frac{1}{h}\mathbf{Mu}_E$

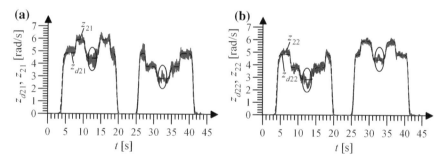

**Fig. 10.25** **a** Values of desired ($z_{d21}$) and executed ($z_{21}$) angular velocity of drive wheel 1 of the WMR, **b** values of desired ($z_{d22}$) and executed ($z_{22}$) angular velocity of drive wheel 2 of the WMR

The values of executed angular velocities of the WMR's drive wheels rotation, shown in Fig. 10.25, are similar to values of the desired angular velocities. During the occurrence of disturbance, at $t_{d1}$, resulting from loading of the WMR with additional mass, the values of angular velocities are decreasing, because the generated control signals are not sufficient enough to execute the motion of the WMR with desired velocity due to increased resistances to motion. The increase in value of singlas compensating for the nonlinearity, generated by the HDP structures, increases the value

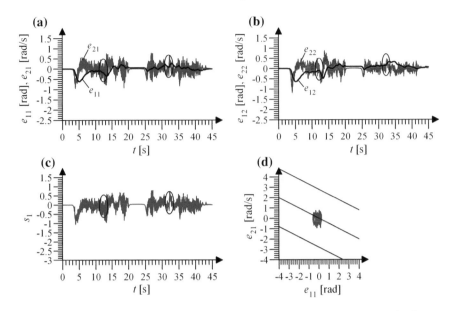

**Fig. 10.26  a** Values of errors of trajectory execution $e_{11}$ and $e_{21}$, **b** values of errors of trajectory execution $e_{12}$ and $e_{22}$, **c** values of filtered tracking error $s_1$, **d** sliding manifold of wheel 1

of angular velocities and results in further execution of the trajectory with desired motion parameters. Similarly, during the second parametric disturbance occurrence, at $t_{d2}$, resulting from the removal of the additional mass from the WMR, the values of generated control signals are too large in relation to required ones, due to smaller resistances to motion of an object with reduced mass. Because of that, the value of executed velocity of point $A$ of the WMR is larger than the desired value. As a result of the adaptation process of NNs output layer weights of HDP structures the values of control signals $u_{A1}$ and $u_{A2}$ are reduced, the velocity of motion execution decreases and reaches the desired values.

Values of errors of the desired trajectory execution are shown in Fig. 10.26. As a result of non-zero initial values of the NN output layer weights application of critic's structures the values of errors of the desired motion parameters execution were reduced, especially in the initial phase of the first stage of the motion - acceleration. The influence of parametric disturbances on the quality of tracking motion execution is small, and increase in the absolute values of errors of the desired motion parameters execution caused by changes in the dynamics of the controlled object is compensated by adaptation of the NNs output layer weights of NDP structures.

Values of actor's and critic's NN output layer weights of first HDP structure are shown in Fig. 10.27. Values of NNs output layer weights of the second HDP structure, which generates signal controlling wheel 2, are similar. Zero initial values of the actor's NNs output layer weights and initial values of the critic's NNs output layer weights, equal to one, were assumed. Values of the NNs output layer weights

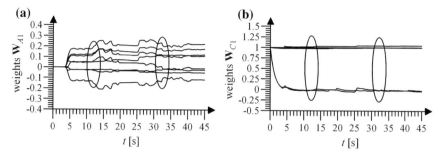

**(a)**

weights $\mathbf{W}_{A1}$

0.4
0.3
0.2
0.1
0
-0.1
-0.2
-0.3
-0.4

0  5  10 15 20 25 30 35 40 45

$t$ [s]

**(b)**

weights $\mathbf{W}_{C1}$

1.5
1.25
1
0.75
0.5
0.25
0
-0.25
-0.5

0  5  10 15 20 25 30 35 40 45

$t$ [s]

**Fig. 10.27** **a** Values of actor 1 $\mathbf{W}_{A1}$ NN output layer weights, **b** values of critic 1 $\mathbf{W}_{C1}$ NN output layer weights

**Fig. 10.28** Desired and executed path of point $A$ of the WMR

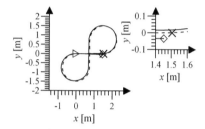

stabilize around certain values and remain limited during the experiment. During the occurrence of parametric disturbances it is possible to notice changes in the actor's NNs output layer weights values, resulting from the adjustments of generated signals compensating for the WMR nonlinearities to the changed parameters of the controlled object. RVFL NNs were applied in actor's structures, while the critic's structures were implemented with the use of the NN with activation functions of Gaussian curves type.

Desired (dashed line) and executed (continuous line) path of selected point $A$ of the WMR is provided in Fig. 10.28. The same markings were used as in the previous experiments. Enlargement of the selected area of the chart (on the left side) is presented on the right side.

Values of individual performance indexes of tracking motion execution with the use of control system with adaptive critic structures in HDP configuration are shown in Tables 10.18 and 10.19.

**Table 10.18** Values of selected performance indexes

| Index | $e_{max1j}$ [rad] | $\varepsilon_{1j}$ [rad] | $e_{max2j}$ [rad/s] | $\varepsilon_{2j}$ [rad/s] | $s_{maxj}$ [rad/s] | $\sigma_j$ [rad/s] |
|---|---|---|---|---|---|---|
| Wheel 1, $j = 1$ | 0.62 | 0.16 | 0.95 | 0.21 | 1.06 | 0.23 |
| Wheel 2, $j = 2$ | 0.64 | 0.2 | 1.18 | 0.19 | 1.33 | 0.21 |

**Table 10.19** Values of selected performance indexes of path execution

| Index | $d_{max}$ [m] | $d_n$ [m] | $\rho_d$ [m] |
|-------|---------------|-----------|--------------|
| Value | 0.089 | 0.063 | 0.054 |

**Table 10.20** Values of selected performance indexes in I and II stage of motion

| Index | $e_{max1j}$ [rad] | $\varepsilon_{1j}$ [rad] | $e_{max2j}$ [rad/s] | $\varepsilon_{2j}$ [rad/s] | $s_{maxj}$ [rad/s] | $\sigma_j$ [rad/s] |
|-------|-------------------|--------------------------|---------------------|----------------------------|--------------------|--------------------|
| Stage I, $k = 1, \dots, 2250$ | | | | | | |
| Wheel 1, $j = 1$ | 0.62 | 0.21 | 0.95 | 0.24 | 1.06 | 0.26 |
| Wheel 2, $j = 2$ | 0.64 | 0.24 | 1.18 | 0.24 | 1.34 | 0.27 |
| Stage II, $k = 2251, \dots, 4500$ | | | | | | |
| Wheel 1, $j = 1$ | 0.26 | 0.09 | 0.84 | 0.19 | 0.79 | 0.19 |
| Wheel 2, $j = 2$ | 0.33 | 0.14 | 0.48 | 0.12 | 0.58 | 0.14 |

Application of non-zero initial values of critic's NNs output layer weights of HDP structures resulted in a significant improvement in the quality of tracking motion execution expressed by lower values of performance indexes in comparison to the implementation of adaptive or neural algorithm and a PD controller. Obtained values of selected performance indexes are similar to those obtained in the numerical test Sect. 7.4.2.2.

Table 10.20 presentes the values of selected performance indexes of first and second stage of motion execution along the path in the shape of digit 8, calculated according to the methodology described in the numerical test Sect. 7.1.2.2.

Similarly as in the numerical tests, during the verification tests a significantly lower values of the performance indexes implemented in the second stage of tracking motion of point $A$ of the WMR Pioneer 2-DX, in comparison to the performance indexes of trajectory execution in the first stage, were obtained.

## 10.5.2  Conclusions

As a result of verification tests carried out on tracking control of point $A$ of the WMR Pioneer 2-DX with the use of NDP structures in HDP configuration, obtained were sequences of individual motion parameters which are similar to the sequences determined in the numerical tests. Application of zero initial values of actor and critic's NNs output layer weights results in slowing down weights adaptation process and a lower quality of tracking motion execution in comparison to the experiment, in which the non-zero initial values of the critic's NNs output layer weights were applied.

## 10.6 Analysis of the DHP Control

Experimental tests were carried out on tracking control system of point $A$ of the WMR Pioneer 2-DX with DHP structure. Its synthesis is shown in Sect. 7.5.1.1. Experimental test were carried out on a laboratory stand described in Sect. 10.1.1. The trajectory with path in the shape of digit 8, described in Sect. 2.1.1.2, was implemented. Two experiments with parametric disturbance, in the form of additional mass carried by the WMR Pioneer 2-DX, and without disturbance were selected. Then the experimental tests were carried out on tracking control system of point $C$ of the end-effector of the RM, which synthesis was presented in Sect. 7.5.1.2. Experimental tests were carried out on a laboratory stand discussed in Sect. 10.1.2. Trajectory with path in the shape of a semi-circle, described in Sect. 2.2.1.2, was implemented. Experiment with parametric disturbance in the form of mass carried by the RM's Scorbot-ER 4pc gripper was selected among other conducted experiments.

In the experimental tests of tracking control systems with DHP algorithm, the same parameter values, as in the numerical tests described in Sect. 7.5.2, were applied. In order to evaluate the quality of tracking motion execution the performance indexes, presented in Sect. 7.4.2, were applied.

The following are selected results of the experimental tests carried out on the algorithm controlling the motion of the WMR with the use of a trajectory with path in the shape of digit 8, with parametric disturbance and with tracking control algorithm of the RM with the use of a trajectory with path in the shape of a semicircle in version with a parametric disturbance.

### 10.6.1 Analysis of the WMR Motion Control

Verification tests were carried out on algorithm controlling the tracking motion of point $A$ of the WMR with the use of NDP structure in DHP configuration assuming zero initial values of actor and critic's NNs output layer weights. The task of the control system was to generate control signals providing the execution of the desired trajectory of point $A$ of the WMR along the trajectory with path in the shape of digit 8, with double parametric disturbance resulting from loading of the robot's frame with $m_{RL} = 4.0$ [kg] mass during $t_{d1} = 12$ [s], and removal of the additional mass during $t_{d2} = 32$ [s].

Values of the overall control signals and control signals of a individual structures of tracking control system of point $A$ of the WMR Pioneer 2-DX are presented in Fig. 10.29 with markings similar to those used in the previous experimental tests.

Adaptation process of the DHP algorithm parameters, analogously as in the numercial test Sect. 7.5.2.2, in the initial phase of the first stage of motion progresses faster in comparison to the implementation of the HDP structures with zero initial values of NNs output layer weights. It triggers the generation of actor's NNs control signals, which presence is significant in the overall control signals from the

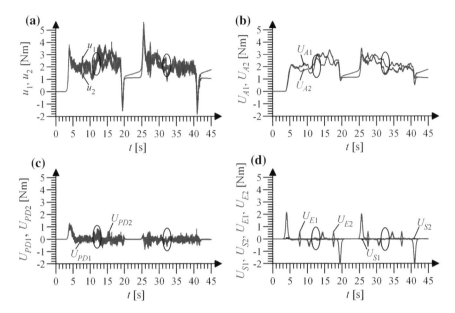

**Fig. 10.29**  **a** Overall control signals $u_1$ and $u_2$ for wheels 1 and 2, **b** actor's NNs control signals $U_{A1}$ and $U_{A2}$, $\mathbf{U}_A = -\frac{1}{h}\mathbf{M}u_A$, **c** PD control signals $U_{PD1}$ and $U_{PD2}$, $\mathbf{U}_{PD} = -\frac{1}{h}\mathbf{M}u_{PD}$, **d** supervisory term's control signals $U_{S1}$ and $U_{S2}$, $\mathbf{U}_S = -\frac{1}{h}\mathbf{M}u_S$, and additional control signals $U_{E1}$ and $U_{E2}$, $\mathbf{U}_E = -\frac{1}{h}\mathbf{M}u_E$

beginning of the WMR motion. Control signals of the PD controller $u_{PD1}$ and $u_{PD2}$ have low values in the initial phase of motion and then are reduced to values close to zero. During the occurrence of disturbances $t_{d1}$ and $t_{d2}$, control signals $u_{A1}$ and $u_{A2}$ compensate for the influence of the additional mass on the changes of the controlled object dynamics. In the second stage of motion the compensation for the nonlinearities of the WMR, implemented with a DHP algorithm, provides a good quality of tracking motion execution because of the information regarding the dynamics of the controlled object contained in the values of NNs output layer weights of actor and critic's structures.

The values of executed angular velocities of drive wheels rotation of the WMR Pioneer 2-DX, presented in Fig. 10.30, are similar to the values of desired angular velocities. During the occurrence of disturbance, at $t_{d1}$, implemented in the form of loading of the WMR with additional mass, the values of angular velocities of motion execution are decreasing, then the DHP algorithm compensates for changes in the dynamics of the controlled object. As a result the values of executed angular velocities of the WMR's drive wheels rotation converge to the desired values. During the occurrence of the second parametric disturbance, at $t_{d2}$, the values of angular velocities are increasing, then the influence of the removal of additional mass on the robot's dynamics is compensated by DHP structure, which generates control signals with reduced values.

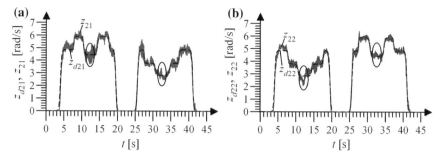

**Fig. 10.30** **a** Values of desired ($z_{d21}$) and executed ($z_{21}$) angular velocity of drive wheel 1 of the WMR, **b** values of desired ($z_{d22}$) and executed ($z_{22}$) angular velocity of drive wheel 2 of the WMR

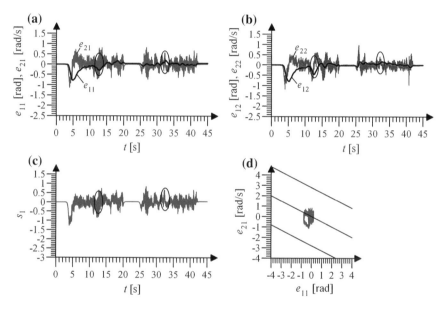

**Fig. 10.31** **a** Values of errors of trajectory execution $e_{11}$ and $e_{21}$, **b** values of errors of trajectory execution $e_{12}$ and $e_{22}$, **c** values of filtered tracking error $s_1$, **d** sliding manifold of wheel 1

Values of errors of the desired trajectory execution of the WMR's wheels 1 and 2 motion are presented in Fig. 10.31a and b, respectively. Figure 10.31c presents the value of filtered tracking error $s_1$. The value of filtered tracking error $s_2$ is similar. Figure 10.31d shows the sliding manifold of wheel 1. The largest values of errors of the trajectory execution occur during the initial phase of motion. They are minimized due to the NNs output layer weights adaptation process of DHP structure. The influence of parametric disturbances on the quality of tracking motion execution is small and an increase in the absolute values of tracking errors caused by changes in the dynamics of the controlled object is compensated by adaptation of the NNs output layer weights of DHP structure.

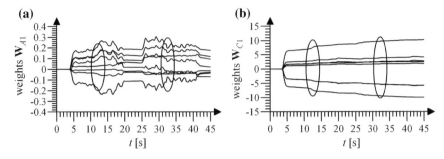

**Fig. 10.32  a** Values of actor 1 $\mathbf{W}_{A1}$ NN output layer weights, **b** values of critic 1 $\mathbf{W}_{C1}$ NN output layer weights

**Fig. 10.33** Desired and executed path of point $A$ of the WMR

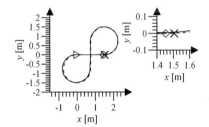

Values of actor and critic's first NN output layer weights of DHP structure are shown in Fig. 10.32. Values of actor and critic's second NN output layer weights are similar. Zero initial values of actor and critic's NNs output layer weights were applied. The values of weights stabilize around certain values and remain bounded during the experiment, and their sequences are similar to those obtained in the numerical test Sect. 7.5.2.2. Changes in the actor's NNs output layer weights, resulting from the adjustments of generated signal compensating for the WMR nonlinearity to the changed parameters of the controlled object can be noticed during the occurrence of parametric disturbances.

The desired (dashed line) and executed (continuous line) path of point $A$ of the WMR Pioneer 2-DX is shown in Fig. 10.33. The same markings as in the previous experiments were applied. A high mapping accuracy of the desired path of point $A$ should be emphasized, whereas the task of the control algorithm was to minimize the errors of execution of the desired angular parameters of wheels motion instead of minimizing the error of the path execution of the robot's point $A$. However, high quality of execution of the desired trajectory in configuration space reflects on the execution of path of the robot's point $A$ similar to the desired one in the working space.

The values of individual performance indexes of tracking motion execution with the use of control system with adaptive critic structure in DHP configuration are provided in Tables 10.21 and 10.22.

The values of selected performance indexes of motion execution are similar to those obtained during the numerical test Sect. 7.5.2.2, and lower in comparison to the

**Table 10.21**  Values of selected performance indexes

| Index | $e_{max1j}$ [rad] | $\varepsilon_{1j}$ [rad] | $e_{max2j}$ [rad/s] | $\varepsilon_{2j}$ [rad/s] | $s_{maxj}$ [rad/s] | $\sigma_j$ [rad/s] |
|---|---|---|---|---|---|---|
| Wheel 1, $j = 1$ | 0.78 | 0.18 | 1.14 | 0.22 | 1.28 | 0.24 |
| Wheel 2, $j = 2$ | 0.8 | 0.18 | 1.17 | 0.21 | 1.34 | 0.23 |

**Table 10.22**  Values of selected performance indexes of path execution

| Index | $d_{max}$ [m] | $d_n$ [m] | $\rho_d$ [m] |
|---|---|---|---|
| Value | 0.068 | 0.053 | 0.041 |

**Table 10.23**  Values of selected performance indexes in I and II stage of motion

| Index | $e_{max1j}$ [rad] | $\varepsilon_{1j}$ [rad] | $e_{max2j}$ [rad/s] | $\varepsilon_{2j}$ [rad/s] | $s_{maxj}$ [rad/s] | $\sigma_j$ [rad/s] |
|---|---|---|---|---|---|---|
| Stage I, $k = 1, \ldots, 2250$ | | | | | | |
| Wheel 1, $j = 1$ | 0.78 | 0.24 | 1.14 | 0.25 | 1.28 | 0.28 |
| Wheel 2, $j = 2$ | 0.8 | 0.25 | 1.17 | 0.26 | 1.34 | 0.29 |
| Stage II, $k = 2251, \ldots, 4500$ | | | | | | |
| Wheel 1, $j = 1$ | 0.2 | 0.06 | 0.83 | 0.18 | 0.85 | 0.18 |
| Wheel 2, $j = 2$ | 0.17 | 0.07 | 0.98 | 0.15 | 1.02 | 0.15 |

values obtained during the verification tests of trajectory execution with the use of tracking control system consisting of PD controller, adaptive control system, neural control system, or control algorithm with HDP structures and zero initial values of actor and critic's NNs output layer weights, described in Sect. 10.5.1.1. The quality of tracking motion execution in the conducted experiment, expressed by values of selected performance indexes, is similar to one obtained with the use of control system with HDP structures and non-zero initial values of the critic's NNs output layer weights in the experiment Sect. 10.5.1.2.

Table 10.23 presents values of selected performance indexes of first and second stage of motion along the path in the shape of digit 8, calculated according to the methodology described in the numerical test Sect. 7.1.2.2.

Obtained values of performance indexes, in the second stage of motion, were lower than in the first stage. This results from the application of non-zero values of the NNs output layer weights in the initial phase of motion of the second stage of the experiment. Knowledge of the controlled object dynamics, included in the NNs output layer weights of DHP structure enables generation of signals compensating for the WMR nonlinearities, which improves the quality of the generated signals controlling the execution of tracking motion. A special attention should be paid to the values of performance index $\varepsilon_{1j}$ of the desired angles of rotation of drive wheels in the second stage of motion, which are lower than those obtained in verification tests of other control algorithms.

### 10.6.2  Analysis of the RM Motion Control

Experimental tests were carried out on tracking control of the RM with the use of NDP algorithm in DHP configuration. The desired trajectory, described in Sect. 2.2.1.2, with path in the shape of a semicircle was implemented. Parametric disturbance in the form of loading of the RM's gripper with $m_{RL} = 1.0$ [kg] mass during $t_{d1} \in\, <21; 27 >$ [s] and $t_{d2} \in\, < 33; 40 >$ [s] was implemented. The mass was attached to the gripper by the operator during stopping of the end-effector at the turning point of executed path corresponding to point $G$ of the desired path and disassembled when the end-effector was at point corresponding to point $S$ of the desired path.

Figure 10.34a presents desired (green line) and executed (red line) path of point $C$ of the RM's end-effector. The initial position (point $S$) is marked on the chart by a triangle. "**X**"represents the point of reversal of the path (point $G$) and the end-position obtained by point $C$ of the RM's end-effector is marked by rhombus. The chart in Fig. 10.34b demonstrates the values of errors of the desired path of point $C$ of the end-effector execution, $e_x = x_{Cd} - x_C$, $e_y = y_{Cd} - y_C$, $e_z = z_{Cd} - z_C$. Increased values of errors of the desired path execution of point $C$ of the RM's arm occur during the periods of motion with additional mass.

The path of point $C$ of the RM's end-effector results from application of the tracking control signals, which sequences are presented in Fig. 10.35, into the actuator systems. The overall control signals $u_1$, $u_2$ and $u_3$ are shown in Fig. 10.35a. They consist of control signals generated by DHP structure $u_{A1}$, $u_{A2}$ and $u_{A3}$ (Fig. 10.35b), control signals of the PD controller $u_{PD1}$, $u_{PD2}$ and $u_{PD3}$ (Fig. 10.35c), supervisory control signals $u_{S1}$, $u_{S2}$ and $u_{S3}$, as well as additional control signals $u_{E1}$, $u_{E2}$ and $u_{E3}$ (Fig. 10.35d). The supervisory control signals are not presented in the charts, because their values, during the experiment, were equal to zero. This results from assumed values of parameter $\rho_j$ of the supervisory control algorithm.

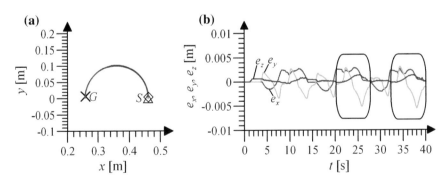

**Fig. 10.34  a** Desired and executed path of point $C$ of the robotic manipulator's end-effector, **b** error values of trajectory $e_x$, $e_y$, $e_z$ execution

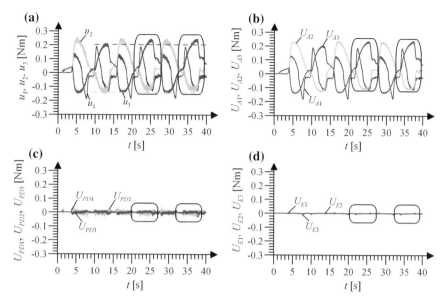

**Fig. 10.35**  **a** Overall control signals $u_1$, $u_2$ oraz $u_3$, **b** actor's NNs control signals $U_{A1}$, $U_{A2}$ i $U_{A3}$, $\mathbf{U}_A = -\frac{1}{h}\mathbf{Mu}_A$, **c** PD control signals $U_{PD1}$, $U_{PD2}$ i $U_{PD3}$, $\mathbf{U}_{PD} = -\frac{1}{h}\mathbf{Mu}_{PD}$, **d** control signals $U_{E1}$, $U_{E2}$ i $U_{E3}$, $\mathbf{U}_E = -\frac{1}{h}\mathbf{Mu}_E$

In the initial phase of the experiment the arm of the RM's was supported in order to prevent its free motion under the influence of gravitational forces and preserve the initial conditions of angles of rotation of individual links, which amount to $z_{11} = 0$, $z_{12} = 0.16533$, $z_{13} = -0.16533$ [rad], respectively. In this position of the RM's arm, point $C$ of the end-effector is at the maximal distance, reached during the execution of the desired trajectory, from the robot's base (point $S$). At $t = 1.5$ [s], the support was removed and the RM's arm was able to move. Links 2 and 3 of the RM's arm, under the influence of torques from the gravitational forces, started to move. This resulted in the occurrence of errors of the desired trajectory execution in configuration space. As a result of the occurrence of errors of the desired trajectory execution the control algorithm generated control signals in order to minimize the errors and prevent further motion of the RM's arm. This is evident in the graph presenting the actor's control signals sequence (Fig. 10.35b) during time $t \in< 1.5; 4 >$ [s]. During the motion the point $C$ of the RM's arm executed three cycles of motion running back and forth from point $S$ and point $G$. During the motion from point $G$ to point $S$ of the second and third cycle the gripper of the RM's was loaded with $m_{RL}$ mass. Loading of the RM's gripper with mass has a small influence on signal $u_1$ which controls the motion of link 1. This link rotates about the $z$ axis of global coordinate system. Considering the transmission ratio of the drive unit of link 1 and inertness of the robot's arm, the influence of additional mass on the motion of this link is negligible small. Figure 10.35a presents the maximum or minimum values of selected control signals, during the motion without loading of the RM's arm, which are marked by dashed lines in appropriate color indicated. It can be noticed that the control signal of link 2 motion, $u_2$, reaches lower values during the motion with load, which results from the configuration of the robot's arm and the specify of the executed motion. Signal $u_3$, which controls the motion of link 3, reaches higher values during the motion with load. Control signals, generated by the actor's NNs of DHP algorithm, are dominant in the values of overall control signals. Other control signals have values close to zero.

Value of desired (red line) and executed (green line) angular velocity of link 2 rotation is shown in Fig. 10.36a. The sequence of the desired absolute value ($|v_{Cd}|$) and the executed value ($|v_C|$) of velocity of point $C$ of the RM's end-effector is presented in Fig. 10.36b. It is difficult to notice the influence of loading of the robot's arm on the aforementioned sequences of executed motion parameters.

Values of errors of the desired angles of rotation execution and angular velocities of RM's links are shown in Fig. 10.37a and b, respectively. Figure 10.37c presents the values of filtered tracking errors $s_1$, $s_2$ and $s_3$. Figure 10.37d demonstarates a sliding manifold of link 1. The largest values of errors of trajectory execution occur during the motion of RM's arm with a gripper loaded with mass. However, the influence of disturbances on the quality of the tracking motion execution is small and an increase in the absolute values of tracking errors, caused by changes in the dynamics of the controlled object, is minimized by adaptation of the NNs output layer weights of NDP structure.

The values of actor and critic's NNs output layer weights of DHP structure are shown in Fig. 10.38. Zero initial values of actor and critic's NNs output layer weights were applied. The values of NNs output layer weights stabilize around certain values and remain bounded during the experiment. Changes in the actor's NNs output layer weights, resulting from the adjustments of signals compensating for the nonlinearities of the RM to the changed parameters of the controlled object, are possible to notice during the occurrence of parametric disturbances. Adaptation of the first NN output layer weights of actor and critic's structures begins when the non-zero value of the

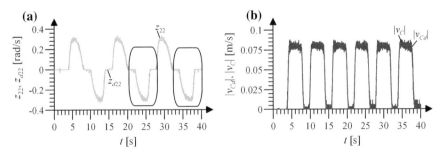

**Fig. 10.36** **a** Values of desired ($z_{d21}$) and executed ($z_{21}$) angular velocity of the robotic manipulator's link 2 rotation, **b** sequence of absolute value of desired ($|v_{Cd}|$) and executed ($|v_C|$) velocity of point $C$ of the manipulator's end-effector

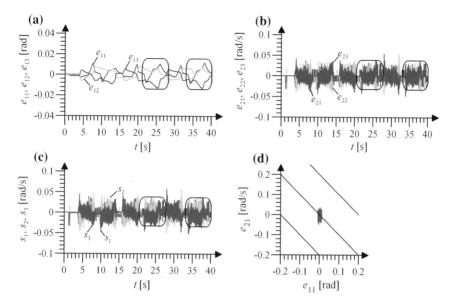

**Fig. 10.37** **a** Values of errors of desired angles of rotation $e_{11}$, $e_{12}$ and $e_{13}$ execution, **b** values of errors of desired angular velocities $e_{21}$, $e_{22}$ and $e_{23}$ execution, **c** values of filtered tracking errors $s_1$, $s_2$ and $s_3$, **d** sliding manifold of link 1

filtered tracking error $s_1$ occurs, over time $t = 3.5$ [s]. In contrast, the output layer weights of remaining NNs of DHP algorithm, connected to motion control of links 2 and 3, are adapted since the removal of the robot's arm support ($t = 1.5$ [s]). This results from the need to generate control signals preventing the motion of the robot's arm under the influence of torques from the gravitational forces and providing the desired position of point $C$ of the end effector.

The values of individual performance indexes of tracking motion execution with the use of control system with the structure of adaptive critic in DHP configuration are presented in Tables 10.24 and 10.25.

As a result of the carried out experiment the obtained values of performance indexes have significantly lower values than in the case of implementing in the algorithm of tracking control of the RM only a PD controller. The values of applied performance indexes are similar to those obtained as a result of the numerical test carried out on control algorithm with DHP structures and described in Sect. 7.5.2.4. During the analysis of performance indexes values, provided in Table 10.25, it should be noted that the used RM is an laboratory project and the obtained accuracy of motion execution cannot be compared to the parameters of industrial robots, which have greater rigidity, smaller clearances and different method of transmitting the propulsion from the actuators to the robot's links.

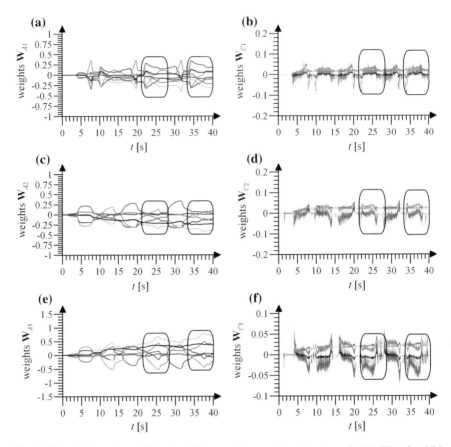

**Fig. 10.38** **a** Values of actor $\mathbf{W}_{A1}$ first NN output layer weights, **b** values of critic $\mathbf{W}_{C1}$ first NN output layer weights, **c** values of actor $\mathbf{W}_{A2}$ second NN output layer weights, **d** values of critic $\mathbf{W}_{C2}$ second NN output layer weights, **e** values of actor $\mathbf{W}_{A3}$ third NN output layer weights, **f** values of critic $\mathbf{W}_{C3}$ third NN output layer weights

**Table 10.24** Values of selected performance indexes

| Index | $e_{max1j}$ [rad] | $\varepsilon_{1j}$ [rad] | $e_{max2j}$ [rad/s] | $\varepsilon_{2j}$ [rad/s] | $s_{maxj}$ [rad/s] | $\sigma_j$ [rad/s] |
|---|---|---|---|---|---|---|
| Link 1, $j = 1$ | 0.0116 | 0.0043 | 0.0478 | 0.0106 | 0.0518 | 0.0114 |
| Link 2, $j = 2$ | 0.0077 | 0.0043 | 0.0542 | 0.0106 | 0.0585 | 0.0116 |
| Link 3, $j = 3$ | 0.013 | 0.0054 | 0.0544 | 0.0099 | 0.0581 | 0.0114 |

**Table 10.25** Values of selected performance indexes of path execution

| Index | $d_{max}$ [m] | $d_n$ [m] | $\rho$ [m] |
|---|---|---|---|
| Value | 0.0056 | 0.0011 | 0.0026 |

### 10.6.3 Conclusions

Tracking control system of the point $A$ of the WMR Pioneer 2-DX, in which the structure of the adaptive critic in DHP configuration was implemented, ensures a high quality of motion execution expressed by low values of performance indexes. In the case of applying the worst case scenario of zero initial values of NNs output layer weights of the actor and critic's structures, the obtained values of performance indexes are lower than those obtained during execution of motion with the use of a PD controller, adaptive control system or neural control system, and comparable to those obtained in the course of carried out verification tests of the control system with the use of HDP structures with non-zero initial values of the critic's NNs output layer weights. What is vital is the low value of tracking errors in the initial phase of motion. The quality of motion is much higher in the case of TCP tracking control of the RM with the use of DHP algorithm than the quality of motion obtained with the implementation of control system built with a PD controller only.

Critic's NNs, in the DHP structure, approximate the derivative of value function with respect to the state vector of the system, which complicates the actor and critic's NNs output layer weights adaptation law, but ensures better quality of tracking control. The knowledge concerning the mathematical model of the controlled object is required in order to perform synthesis of the NNs weights adaptation algorithms of the DHP structure.

## 10.7 Analysis of the GDHP Control

The experimental tests were carried out on tracking control system of point $A$ of the WMR Pioneer 2-DX with GDHP structure, which synthesis is shown in subchapter 7.6.1.1. Experimental tests were carried out on a laboratory stand described in subchapter 10.1.1. The applied trajectory with path in the shape of digit 8 was described in subchapter 2.1.1.2. From the conducted analyses, two experiments in version with parametric disturbance, in the form of additional load carried by the WMR Pioneer 2-DX, and without disturbance were selected.

In experimental tests of tracking control system with the use of GDHP structure, the same parameter values were applied as in the numerical tests described in Sect. 7.6.2.2. In order to evaluate the quality of the generated control and the execution of tracking motion the performance indexes, presented in subchapter 7.4.2, were applied.

The following are selected results of the experimental test carried out with trajectory with path in the shape of digit 8, with parametric disturbance.

### 10.7.1  Analysis of the WMR Motion Control

The verification analyses were carried out on tracking control algorithm of point $A$ of the WMR with NDP structure in GDHP configuration with zero initial values of the actor and critic's output layer weights. The task of the control system was to generate control signals providing the execution of the desired trajectory of point $A$ of the WMR with path in the shape of digit 8, with double parametric disturbance implemented in the form of loading of the robot's frame with $m_{RL} = 4.0$ [kg] mass at time $t_{d1} = 12.5$ [s], and removal of the additional mass at time $t_{d2} = 33$ [s].

The values of overall tracking control signals of the WMR Pioneer 2-DX and control signals of individual structures included in the control system are shown in Fig. 10.39. Similar markings were used as in the previous experimental tests.

Adaptation process of the GDHP structure parameters runs faster in the initial phase of the first stage of motion than in the case of HDP structures when zero initial values of actor and critic's NNs output layer weights are applied, but is slower, when compared to the use of DHP algorithm. The GDHP algorithm generates control signals, which percentage in the overall control signals is high from the beginning of the WMR's point $A$ motion. Control signals of the PD controller reach low values in the initial phase of motion, and then they are minimized to values close to zero.

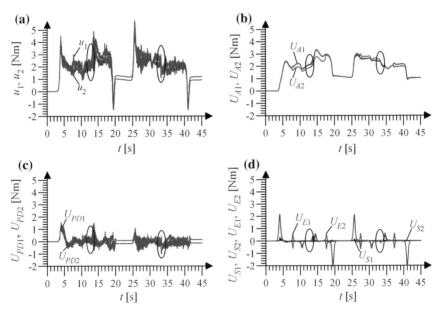

**Fig. 10.39** **a** Overall control signals $u_1$ and $u_2$ for wheels 1 and 2, **b** actor's NNs control signals $U_{A1}$ and $U_{A2}$, $\mathbf{U}_A = -\frac{1}{h}\mathbf{Mu}_A$, **c** PD control signals $U_{PD1}$ and $U_{PD2}$, $\mathbf{U}_{PD} = -\frac{1}{h}\mathbf{Mu}_{PD}$, **d** supervisory term's control signals $U_{S1}$ and $U_{S2}$, $\mathbf{U}_S = -\frac{1}{h}\mathbf{Mu}_S$, and additional control signals $U_{E1}$ and $U_{E2}$, $\mathbf{U}_E = -\frac{1}{h}\mathbf{Mu}_E$

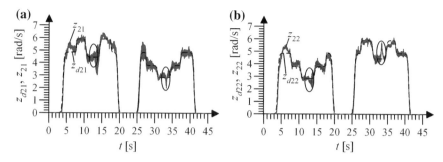

**Fig. 10.40  a** Values of desired ($z_{d21}$) and executed ($z_{21}$) angular velocity of drive wheel 1 of the WMR, **b** values of desired ($z_{d22}$) and executed ($z_{22}$) angular velocity of drive wheel 2 of the WMR

During the occurrence of disturbance $t_{d1}$ and $t_{d2}$, control signals generated by the GDHP structure compensate for the influence of changes in the parameters of the controlled object.

The values of executed angular velocities of the WMR drive wheels rotation are shown in Fig. 10.40. A clear differences between $z_{d21}$ and $z_{d22}$, and $z_{21}$ and $z_{22}$ values, resulting from the implementation of zero initial values of NNs output layer weights of the GDHP structure, are noticeable in the initial phase of the first stage of motion. Similarly, the differences between the desired and executed angular velocities are clearly visible during and after the occurrence of disturbances.

The values of errors of the desired trajectory of wheels 1 and 2 of the WMR execution are shown in Fig. 10.41a and b, respectively. Figure 10.41c presents values of filtered tracking error $s_1$, the values of filtered tracking error $s_2$ are similar. Figure 10.41d presents a sliding manifold of wheel 1. The largest values of tracking errors occur in the initial phase of motion and during the occurrence of disturbances, and are minimized due to the adaptation process of the NNs output layer weights of the GDHP structure. The observed influence of disturbances on the execution of the WMR motion is greater than in the case of other control systems with NDP algorithms.

The values of output layer weights of the actor's NN, generating a signal controlling the motion of wheel 1, and critic's NN output layer weights of GDHP algorithm are shown in Fig. 10.42. The values of the actor's NN output layer weights generating a signal controlling the motion of wheel 2 are similar to those presented in Fig. 10.42a. Zero initial values of the actor and critic's NNs output layer weights were assumed. Weight values are stabilizing around certain values and remain bounded during the experiment. During the parametric disturbance, changes in the values of actor's NNs output layer weights resulting from the adjustments of the generated signal compensating for the nonlinearity of the WMR to the changed parameters of the controlled object and changes in the value of the critic's NN output layer weights, are noticeable.

The desired (dashed line) and executed (continuous line) path of point $A$ of the WMR is shown in Fig. 10.43. The lower quality of tracking motion, resulting from

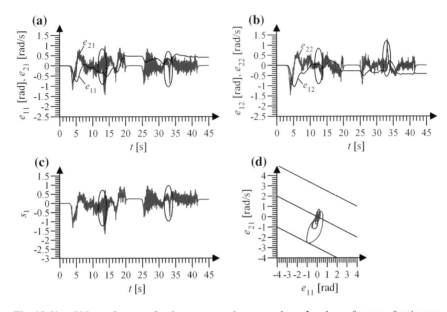

**Fig. 10.41** **a** Values of errors of trajectory execution $e_{11}$ and $e_{21}$, **b** values of errors of trajectory execution $e_{12}$ and $e_{22}$, **c** values of filtered tracking error $s_1$, **d** sliding manifold of wheel 1

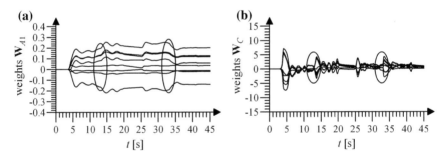

**Fig. 10.42** **a** Values of actor's 1 $\mathbf{W}_{A1}$ NN output layer weights, **b** values of critic's $\mathbf{W}_C$ NN output layer weights

the larger values of errors of the desired angular motion parameters execution, results in mapping of the desired path of point $A$ of the WMR with a lower accuracy.

The values of individual performance indexes of tracking motion execution with the use of a control system with the adaptive critic structure in GDHP configuration are shown in the Tables 10.26 and 10.27.

As a result of conducted verification tests, the obtained values of performance indexes were lower than those obtained with the use of a PD controller, neural control algorithm or control system with HDP algorithms, in which the zero initial values of the actor and critic's NNs output layer weights were applied. The values of performance indexes of the tracking motion execution with the use of GDHP

**Fig. 10.43** Desired and executed path of point $A$ of the WMR

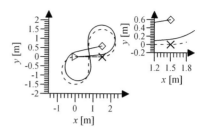

**Table 10.26** Values of selected performance indexes

| Index | $e_{max1j}$ [rad] | $\varepsilon_{1j}$ [rad] | $e_{max2j}$ [rad/s] | $\varepsilon_{2j}$ [rad/s] | $s_{maxj}$ [rad/s] | $\sigma_j$ [rad/s] |
|---|---|---|---|---|---|---|
| Wheel 1, $j = 1$ | 0.77 | 0.36 | 1.43 | 0.26 | 1.66 | 0.33 |
| Wheel 2, $j = 2$ | 1.05 | 0.35 | 1.47 | 0.25 | 1.66 | 0.31 |

**Table 10.27** Values of selected performance indexes of path execution

| Index | $d_{max}$ [m] | $d_n$ [m] | $\rho_d$ [m] |
|---|---|---|---|
| Value | 0.599 | 0.599 | 0.313 |

**Table 10.28** Values of selected performance indexes in I and II stage of motion

| Index | $e_{max1j}$ [rad] | $\varepsilon_{1j}$ [rad] | $e_{max2j}$ [rad/s] | $\varepsilon_{2j}$ [rad/s] | $s_{maxj}$ [rad/s] | $\sigma_j$ [rad/s] |
|---|---|---|---|---|---|---|
| Stage I, $k = 1, \ldots, 2250$ | | | | | | |
| Wheel 1, $j = 1$ | 0.77 | 0.33 | 1.43 | 0.31 | 1.66 | 0.36 |
| Wheel 2, $j = 2$ | 1.05 | 0.34 | 1.47 | 0.28 | 1.66 | 0.33 |
| Stage II, $k = 2251, \ldots, 4500$ | | | | | | |
| Wheel 1, $j = 1$ | 0.57 | 0.4 | 0.9 | 0.21 | 0.96 | 0.28 |
| Wheel 2, $j = 2$ | 0.52 | 0.36 | 1.31 | 0.22 | 1.27 | 0.29 |

algorithm are comparable to those obtained with the use of adaptive algorithm, and higher than those obtained with DHP algorithm and HDP algorithms, in the case of applying non-zero initial values of the critic's NNs output layer weights.

Table 10.28 provides the values of selected performance indexes of the first and second stage of motion execution along the path in the shape of digit 8, calculated according to the methodology described in the numerical test in Sect. 7.1.2.2.

The obtained values of performance indexes, in the first stage of motion execution, are comparable to those obtained in the numerical test. In contrast, the values of performance indexes of the second stage of motion execution are significantly higher than those obtained in the simulation tests.

## 10.7.2  Conclusions

Implementation of the NDP algorithm in GDHP configuration in the tracking control system of the WMR Pioneer 2-DX ensures the quality of tracking motion execution similar to one achieved with the use of the adaptive control algorithm, but lower than one achieved in the case of the implemented control system with DHP structure. A slight increase in the quality of tracking motion execution in the second stage of motion, in relation to the first stage, which results in obtaining similar values of the selected performance indexes, was noted in the carried out experiment. This may result from the greater sensitivity of the GDHP algorithm to changes in parameters of the controlled object resulting from the occurrence of disturbances.

## 10.8  Analysis of the ADHDP Control

The experimental tests were carried out on tracking control system with the use of NDP algorithms in ADHDP configuration, which synthesis is shown in Sect. 7.7.1. Experimental tests were carried out on a laboratory stand described in Sect. 10.1.1. From results of experimental tests that were carried out two implemented sequences with trajectory in the shape of digit 8, with parametric disturbance in the form of additional mass carried by the WMR Pioneer 2-DX and without disturbance, were selected.

In all tests of tracking control system with ADHDP structures the same parameters values, as in the numerical tests presented in Sect. 7.7.2, were applied. Performance indexes, presented in Sect. 7.4.2, were applied in order to evaluate the quality of the tracking motion execution.

The following are selected results of experimental tests that were carried out in the version with trajectory with path in the shape of digit 8 and with parametric disturbance.

## 10.8.1  Analysis of the WMR Motion Control

Verification tests were carried out on tracking control of the WMR Pioneer 2-DX with adaptive critic structures in ADHDP configuration during the execution of desired trajectory with path in the shape of digit 8. Parametric disturbance implemented in the form of loading the robot's frame with $m_{RL} = 4.0$ [kg] mass at time $t_{d1} = 12$ [s] and removing the additional mass at time $t_{d2} = 33$ [s] were introduced twice during the motion of point $A$ of the WMR.

The values of overall control signals of the WMR Pioneer 2-DX and control signals of individual structures included in the control system composition are presented in Fig. 10.44, and are marked similarly as in the previous experimental tests.

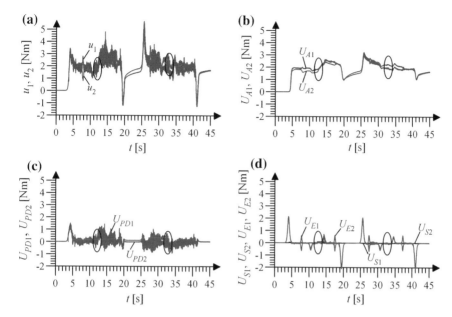

**Fig. 10.44**  **a** Overall control signals $u_1$ and $u_2$ for wheels 1 and 2, **b** actor's NNs control signals $U_{A1}$ and $U_{A2}$, $\mathbf{U}_A = -\frac{1}{h}\mathbf{Mu}_A$, **c** PD control signals $U_{PD1}$ and $U_{PD2}$, $\mathbf{U}_{PD} = -\frac{1}{h}\mathbf{Mu}_{PD}$, **d** supervisory term's control signals $U_{S1}$ and $U_{S2}$, $\mathbf{U}_S = -\frac{1}{h}\mathbf{Mu}_S$, and additional control signals $U_{E1}$ and $U_{E2}$, $\mathbf{U}_E = -\frac{1}{h}\mathbf{Mu}_E$

The control signals generated by the PD controller $\mathbf{u}_{PD}$ and signals $\mathbf{u}_E$ play a very important role in the overall control signals $\mathbf{u}$ in the initial phase of the first stage of motion - acceleration. Progress of the NNs output layer weights adaptation process of ADHDP structures causes a rapid increase of the $\mathbf{u}_A$ control signals presence in the overall control signals. During the occurrence of disturbances $t_{d1}$ and $t_{d2}$, it can be noticed that the values of the generated control signals of the PD controller are changed and the control signals of ADHDP structures react slowly to changes in the dynamics of the controlled object. The parametric disturbances causing changes in the dynamics of the WMR have small influence on the control signals $\mathbf{u}_S$ and $\mathbf{u}_E$.

The values of desired and executed angular velocities of the WMR's drive wheels 1 and 2 rotation are shown in Fig. 10.45a, b. The values of executed angular velocities of drive wheels of the WMR are similar to the desired parameters. At time $t_{d1}$, during the occurrence of disturbance implemented in the form of loading the WMR with additional mass, the values of angular velocities of motion decline and at time $t_{d2}$, during the second parametric disturbance occurrence, the values of angular velocities are increasing. However, after some time, they intend to reach the desired values, which results from the NNs output layer weights adaptation of the ADHDP algorithms.

Error values of desired trajectory execution of the WMR's wheels 1 and 2 are presented in Fig. 10.46a, b. Figure 10.46c presents the values of filtered tracking error $s_1$, and Fig. 10.46d a sliding manifold of wheel 1. The largest absolute values

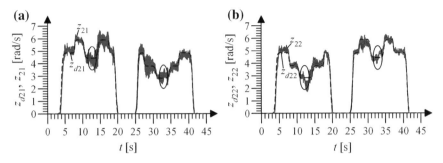

**Fig. 10.45  a** Values of desired ($z_{d21}$) and executed ($z_{21}$) angular velocity of drive wheel 1 of the WMR, **b** values of desired ($z_{d22}$) and executed ($z_{22}$) angular velocity of drive wheel 2 of the WMR

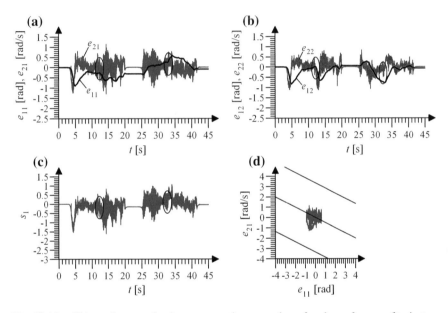

**Fig. 10.46  a** Values of errors of trajectory execution $e_{11}$ and $e_{21}$, **b** values of errors of trajectory execution $e_{12}$ and $e_{22}$, **c** values of filtered tracking error $s_1$, **d** sliding manifold of wheel 1

of tracking errors occur in the initial phase of the first stage of motion - acceleration, and at the time when the first disturbance, $t_{d1}$, occurs.

Figure 10.47 presents the values of actor and critic's NNs output layer weights of the first ADHDP structure. The NNs with neurons activation functions of Gaussian curve type were applied in the critic's structures; actor's structures were implemented as RVFL NNs. Because the adaptation process of ADHDP structures in relation to the WMR's dynamics changes is slow, it is possible to notice a small changes in the actor's NN output layer weights values during the occurrence of the parametric disturbance. The NNs output layer weights values are stabilizing and remain bounded during the experimental tests.

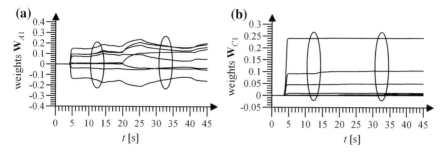

**Fig. 10.47   a** Values of actor 1 $\mathbf{W}_{A1}$ NN output layer weights, **b** values of critic 1 $\mathbf{W}_{C1}$ NN output layer weights

**Fig. 10.48**   Desired and executed path of point $A$ of the WMR

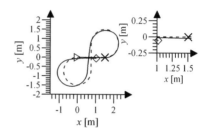

**Table 10.29**   Values of selected performance indexes

| Index | $e_{max1j}$ [rad] | $\varepsilon_{1j}$ [rad] | $e_{max2j}$ [rad/s] | $\varepsilon_{2j}$ [rad/s] | $s_{maxj}$ [rad/s] | $\sigma_j$ [rad/s] |
|---|---|---|---|---|---|---|
| Wheel 1, $j = 1$ | 0.89 | 0.36 | 1.29 | 0.29 | 1.62 | 0.34 |
| Wheel 2, $j = 2$ | 0.83 | 0.3 | 1.12 | 0.25 | 1.39 | 0.29 |

**Table 10.30**   Values of selected performance indexes of path execution

| Index | $d_{max}$ [m] | $d_n$ [m] | $\rho_d$ [m] |
|---|---|---|---|
| Value | 0.491 | 0.489 | 0.262 |

The desired (dashed line) and executed (continuous line) path of point $A$ of the WMR is shown in Fig. 10.48. The same markings were applied as in the descriptions of results of the previous experiments.

The values of individual performance indexes of the tracking motion execution with the use of control system with adaptive critic structures in ADHDP configuration are shown in Tables 10.29 and 10.30.

The obtained values of performance indexes are similar to the values obtained in the numerical test Sect. 7.7.2.1.

Table 10.31 presents the values of selected performance indexes of first and second stage of motion execution along the path in the shape of digit 8, calculated according to the methodology described in the numerical test Sect. 7.1.2.2.

**Table 10.31**  Values of selected performance indexes in I and II stage of motion

| Index | $e_{maxj}$ [rad] | $\varepsilon_j$ [rad] | $\dot{e}_{maxj}$ [rad/s] | $\dot{\varepsilon}_j$ [rad/s] | $s_{maxj}$ [rad/s] | $\sigma_j$ [rad/s] |
|---|---|---|---|---|---|---|
| Stage I, $k = 1, \ldots, 2250$ | | | | | | |
| Wheel 1, $j = 1$ | 0.89 | 0.43 | 1.29 | 0.31 | 1.62 | 0.38 |
| Wheel 2, $j = 2$ | 0.83 | 0.31 | 1.12 | 0.29 | 1.59 | 0.33 |
| Stage II, $k = 2251, \ldots, 4500$ | | | | | | |
| Wheel 1, $j = 1$ | 0.59 | 0.26 | 0.99 | 0.26 | 1.08 | 0.28 |
| Wheel 2, $j = 2$ | 0.78 | 0.3 | 1.07 | 0.2 | 1.07 | 0.25 |

The obtained values of performance indexes, in the second stage of motion, were lower than in the first stage, but the improvement in motion execution of point $A$ of the WMR is not as significant as when the DHP algorithm is implemented.

### 10.8.2   Conclusions

Application of NDP structures in ADHDP configuration in the tracking control system of point $A$ of the WMR Pioneer 2-DX ensures the similar quality of tracking motion execution to one obtained in the experimental tests of neural control system. The assumed performance indexes have larger values than those obtained in the analyses of tracking control system with DHP algorithm or HDP algorithms with non-zero initial values of critic's NN output layer weights. This may result from the lack of the predictive model of the controlled object in the ADHDP algorithm. It was noticed that the process of NNs output layer weights adaptation of the ADHDP structures during disturbance occurrence, in comparison to the HDP or DHP algorithms, was slower. Compensation for the disturbance influence in the verification tests runs faster than in the numerical tests. This may result from the better system stimulation.

## 10.9   Analysis of Behavioral Control

The experimental tests of the control layer generating collision-free trajectory of point $A$ of the WMR Pioneer 2-DX in the task of CB type was carried out with designed neural trajectory generator composed of two NDP structures in ADHDP configuration and a PD controller. The hierarchical control system was discussed in Sect. 7.8. The same parameters of control system and markings were applied in the experimental tests and in the numerical tests described in Sect. 7.8.2. The experiments were performed on a laboratory stand discussed in Sect. 10.1.1, with the measurement track mapping a virtual measurement track from the numerical test presented in Sect. 7.8.2.1.

The task of the hierarchical control system, in the carried out verification tests, was to generate and execute in real-time a collision-free trajectory of point $A$ of the WMR to the goal located in points $G_A$ (4.9, 3.5), $G_B$ (10.3, 3.0) and $G_C$ (7.6, 1.5) of the actual measurement track. The paths of point $A$ of the WMR heading towards the goal $G$, which is placed in different locations, are presented in Fig. 10.49. The assumed markings of individual elements of the figure are similar to those in the numerical test and are presented in Sect. 7.8.2.1.

Obstacles, presented in the Fig. 10.49, localized with the use of ultrasonic sensors and a laser range finder were marked by red points (motion towards $G_A$ goal), green points (motion towards $G_B$ goal) or blue points (motion towards $G_C$ goal). The outline of the laboratory measurement environment, in certain areas, does not coincide with the outline of the obstacles detected by the sensory system of the WMR Pioneer 2-DX. This condition may results from several factors, such as:

- inaccurate setting of the WMR Pioneer 2-DX in initial position,
- inaccurate counting of pulses from encoders during motion,
- inaccurate reading of measurement data from the proximity sensors, additionally disturbed by the sensor motion,
- clearance occurring in the drive units of the WMR,
- the difference in dimension of the test track in relation to its virtual model.

The influence of these factors, which may adversely affect the process of generating collision-free trajectories, accumulates and introduces inaccuracies in generating and executing the desired trajectory of point $A$ of the WMR Pioneer 2-DX. Erroneous measurements, which were used as an information for the control system on the obstacles distance and had influence on the process of trajectory generation, could occur during the motion. Figure 10.49 presents all captured locations of obstacles. It should be noted, that despite the fact that the actual location of obstacles detected by the

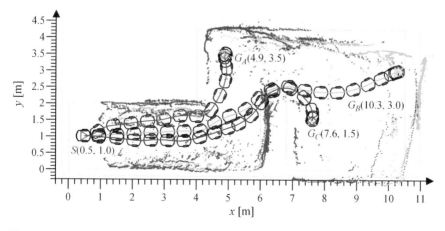

**Fig. 10.49** Paths of point $A$ of the WMR Pioneer 2-DX heading towards the individual location of $G$ goal

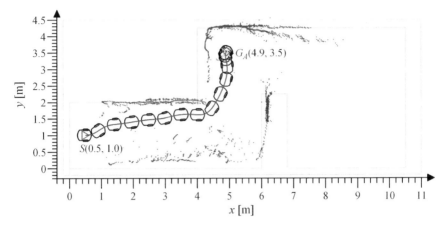

**Fig. 10.50** The path of point $A$ of the WMR Pioneer 2-DX heading towards the $G_A$ (4.9, 3.5) goal

WMR's sensory system are shifted in regard to the map of the measurement track created within the measurement needed to build an environment simulation model, but this does not influence the correctness of the trajectory generation process in the carried out task, where the control signals are determined on the basis of the environment condition. The WMR Pioneer 2-DX had no collision with an obstacles during verification tests.

The same values of system parameters, as in the numerical test presented in Sect. 7.8.2, were applied in experimental tests of the hierarchical control system with layer generating collision-free trajectories implemented in the form of a neural trajectory generator, designed with the use of the adaptive critic structures in ADHDP configuration and a layer of motion execution with NDP structures in DHP configuration.

### 10.9.1  Analysis of the WMR Motion Control

Verification tests were carried out on algorithm generating collision-free trajectory of the WMR in unknown environment with the use of NDP algorithms in ADHDP configuration. The task of the hierarchical control system was to generate the trajectory of point $A$ of the WMR on-line, on the basis of signals from the sensory system of the robot, which included information regarding the environment condition in the form of a measurement data from the proximity sensors and, at the same time, execution of the trajectory. The motion of point $A$ of the WMR Pioneer 2-DX started at the predetermined initial position $S$ (0.5, 1.0) and headed towards the goal located at point $G_A$ (4.9, 3.5). The path of point $A$ of the WMR is presented in Fig. 10.50.

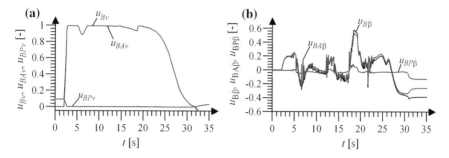

**Fig. 10.51** **a** Control signals $u_{Bv}$, $u_{BAv}$, $u_{BPv}$, **b** control signals $u_{B\dot{\beta}}$, $u_{BA\dot{\beta}}$, $u_{BP\dot{\beta}}$

In the initial phase of the WMR motion the dominant influence of the execution of the GS task can be noticed, when point $A$ of the WMR "is pulled by the $G_A$ goal" and the robot is moving towards wall. Next executed behavior of the robot results from the influence of two behavioral control signals on the generated trajectory, which results in the execution of motion along the wall of the test track. Point $A$ of the WMR, in the final phase of motion, is pulled by the goal. Implementation of the GS tasks is dominant in this phase of motion.

The path of point $A$ of the WMR Pioneer 2-DX results from the execution of the trajectory generated on the basis of the control signals of neural trajectory generator. The overall control signals of the trajectory planning layer ($u_{Bv}$ and $u_{B\dot{\beta}}$) consist of control signals generated by the actor's NNs ($u_{BAv}$ and $u_{BA\dot{\beta}}$) and a proportional controller ($u_{BPv}$ and $u_{BP\dot{\beta}}$). Their values are shown in Fig. 10.51. The control signals generated by the P controller assume small values in the initial stage of the trajectory generation process. They are minimized from $t = 2$ [s], because of the NNs output layer weights adaptation process of the adaptive critic structures. During the NNs output layer weights adaptation process progression the $u_{BAv}$ and $u_{BA\dot{\beta}}$ control signals are dominant in the overall control signals of the layer generating the trajectory and the control signals of the P control assume values close to zero.

The worst case scenario, in which the values of NNs output layer weights are equal to zero, was assumed in the initialization process of the ADHDP structures, what corresponds to the lack of preliminary knowledge. The NNs output layer weights adaptation process is executed in real time, in accordance with the assumed adaptation algorithm. The values of NNs output layer weights of actors and critic's structures of ADHDP algorithms are shown in Fig. 10.52. The values of weights are bounded.

The distance of point $A$ of the WMR to the goal ($d_G$) is shown in Fig. 10.53a and the value of $\psi_G$ angle in Fig. 10.53b. The $d_G$ distance is reduced during the WMR motion. The $\psi_G$ angle is reduced in the initial stage of motion when $t \in \leq 4, 6 \geq$, which results from the dominant influence of the behavioral control of GS type on the overall control signals of the trajectory generation layer. Then the value of $\psi_G$ angle is increasing, which results from the influence of the OA task execution and translates into WMR motion along the wall of the test track. In the final stage of motion the $\psi_G$ angle is reduced to zero and then assumes $\psi_G \approx -\pi$ value, due to

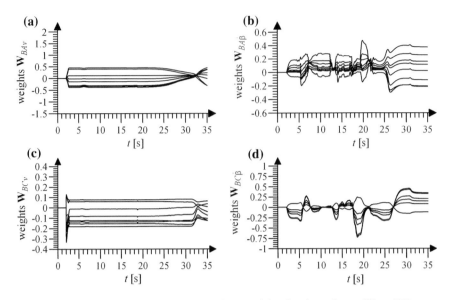

**Fig. 10.52** **a** Values of actor $\mathbf{W}_{BAv}$ NN output layer weights, **b** values of actor $\mathbf{W}_{BA\dot\beta}$ NN output layer weights, **c** values of critic $\mathbf{W}_{BCv}$ NN output layer weights, **d** values of critic $\mathbf{W}_{BC\dot\beta}$ NN output layer weights

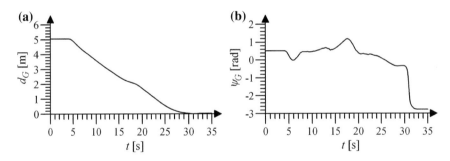

**Fig. 10.53** **a** The distance of point $A$ of the WMR from goal $G_A$, $d_G$, **b** $\psi_G$ angle

stopping of the WMR in a position where point $G_A$ is behind point $A$ (its coordinate in the direction of $x_1$ axis of the movable coordinate system associated with the WMR's frame is negative).

The path of point $A$ of the WMR Pioneer 2-DX with obstacles located within the individual proximity sensors of the robot's sensory system are shown in Fig. 10.54a, b, c, d and e, respectively. In the presented experiment, the best quality of measurements was obtained with the use of laser range finder $s_L$. A lot of erroneous measurements were obtained in the case of distance measuring carried out with the use of ultrasonic sensors $s_{u6}$ and $s_{u7}$.

On the basis of overall control signals of the trajectory generating layer during the motion the values of desired angular velocities of the WMR drive wheels rotation,

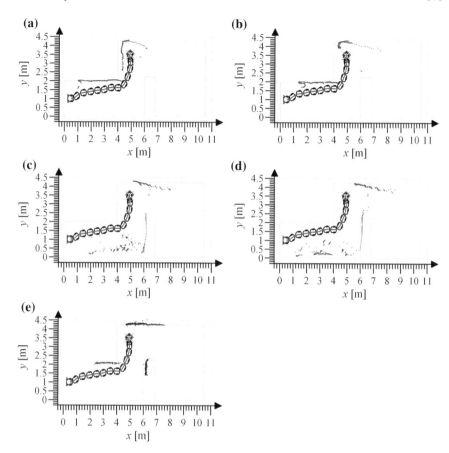

**Fig. 10.54** Path of point $A$ of the WMR and location of obstacles: **a** detected by the sensor $s_{u2}$, **b** detected by the sensor $s_{u3}$, **c** detected by the sensor $s_{u6}$, **d** detected by the sensor $s_{u7}$ **e** detected by the laser range finder $s_L$

shown in Fig. 10.55a, were generated. Execution of the desired parameters of motion required generation of control signals of the lower layer of hierarchical control system. The overall signal controlling the motion of wheel 1 of the motion execution layer $u_1$, presented in Fig. 10.55b, consists of a control signal generated by the actor NN $u_{A1}$, the control signal of the PD controller $u_{PD1}$, the control signal of the supervisory controller $u_{S1}$, and additional control signal $u_{E1}$. Tracking errors of wheel 1 of the WMR Pioneer 2-DX are shown in Fig. 10.55c. The values of actor's 1, $\mathbf{W}_{A1}$ NN weights of NDP algorithm in DHP configuration are presented in Fig. 10.55d. The same zero initial values of NN output layer weights of DHP algorithm, as in the numerical test Sect. 7.8.2.1, were assumed. The values of NNs output layer weights remained bounded during the experimental tests. The values of signals controlling the motion of wheel 2, the values of errors of desired motion parameters execution of wheel 2, and the values of weights $\mathbf{W}_{A2}$ are similar.

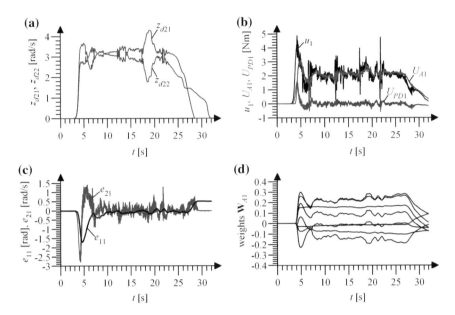

**Fig. 10.55** **a** Values of desired angular velocities of the driving wheels' own rotation, $z_{d21}$ and $z_{d22}$, **b** control signal $u_1$, scaled control signal generated by the actor's NN $U_{A1}$, $\mathbf{U}_A = -\frac{1}{h}\mathbf{Mu}_A$, scaled control signal generated by PD controller $U_{PD1}$, $\mathbf{U}_{PD} = -\frac{1}{h}\mathbf{Mu}_{PD}$, **c** values of tracking errors $e_{11}$ and $e_{21}$, **d** values of the weights of the actor's NN output layer $\mathbf{W}_{A1}$

The proposed hierarchical control system, in which the algorithm of neural trajectory generator with NDP structures in ADHDP configuration was applied, generates a trajectory which execution will ensure reaching by the point $A$ of the robot the desired position and the measurement of the traveled distance, based on odometry, may introduce errors into the process of trajectory generation. The reached, by point $A$ of the WMR, end position in real conditions may differ from the assumed position of $G$ goal in map coordinates. The values of individual signals obtained in the experimental test of hierarchical control system with neural trajectory generator are similar to those obtained in the numerical test Sect. 7.8.2.1.

## 10.9.2   Conclusions

The designed algorithm of collision-free trajectory generating of the WMR in an unknown environment with static obstacles generates a trajectory, which execution is carried out by the selected point $A$ of the WMR Pioneer 2-DX from the desired initial position towards the $G$ goal without colliding with obstacles. The process of trajectory generation is based on the information from the robot's sensory system and measurement of the angles of rotation of the drive wheels implemented by counting the impulses from the incremental encoders included in the drive units. The carried out experimental tests confirm the influence of the measurements accuracy conducted with the use of proximity sensors on the progression of the trajectory planning process. The measurement of an angle of rotation of the drive wheels, implemented with the use of incremental encoders, in conjunction with the clearances in the transmissions of the drive units, results in determining, with an error, the executed path of point $A$ of the WMR Pioneer 2-DX along the test track. This results in generating a trajectory of point $A$ of the WMR towards the point shifted, with respect to the actual location of the $G$ goal, by the distance resulting from an erroneous measurement of the covered distance. Implementation of an external system measuring position of point $A$ of the WMR in the environment, for example, an external local positioning system, could be the solution to this problem.

## 10.10  Summary

Experimental tests were carried out of the presented tracking control systems of point $A$ of the WMR Pioneer 2-DX end enable comparison of the quality of motion execution on the basis of obtained values of the selected performance indexes. Experimental tests of tracking control systems were carried out under the same laboratory conditions and with the trajectory with path in the shape of digit 8.

The values of averaged performance indexes of motion execution, described in Sect. 7.9, were determined.

Comparison of performance indexes of the following algorithms of tracking control was performed:

1. PD controller,
2. adaptive control system,
3. neural control system,
4. control system with ADHDP algorithms,
5. control system with HDP algorithms with zero initial values of critic's NNs output layer weights $\mathbf{W}_{C\{0\}} = \mathbf{0}$ (HDP$_{W0}$),
6. control system with HDP algorithms with non-zero initial values of critic's NNs output layer weights $\mathbf{W}_{C\{0\}} = \mathbf{1}$ (HDP$_{W1}$),
7. control system with GDHP algorithm,
8. control system with DHP algorithm.

The values of selected performance indexes of the WMR's point $A$ motion execution with the use of trajectory with path in the shape of digit 8 were compared during experiments without parametric disturbance. The values of performance indexes are presented in Table 10.32 with numbering that places in order the control systems in tables and comparative charts applied later in this section.

The graphic interpretation of the $s_{maxa}$ and $\sigma_{av}$ performance indexes are presented in bar charts (Fig. 10.56). White color (1) represents the values of the PD controller performance indexes, gray color (2) the values of performance indexes of the adaptive

**Table 10.32** Values of selected performance indexes of trajectory with path in the shape of digit 8 execution, experiments in version without disturbance

| No. | Algorithm | $e_{maxa}$ [rad] | $\varepsilon_{av}$ [rad] | $\dot{e}_{maxa}$ [rad/s] | $\dot{\varepsilon}_{av}$ [rad/s] | $s_{maxa}$ [rad/s] | $\sigma_{av}$ [rad/s] |
|---|---|---|---|---|---|---|---|
| 1 | PD | 4.54 | 3.34 | 2.79 | 0.52 | 3.6 | 1.77 |
| 2 | Adaptive | 1.25 | 0.33 | 1.84 | 0.25 | 2.1 | 0.3 |
| 3 | Neural | 1.02 | 0.22 | 2.02 | 0.31 | 2.3 | 0.33 |
| 4 | ADHDP | 0.86 | 0.33 | 1.31 | 0.23 | 1.52 | 0.28 |
| 5 | $HDP_{W0}$ | 1.93 | 0.39 | 1.7 | 0.35 | 2.21 | 0.41 |
| 6 | $HDP_{W1}$ | 0.65 | 0.17 | 1.07 | 0.18 | 1.21 | 0.2 |
| 7 | GDHP | 0.89 | 0.24 | 1.3 | 0.2 | 1.49 | 0.24 |
| 8 | DHP | 0.78 | 0.18 | 1.14 | 0.2 | 1.32 | 0.22 |

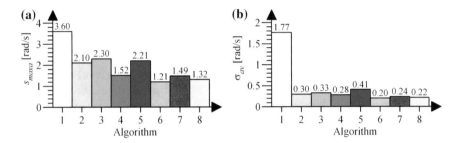

**Fig. 10.56** Values of selected performance indexes of trajectory with path in the shape of digit 8 execution, experiments in version without disturbance

control system, green color (3) the values of performance indexes of the neural control algorithm, purple color (4) represents the values of control system with NDP structures in ADHDP configuration, dark blue (5) represents the values of control system with HDP algorithms and zero initial values of the critic's NNs output layer weights, light blue (6) represents the values of performance indexes of the control system with HDP structures and non-zero initial values of the critic's NNs output layer weights, red color (7) the values obtained with the use of the control system with adaptive critic algorithm of GDHP type, and yellow color (8) represents the values obtained with the use of the DHP algorithm.

From the point of view of presented quality analysis of the WMR's desired trajectory execution in real conditions, the most important is the value of the $\sigma_{av}$ performance index, which depends on the error values of the desired angles of rotation and the angular velocities of the WMR's drive wheels rotation in each step of the experiment. Under certain conditions, the obtained maximum values of errors of the desired trajectory execution, $e_{maxa}$, $\dot{e}_{maxa}$ and $s_{maxa}$, may not be representative and result from the occurrence of a random disturbance acting on the robot, for example, driving along the uneven surface, contact with an obstacle, or change in the dynamics of the WMR caused by change of trasported mass.

The point of reference in the comparative analysis of the quality of the WMR's point $A$ tracking motion execution of individual control systems are the values of selected performance indexes obtained in the verification tests of the PD controller, marked in tables as (1). Other control algorithms consist of a PD controller and additional links, e.g., the link approximating the nonlinearities of the controlled object, which has an influence on the improvement on the quality of the WMR tracking motion execution.

Similarly as in numerical tests, in terms of quality of motion execution expressed by the value of $\sigma_{av}$ performance index (Fig. 10.56b), the analyzed control algorithms can be divided into four groups: $A$, $B$, $C$ and $D$, where the algorithms ensuring the highest quality of tracking motion execution expressed by the lowest values of applied performance indexes were assigned to group $A$. The largest value of the performance index occurs in the case of the PD controller (1), which is assigned to group $D$. It is much larger than the value of the second in order HDP$_{W0}$ algorithm (5) assigned to group $C$. The value of the performance index $\sigma_{av}$ of quality of motion execution of the control system with the use of HDP algorithms with zero initial values of actor and critic's NNs output layer weights results from the large values of errors of desired trajectory execution in the initial stage of weights adaptation. Introduction of non-zero initial values of the critic's NNs output layer weights resulted in improving the quality of motion execution. Group $B$ may include an adaptive control algorithm (2), neural control algorithm (3), and a control system with NDP algorithms in ADHDP configuration (4), where the $\sigma_{av}$ values amount between 0.28–0.33. Tracking control systems using HDP$_{W1}$ (6), GDHP (7), or DHP (8) algorithms are assigned to group $A$. In this case, the values of $\sigma_{av}$ performance index are the lowest and amount between 0.2–0.24. The implemented division into specified groups of algorithms can be noticed by analyzing the values of other performance indexes. The best quality of tracking control is achieved by the control systems with NDP algorithms. It should be noted that in the case of the HDP$_{W1}$ algorithm (6), the applied non-zero initial values of the critic's NNs output layer weights improved significantly the quality of the desired trajectory execution, especially during the initial phase of the robot's motion. The worst case scenario of zero initial values of applied parameters was implemented in other control algorithms. Therefore, the results of the HDP$_{W1}$ algorithm (6) should be considered only as information and should be compared against the results of the HDP$_{W0}$ algorithm (5), in the context of differences in the quality of control caused by the change of the initial conditions of the applied parameters (NN weights).

Table 10.33 presents the values of selected performance indexes of the desired trajectory execution during individual stages of verification tests.

Charts with a graphical comparison of values of selected performance indexes obtained during individual stages of motion are shown in Fig. 10.57 with previously introduced markings and color scheme. Bar charts corresponding to the values of performance indexes of the second stage of motion are marked by oblique lines.

Continuation of the parameters adaptation process has a positive influence on improving the quality of the desired trajectory execution, which can be noticed by comparing the values of selected performance indexes of first and second stage of motion. A big difference can be noticed in the $\sigma_{av}$ values of the first and second stage

**Table 10.33** Values of selected performance indexes in individual stages of trajectory execution with path in the shape of digit 8, during experiments without parametric disturbance

|      |           | Stage I | | | | Stage II | | | |
|------|-----------|---------|---|---|---|----------|---|---|---|
|      |           | $k = 1, \ldots, 2250$ | | | | $k = 2251, \ldots, 4500$ | | | |
| No.  | Algorithm | $e_{maxa}$ [rad] | $\varepsilon_{av}$ [rad] | $s_{maxa}$ [rad/s] | $\sigma_{av}$ [rad/s] | $e_{maxa}$ [rad] | $\varepsilon_{av}$ [rad] | $s_{maxa}$ [rad/s] | $\sigma_{av}$ [rad/s] |
| 1    | PD        | 4.28 | 3.12 | 3.22 | 1.71 | 4.27 | 3.44 | 3.6  | 1.76 |
| 2    | Adaptive  | 1.25 | 0.45 | 2.1  | 0.38 | 0.22 | 0.1  | 0.9  | 0.17 |
| 3    | Neural    | 1.02 | 0.25 | 2.3  | 0.35 | 0.46 | 0.16 | 1.52 | 0.27 |
| 4    | ADHDP     | 0.86 | 0.37 | 1.52 | 0.31 | 0.67 | 0.29 | 1.19 | 0.24 |
| 5    | HDP$_{W0}$ | 1.93 | 0.55 | 2.21 | 0.55 | 0.25 | 0.08 | 0.87 | 0.17 |
| 6    | HDP$_{W1}$ | 0.65 | 0.21 | 1.21 | 0.24 | 0.2  | 0.07 | 0.96 | 0.17 |
| 7    | GDHP      | 0.89 | 0.26 | 1.49 | 0.28 | 0.46 | 0.21 | 0.91 | 0.18 |
| 8    | DHP       | 0.78 | 0.23 | 1.32 | 0.25 | 0.29 | 0.08 | 1.05 | 0.17 |

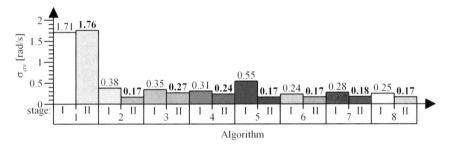

**Fig. 10.57** Values of selected performance indexes in individual stages of trajectory execution with path in the shape of digit 8, during experiments without parametric disturbance

of the WMR's point $A$ motion, implemented with the use of the adaptive algorithm and control system with HDP$_{W0}$ structures, which indicates the correct adjustment of systems' parameters in adaptation process in the first stage of motion. The lowest value of the performance index, $\sigma_{av} = 0.17$, in the second stage of motion was obtained by applying the adaptive control system and control system with HDP and DHP algorithms.

Values of selected performance indexes obtained during verification tests of algorithms controlling the tracking motion of point $A$ of the WMR with the use of trajectory with path in the shape of digit 8, with parametric disturbance, are shown in Table 10.34.

Charts with a graphical comparison of selected performance indexes, with previously introduced markings and color scheme, are presented in Fig. 10.58.

Introduction of double parametric disturbance during the WMR motion increased the value of the performance indexes of all analyzed algorithms. Attention should be drawn to the increase in the value of the $\sigma_{av}$ performance index of motion execution with the use of control system with GDHP algorithm, in relation to the value of the

**Table 10.34** Values of selected performance indexes of trajectory with path in the shape of digit 8 execution, experiments in version with disturbance

| No. | Algorithm | $e_{maxa}$ [rad] | $\varepsilon_{av}$ [rad] | $\dot{e}_{maxa}$ [rad/s] | $\dot{\varepsilon}_{av}$ [rad/s] | $s_{maxa}$ [rad/s] | $\sigma_{av}$ [rad/s] |
|-----|-----------|------|------|------|------|------|------|
| 1 | PD | 5.51 | 3.79 | 2.72 | 0.55 | 4.08 | 1.99 |
| 2 | Adapive | 1.24 | 0.37 | 1.78 | 0.26 | 2.07 | 0.32 |
| 3 | Neural | 0.99 | 0.26 | 2.17 | 0.36 | 2.3 | 0.39 |
| 4 | ADHDP | 0.86 | 0.33 | 1.21 | 0.27 | 1.51 | 0.32 |
| 5 | HDP$_{W0}$ | 1.88 | 0.38 | 1.71 | 0.38 | 2.24 | 0.43 |
| 6 | HDP$_{W1}$ | 0.63 | 0.18 | 1.07 | 0.2 | 1.2 | 0.22 |
| 7 | GDHP | 0.91 | 0.36 | 1.45 | 0.26 | 1.66 | 0.32 |
| 8 | DHP | 0.79 | 0.18 | 1.16 | 0.22 | 1.31 | 0.24 |

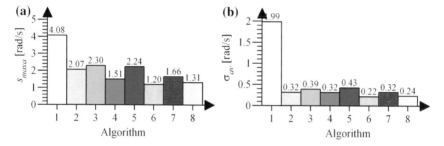

**Fig. 10.58** Values of selected performance indexes of trajectory with path in the shape of digit 8 execution, experiments in version with disturbance

performance index from the experiment without a disturbance. This may indicate a lower resistance of this algorithm to changes in the dynamics of the controlled object, in comparison to other NDP algorithms. The obtained value of the $\sigma_{av}$ performance index of control system with GDHP algorithm is the same as in the case of the adaptive control algorithm, or the control systm with ADHDP algorithms. The highest quality of tracking motion execution is provided by control systems with NDP algorithms.

Table 10.35 provides values of selected performance indexes of verification tests of the WMR's point $A$ tracking control systems, during individual stages of execution of trajectory with path in the shape of digit 8, with parametric disturbance.

Charts with a graphical comparison of the $\sigma_{av}$ performance index values obtained during the individual stages of trajectory execution with path in the shape of digit 8, and occurrence of the parametric disturbance, are shown in Fig. 10.59.

Introduction of double parametric disturbance caused a deterioration of quality of tracking motion execution with adaptive control system and a control system with NDP structures in HDP$_{W0}$ configuration in the second stage of the experiment, in comparison to sequences without disturbances. In the case of control systems with the use of NDP structures in DHP and HDP$_{W1}$ configurations, in spite of parametric disturbances, the lowest value of the performance index, $\sigma_{av} = 0.17$, in the second

**Table 10.35** Values of selected performance indexes in individual stages of trajectory execution with path in the shape of digit 8, during experiments with parametric disturbance

| No. | Algorithm | Stage I $k = 1, \ldots, 2250$ | | | | Stage II $k = 2251, \ldots, 4500$ | | | |
|---|---|---|---|---|---|---|---|---|---|
| | | $e_{maxa}$ [rad] | $\varepsilon_{av}$ [rad] | $s_{maxa}$ [rad/s] | $\sigma_{av}$ [rad/s] | $e_{maxa}$ [rad] | $\varepsilon_{av}$ [rad] | $s_{maxa}$ [rad/s] | $\sigma_{av}$ [rad/s] |
| 1 | PD | 5.17 | 3.53 | 3.47 | 1.93 | 5.3 | 3.91 | 4.08 | 1.98 |
| 2 | Adaptive | 1.24 | 0.48 | 2.07 | 0.4 | 0.47 | 0.2 | 0.98 | 0.19 |
| 3 | Neural | 0.99 | 0.28 | 2.18 | 0.4 | 0.79 | 0.22 | 2.3 | 0.37 |
| 4 | ADHDP | 0.86 | 0.37 | 1.51 | 0.36 | 0.69 | 0.28 | 1.08 | 0.27 |
| 5 | HDP$_{W0}$ | 1.88 | 0.51 | 2.24 | 0.55 | 0.4 | 0.14 | 1.44 | 0.25 |
| 6 | HDP$_{W1}$ | 0.63 | 0.23 | 1.2 | 0.27 | 0.3 | 0.12 | 0.69 | 0.17 |
| 7 | GDHP | 0.91 | 0.34 | 1.66 | 0.35 | 0.55 | 0.38 | 1.12 | 0.29 |
| 8 | DHP | 0.79 | 0.25 | 1.31 | 0.29 | 0.19 | 0.07 | 0.94 | 0.17 |

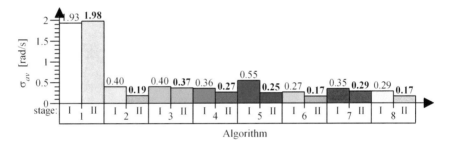

**Fig. 10.59** Values of selected performance indexes in individual stages of trajectory execution with path in the shape of digit 8, during experiments with parametric disturbance

stage of motion was obtained. The process of neural control algorithm and control system with the use of ADHDP and GDHP structures NNs output layer weights adaptation is slow, as evidenced by a small reduction in value of the $\sigma_{av}$ performance index in the second stage of the experiment.

The highest quality of execution of the desired trajectory of point $A$ of the WMR Pioneer 2-DX was obtained with the use of control systems with NDP algorithms in DHP configuration and HDP configuration, in the case of application of non-zero initial values of the critic's NNs output layer weights. The values of control system performance indexes with the use of ADHDP and GDHP algorithms are comparable to values obtained with the use of an adaptive and neural control system, but larger than in the case of applying the algorithm with DHP structure.

# References

1. dSPACE: DS1102 User's Guide. Document Version 3.0, dSpace GmbH, Padeborn (1996)
2. dSPACE: DS1102 RTLib Reference. Document Version 4.0, dSpace GmbH, Paderborn (1998)
3. dSPACE: Control Desk Experiment Guide For Control Desk Version 1.1. dSpace GmbH, Paderborn (1999)
4. dSPACE: DS1102 DSP Controller Board. RTI Reference. dSpace GmbH, Paderborn (1999)
5. dSPACE: Real-Time Interface. Implementation Guide For RTI 3.3. dSpace GmbH, Paderborn (1999)
6. dSPACE: Control Desk Experiment Guide For Release 4.0. dSpace GmbH, Paderborn (2003)
7. dSPACE: DS1104 R&D Controller Board. Installation and Configuration For Release 4.0, dSpace GmbH, Paderborn (2003)
8. dSPACE: Real-Time Interface (RTI and RTI-MP). Implementation Guide For Release 4.0. dSpace GmbH, Paderborn (2003)
9. SIMULINK Dynamic System Simulation for MATLAB. Using Simulink. Version 2.2, on-line version (1998)

# Chapter 11
# Summary

This work presents issues related to vital and up-to-date problems regarding intelligent control for dynamic systems which operate in variable conditions, based on the example of the WMR Pioneer 2-DX and RM Scorbot-ER 4pc. The work also presents the analysis and synthesis of the tracking control systems of selected robot points, in which advanced AI techniques had been used, such as NNs or NDP algorithms. Tracking control system may be perceived as a motion execution layer of hierarchical control system.

Numerical and experimental analyses of tracking control systems of point $A$ for the WMR Pioneer 2-DX have been conducted, such as: PD controller, adaptive control system, neural control system, and tracking control systems with the use of NDP structures in HDP, GDHP, DHP and ADHDP configurations. Numerical and verification analyses of control systems for RM Scorbot-ER 4pc have also been carried out, such as: PD controller and tracking control systems with the use of NDP structures in DHP and GDHP configurations. Similar values have been obtained from numerical and experimental tests of particular control algorithms, which proves the correctness of the mathematical models used, that is WMR Pioneer 2-DX, RM Scorbot-ER 4pc and the virtual measuring environment. The quality of trajectory execution was compared, in the form of assumed performance indexes. The least favorable example for zero initial values of the adopted parameters of structures such as NN or adaptive algorithm was assumed. The best control quality in the analyzed algorithms class is ensured by the control system with DHP algorithm. Similar quality of motion execution was observed while using HDP algorithms, but only in case of non-zero initial values of the critic's NNs output layer weights. The presented systems of motion execution operate in real-time and do not require the initial training process of NN weights.

The inconvenience in the process of designing tracking control system with the use of adaptive critic structures in HDP, DHP and GDHP configurations is that it is necessary to know the mathematical model of the controlled object, essential for the synthesis of the critic and actor's NNs weights adaptation law. In case of

© Springer International Publishing AG 2018  
M. Szuster and Z. Hendzel, *Intelligent Optimal Adaptive Control for Mechatronic Systems*, Studies in Systems, Decision and Control 120,  
https://doi.org/10.1007/978-3-319-68826-8_11

ADHDP structure, such inconvenience does not occur. Mathematical descriptions of mechanical objects, such as WMRs or RMs, are well known, and the models used in NDP structure may be, to some extent, characterized by the mapping inaccuracy of system operation, what does not influence the quality of control. Due to the complex structure and the algorithm of NNs weights adaptation, the control algorithms with NDP structures are characterized by a greater computational cost in comparison to, for example, neural control system.

The work also includes considerations regarding the design and tests on operation of algorithms, in real conditions, generating collision-free trajectories of point $A$ of the WMR in an unknown environment with static obstacles and with the use of the concept of reactive navigation. A hierarchical system generating trajectory and controlling motion of the WMR performing a complex task of CB type with neural trajectory generator built with the use of actor-critic structures in ADHDP configuration was developed. An algorithm of neural trajectory generator in combination with the kinematics module, which converts the generated control signals to motion parameters of WMR's drive wheels, can be seen as a trajectory generating layer of the hierarchical control system.

The problem in the task of synthesis of the algorithm generating the trajectory of the WMR in an unknown environment with immobile obstacles is the lack of knowledge of the environment model, making the design process of control algorithm difficult. In the presented system generating the trajectory with the use of adaptive critic algorithms, an innovative approach was implemented, in which the structure of the control system was extended by the proportional controller similarly as in the solutions of nonlinear control systems. This approach limits the exploration process of reinforcement learning algorithm in favor of the adopted correct control strategy "indicated" by the control signal of the proportional controller. The aim of the proposed solution is to reduce the "trial and error" learning process, because such an approach is not suitable for the control of the WMR in an unknown environment in real time.

Printed in the United States
By Bookmasters